LONDON MATHEMATICAL SOCIETY LECTURE NOTE SERIES

Managing Editor: Professor M. Reid, Mathematics Institute, University of Warwick, Coventry CV4 7AL, United Kingdom

The titles below are available from booksellers, or from Cambridge University Press at www.cambridge.org/mathematics

London Mathematical Society Lecture Notes Series: 388

Groups St Andrews 2009 in Bath

Volume 2

Edited by

C. M. CAMPBELL
University of St Andrews

M. R. QUICK
University of St Andrews

E. F. ROBERTSON
University of St Andrews

C. M. RONEY-DOUGAL
University of St Andrews

G. C. SMITH
University of Bath

G. TRAUSTASON
University of Bath

CAMBRIDGE
UNIVERSITY PRESS

CAMBRIDGE
UNIVERSITY PRESS

University Printing House, Cambridge CB2 8BS, United Kingdom

One Liberty Plaza, 20th Floor, New York, NY 10006, USA

477 Williamstown Road, Port Melbourne, VIC 3207, Australia

314-321, 3rd Floor, Plot 3, Splendor Forum, Jasola District Centre, New Delhi - 110025, India

103 Penang Road, #05-06/07, Visioncrest Commercial, Singapore 238467

Cambridge University Press is part of the University of Cambridge.

It furthers the University's mission by disseminating knowledge in the pursuit of education, learning and research at the highest international levels of excellence.

www.cambridge.org
Information on this title: www.cambridge.org/9780521279048

© Cambridge University Press 2011

First published 2011

A catalogue record for this publication is available from the British Library

ISBN 978-0-521-27904-8 Paperback

CONTENTS

Volume 2

Contents of Volume 1

INTRODUCTION

Groups St Andrews 2009 was held in the University of Bath from 1 August to 15 August 2009. This was the eighth in the series of Groups St Andrews group theory conferences organised by Colin Campbell and Edmund Robertson of the University of St Andrews. The first three were held in St Andrews, and subsequent conferences held in Galway, Bath and Oxford, before returning to St Andrews in 2005 and to Bath in 2009. There were about 200 mathematicians from 30 countries involved in the meeting as well as some family members and partners. The Scientific Organising Committee of Groups St Andrews 2009 was: Colin Campbell (St Andrews), Martyn Quick (St Andrews), Edmund Robertson (St Andrews), Colva Roney-Dougal (St Andrews), Geoff Smith (Bath), Gunnar Traustason (Bath).

The shape of the conference was similar to the previous conferences (with the exception of Groups St Andrews 1981 and 2005) in that the first week was dominated by five series of talks, each surveying an area of rapid contemporary development in group theory and related areas. The main speakers were Gerhard Hiss (RWTH Aachen), Volodymyr Nekrashevych (Texas A&M), Eamonn O'Brien (Auckland), Mark Sapir (Vanderbilt) and Dan Segal (Oxford). The second week featured three special days, a Cannon/Holt Day, a B H Neumann Day and an Engel Day. The invited speakers at the Cannon/Holt Day included George Havas (Queensland), Claas Roever (Galway) and Marston Conder (Auckland). For the B H Neumann Day, two of his sons, Peter Neumann (Oxford) and Walter Neumann (Columbia, New York), were invited speakers as were Michael Vaughan-Lee (Oxford), Cheryl Praeger (Western Australia) and Gilbert Baumslag (CUNY). For the Engel Day invited speakers included Gunnar Traustason (Bath), Olga Macedońska (Katowice) and Patrizia Longobardi (Salerno). Our thanks are due to Charles Leedham-Green (QMWC, London), Roger Bryant (Manchester) and Gunnar Traustason (Bath) for helping organise the programmes for the special days, and to the speakers on these special days.

Each week contained an extensive programme of research seminars and one-hour invited talks. In the evenings throughout the conference, and during the rest periods, there was an extensive social programme. There were two conference outings. The first was to Stonehenge and Salisbury, and the second was to Stourhead Gardens and Wells. In the first week there was a conference banquet at Cumberwell Golf Club. In the second week there was a wine reception at the American Museum in Britain for the B H Neumann Day and a conference banquet at Bath Racecourse on the Cannon/Holt Day. We wish to thank Charles Leedham-Green for allowing us to publish the after-dinner address that he gave at the banquet. Once again the 𝕯𝖆𝖎𝖑𝖞 𝕲𝖗𝖔𝖚𝖕 𝕿𝖍𝖊𝖔𝖗𝖎𝖘𝖙 was a nice feature of the conference. We thank the various editors of this, by now traditional, publication.

Once again, we believe that the support of the two main British mathematics societies, the Edinburgh Mathematical Society and the London Mathematical Society has been an important factor in the success of these conferences. As well as supporting some of the expenses of the main speakers, the grants from these societies were used to support postgraduate students and also participants from Scheme 5 and fSU countries.

As has become the tradition, all the main speakers have written substantial articles for these Proceedings. These articles along with the majority of the other papers are of a survey nature. All papers have been subjected to a formal refereeing process comparable to that of a major international journal. Publishing constraints have forced the editors to exclude some very worthwhile papers, and this is of course a matter of regret. Volume 1 begins with the papers by the main speakers Gerhard Hiss and Volodymyr Nekrashevych. These are followed by those papers whose first-named author begins with a letter in the range A to E. Volume 2 begins with the papers by the main speakers Eamonn O'Brien, Mark Sapir and Dan Segal. These are followed by those papers whose first-named author begins with a letter in the range F to Z.

The next conference in this series will be held in St Andrews in 2013. We are confident that this will be, as usual, a chance to meet many old friends and to make many new friends.

We would like to thank Martyn Quick, Colva Roney-Dougal, Geoff Smith and Gunnar Traustason both for their editorial assistance with these Proceedings and for all their hard work in organising the conference. Our final thanks go not only to the authors of the articles but also to Roger Astley and the rest of the Cambridge University Press team for their assistance and friendly advice throughout the production of these Proceedings.

CMC, EFR

ALGORITHMS FOR MATRIX GROUPS

E.A. O'BRIEN

Department of Mathematics, University of Auckland, Auckland, New Zealand
Email: obrien@math.auckland.ac.nz

Abstract

Existing algorithms have only limited ability to answer structural questions about subgroups G of $GL(d, F)$, where F is a finite field. We discuss new and promising algorithmic approaches, both theoretical and practical, which as a first step construct a chief series for G.

1 Introduction

Research in Computational Group Theory has concentrated on four primary areas: permutation groups, finitely-presented groups, soluble groups, and matrix groups. It is now possible to study the structure of permutation groups having degrees up to about ten million; Seress [97] describes in detail the relevant algorithms. We can compute *useful* descriptions for quotients of finitely-presented groups; as one example, O'Brien & Vaughan-Lee [90] computed a power-conjugate presentation for the largest finite 2-generator group of exponent 7, showing that it has order 7^{20416}. Practical algorithms for the study of polycyclic groups are described in [59, Chapter 8].

We contrast the success in these areas with the paucity of algorithms to investigate the structure of matrix groups. Let $G = \langle X \rangle \leq GL(d, F)$ where $F = GF(q)$. Natural questions of interest to group-theorists include: What is the order of G? What are its composition factors? How many conjugacy classes of elements does it have? Such questions about a subgroup of S_n, the symmetric group of degree n, are answered both theoretically and practically using highly effective polynomial-time algorithms. However, for linear groups these can be answered only in certain limited contexts. As one indicator, it is difficult (using standard functions) to answer such questions about $GL(8, 7)$ using either of the major computational algebra systems, GAP [46] and MAGMA [16].

A major topic of research over the past 15 years, the so-called "matrix recognition" project, has sought to address these limitations by developing effective well-understood algorithms for the study of such groups. A secondary goal is to realise the performance of these algorithms in practice, via publicly available implementations.

Thanks to Peter Brooksbank, Heiko Dietrich, Stephen Glasby, Derek Holt, Colva Roney-Dougal and Ákos Seress for their comments and corrections to the paper. This work was partially supported by the Marsden Fund of New Zealand via grant UOA721.

Two approaches dominate. The *black-box approach*, discussed in Section 4, aims to construct a characteristic series $\mathcal{C}$ of subgroups for G which can be readily refined to provide a chief series; the associated algorithms are independent of the given representation. The *geometric approach*, discussed in Section 5, aims to exploit the natural linear action of G on its underlying vector space to construct a composition series for G; the associated algorithms exploit the linear representation of G. Both approaches rely on the solution of certain key tasks for simple groups which we discuss in Section 3; we survey their solutions in Sections 6–9. Presentations for the groups of Lie type on certain *standard generators* are used to ensure correctness; these are discussed in Section 10.

As we demonstrate in Section 11, the geometric approach is realised via a *composition tree*. In practice, the composition series produced from the geometric approach is readily modified to produce a chief series of G exhibiting $\mathcal{C}$. In Section 12 we consider briefly algorithms which exploit the chief series and its associated *Trivial Fitting* paradigm to answer structural questions about G. While it is not yet possible to make definitive statements about the outcome of this project, a realistic and achievable goal is to provide algorithms to answer many questions for linear groups of "small" degree, say up to degree 20 defined over moderate-sized fields.

In this paper, we aim to supplement and update the related surveys [65], [72] and [91]. Its length precludes comprehensiveness. For example, we consider neither nilpotent nor solvable linear groups. Nor do we discuss the algorithms of Detinko and Flannery and others to study finitely generated matrix groups defined over infinite fields. The excellent survey [43] addresses both omissions.

2 Basic concepts

We commence with a review of basic concepts.

2.1 Complexity

If f and g are real-valued functions defined on the positive integers, then $f(n) = O(g(n))$ means $|f(n)| < C|g(n)|$ for some positive constant C and all sufficiently large n.

One measure of performance is that an algorithm is *polynomial in the size of the input*. If $G = \langle X \rangle \leq \mathrm{GL}(d, q)$, then the size of the input is $|X| d^2 \log q$, since each of the d^2 entries in a matrix requires $\log q$ bits.

2.2 Black-box groups

The concept of a *black-box group* was introduced in [6]. In this model, group elements are represented by bit-strings of uniform length; the only group operations permissible are multiplication, inversion, and checking for equality with the identity element. Permutation groups and matrix groups defined over finite fields are covered by this model.

Seress [97, p. 17] defines a *black-box algorithm* as one which does not use specific features of the group representation, nor particulars of how group operations are performed; it can only use the operations listed above. However, a common assumption is that *oracles* are available to perform certain tasks – usually those not known to be solvable in polynomial time.

One such is a *discrete log oracle*: for a given non-zero $\mu \in \mathrm{GF}(q)$ and a fixed primitive element ω of $\mathrm{GF}(q)$, it returns the unique integer k in the range $1 \leq k < q$ for which $\mu = \omega^k$. The most efficient algorithms for this task run in sub-exponential time (see [98, Chapter 4]).

If the elements of a black-box group G are represented by bit-strings of uniform length n, then n is the *encoding length* of G and $|G| \leq 2^n$. If G is described by a bounded list of generators, then the size of the input to a black-box algorithm is $O(n)$. If G also has Lie rank r and is defined over a field of size q, then $|G| \geq (q-1)^r$, so both r and $\log q$ are $O(n)$.

2.3 Algorithm types and random elements

Most algorithms for linear groups are *randomised*: they rely on random selections. A *Monte Carlo* algorithm is a randomised algorithm that, with prescribed probability less than $1/2$, may return an incorrect answer to a decision question. A *Las Vegas* algorithm is one that never returns an incorrect answer, but may report failure with probability less than some specified $\epsilon \in (0, 1)$. At the cost of n iterations, the probability of a correct answer can be increased to $1 - \epsilon^n$. We refer the reader to [5] for a discussion of these concepts.

Monte Carlo algorithms to construct the normal closure of a subgroup and the derived group of a black-box group are described in [97, Chapter 2].

Many algorithms use random search in a group $G \leq \mathrm{GL}(d, q)$ to find elements having prescribed property $\mathcal{P}$. Examples of $\mathcal{P}$ are having a characteristic polynomial with a factor of degree greater than $d/2$, or order divisible by a prescribed prime.

A common feature is that these algorithms depend on detailed analysis of the *proportion* of elements of finite simple groups satisfying $\mathcal{P}$. Assume we determine a lower bound, say $1/k$, for the proportion of elements in G satisfying $\mathcal{P}$. To find an element satisfying $\mathcal{P}$ by random search with probability of failure less than a given $\epsilon \in (0, 1)$, we choose a sample of uniformly distributed random elements in G of size at least $\lceil \log_e(1/\epsilon) \rceil k$.

Following [97, p. 24], an algorithm constructs an ϵ-uniformly distributed random element x of a finite group G if $(1-\epsilon)/|G| < \mathrm{Prob}(x = g) < (1+\epsilon)/|G|$ for all $g \in G$; if $\epsilon < 1/2$, then the algorithm constructs *nearly uniformly distributed* random elements of G. Babai [4] presents a black-box Monte Carlo algorithm to construct such elements in polynomial time. An alternative is the *product replacement algorithm* of Celler *et al.* [34]. That this runs in polynomial time was established by Pak [92]. Its implementations in GAP and MAGMA are widely used. For a discussion of both algorithms, see [97, pp. 26–30]. Another algorithm, proposed by Cooperman [39], was analysed by Dixon [44].

2.4 Some basic operations

Consider the task of multiplying two $d \times d$ matrices. Its complexity is $O(d^\omega)$ field operations, where $\omega = 3$ if we employ the traditional algorithm. Strassen's divide-and-conquer algorithm [100] reduces ω to $\log_2 7$ but at a cost: namely, the additional intricacy of an implementation and larger memory demands. Coppersmith & Winograd's result [40] that ω can be smaller than 2.376 remains of limited practical significance.

We can compute large powers m of a matrix g in at most $2 \lfloor \log_2 m \rfloor$ multiplications by the standard doubling algorithm: $g^m = g^{m-1}g$ if m is odd and $g^m = g^{(m/2)2}$ if m is even.

Lemma 2.1

(i) *Multiplication and division operations for polynomials of degree d defined over $\mathrm{GF}(q)$ can be performed deterministically in $O(d \log d \log \log d)$ field operations. Using a Las Vegas algorithm, such a polynomial can be factored into its irreducible factors in $O(d^2 \log d \log \log d \log(qd))$ field operations.*

(ii) *Using Las Vegas algorithms, both the characteristic and minimal polynomial of $g \in \mathrm{GL}(d, q)$ can be computed in $O(d^3 \log d)$ field operations.*

For the cost of polynomial operations, see [101, §8.3, §9.1, Theorem 14.14]. Characteristic and minimal polynomials can be computed in the claimed time using the Las Vegas algorithms of [2, 69] and [47] respectively. Neunhöffer & Praeger [87] describe Monte Carlo and deterministic algorithms to construct the minimal polynomial; these have complexity $O(d^3)$ and $O(d^4)$ respectively and are implemented in GAP.

2.5 The pseudo-order of a matrix

To determine the order of $g \in \mathrm{GL}(d, q)$ currently requires factorisation of numbers of the form $q^i - 1$, a problem generally believed not to be solvable in polynomial time. Since $\mathrm{GL}(d, q)$ has elements of order $q^d - 1$ (namely, Singer cycles), it is not practical to compute powers of g until we obtain the identity.

Celler & Leedham-Green [35] present the following algorithm to compute the order of $g \in \mathrm{GL}(d, q)$.

- Compute a "good" multiplicative upper bound B for $|g|$.
- Factorise $B = \prod_{i=1}^{m} p_i^{\alpha_i}$ where the primes p_i are distinct.
- If $m = 1$, then calculate $g^{p_1^j}$ for $j = 1, 2, \ldots, \alpha_1 - 1$ until the identity is constructed.
- If $m > 1$ then express $B = uv$, where u, v are coprime and have approximately the same number of distinct prime factors. Now g^u has order k dividing v and g^k has order ℓ say dividing u, and the order of g is $k\ell$. Hence the algorithm proceeds by recursion on m.

They prove the following:

Theorem 2.2 *If we know a factorisation of B, then the cost of the algorithm is $O(d^4 \log q \log \log q^d)$ field operations.*

We can readily compute in polynomial time a "good" multiplicative upper bound for $|g|$. Let the factorisation over $GF(q)$ of the minimal polynomial $f(x)$ of g into powers of distinct irreducible monic polynomials be given by $f(x) = \prod_{i=1}^{t} f_i(x)^{n_i}$, where $\deg(f_i) = e_i$. Then $|g|$ divides $B := \mathrm{lcm}(q^{e_1} - 1, \ldots, q^{e_t} - 1) \times p^{\beta}$, where $\beta = \lceil \log_p \max n_i \rceil$ and $GF(q)$ has characteristic p.

The GAP and MAGMA implementations of the order algorithm are very efficient, and use databases of factorisations of numbers of the form $q^i - 1$, prepared as part of the Cunningham Project [20].

From B, we can learn in polynomial time the *exact* power of 2 (or of any specified prime) which divides $|g|$. By repeated division by 2, we write $B = 2^m b$ where b is odd. Now we compute $h = g^b$, and determine (by powering) its order, which divides 2^m. In particular, we can deduce if g has *even order*.

For most applications, it suffices to know the *pseudo-order* of $g \in GL(d, q)$, a refined version of B. Leedham-Green & O'Brien [73, Section 2] define this formally and show that it can be computed in $O(d^3 \log d + d^2 \log d \log \log d \log q)$ field operations.

2.6   Straight-line programs

One may intuitively think of a *straight-line program* (SLP) for $g \in G = \langle X \rangle$ as an efficiently stored word in X that evaluates to g; for a formal definition and discussion of their significance, see [97, p. 10]. While the length of a word in a given generating set constructed in n multiplications and inversions can increase exponentially with n, the length of the corresponding SLP is *linear* in n. Babai & Szemerédi [6] prove that every element of a finite group G has an SLP of length $O(\log^2 |G|)$ in every generating set. Both MAGMA and GAP use SLPs.

3   The major tasks

We identify three major problems for a (quasi)simple group $G = \langle X \rangle$. (Recall that G is *quasisimple* if G is perfect and $G/Z(G)$ is simple.)

  (i) The *naming problem*: determine the name of G.

 (ii) The *constructive recognition problem*: construct an isomorphism (possibly modulo scalars) between G and a "standard copy" of G.

(iii) The *constructive membership problem*: if $x \in G$, then write x as an SLP in X.

An algorithm to solve (i) may simply establish that G *contains* a named group as its unique non-abelian composition factor. Such information is useful: if we learn that G is a member of a particular family of finite simple groups, then we can apply algorithms to G which are specific to this family.

For each finite (quasi)simple group, we designate one explicit representation as its *standard copy* and designate a particular generating set as its *standard generators*.

For example, the standard copy of A_n is on n points; its standard generators are $(1, 2, 3)$ and either of $(3, \ldots, n)$ or $(1, 2)(3, \ldots, n)$ according to the parity of n.

To aid exposition, we focus on one common situation. Consider the classical groups, where the standard copy is the natural representation. Let $H \leq \mathrm{GL}(d, q)$ denote the natural representation of a classical group. Given as input an arbitrary permutation or projective matrix representation $G = \langle X \rangle$, a constructive recognition algorithm sets up an isomorphism between G and $H/Z(H)$.

To enable this construction, we define standard generators S for H. Assume we can construct the image $\bar{S}$ of these standard generators in G as SLPs in X. We may now define the isomorphism $\phi : H/Z(H) \to G$. If we can solve the constructive membership problem in H, then the image in G of an arbitrary element of H can be constructed: if h has a known SLP in S then $\phi(h)$ is the SLP evaluated in $\bar{S}$. Similarly if we can solve the constructive membership problem in G, then we can define $\tau : G \to H/Z(H)$. We say that these isomorphisms are *constructive*.

4   The black-box approach

The *black-box group approach*, initiated and pioneered by Babai and Beals (see [7] for an excellent account), focuses on the abstract structure of a finite group G. Recall, for example from [59, pp. 31–32], that G has a characteristic series of subgroups:

$$1 \leq O_\infty(G) \leq S^*(G) \leq P(G) \leq G$$

where

- $O_\infty(G)$ is the largest soluble normal subgroup of G, the *soluble radical*;
- $S^*(G)/O_\infty(G)$ is the socle of $G/O_\infty(G)$ and equals $T_1 \times \cdots \times T_k$, where each T_i is non-abelian simple;
- $\phi : G \to \mathrm{Sym}(k)$ is the representation of G induced by conjugation on $\{T_1, \ldots, T_k\}$, and $P(G) = \ker \phi$;
- $P(G)/S^*(G) \leq \mathrm{Out}(T_1) \times \cdots \times \mathrm{Out}(T_k)$ and so is soluble (by the proof of the Schreier conjecture);
- $G/P(G) \leq \mathrm{Sym}(k)$ where $k \leq \log |G| / \log 60$.

In summary, the black-box approach aims to construct this characteristic series $\mathcal{C}$ for $G \leq \mathrm{GL}(d, q)$ using black-box algorithms. In 2009, as a culmination of 25 years of work, Babai, Beals & Seress [10] proved that, subject to the existence of a discrete log oracle and the ability to factorise integers of the form $q^i - 1$ for $1 \leq i \leq d$, there exist black-box polynomial-time Las Vegas algorithms to construct $\mathcal{C}$ for a large class of matrix groups. Building on results of [9], [56], [81] and [93], they solve the major tasks identified in Section 3 (and others) for groups in this class. We refer the reader to [7] and [10] for details.

In Section 12 we consider how the black-box approach underpins various practical algorithms for matrix groups.

5   Geometry following Aschbacher

By contrast, the *geometric approach* investigates whether a linear group satisfies natural and inherent geometric properties in *its action on the underlying space*. A classification of the maximal subgroups of classical groups by Aschbacher [3] underpins this approach. Let Z denote the subgroup of scalar matrices of $G \leq \mathrm{GL}(d,q)$. Then G is *almost simple modulo scalars* if there is a non-abelian simple group T such that $T \leq G/Z \leq \mathrm{Aut}(T)$, the automorphism group of T. We paraphrase Aschbacher's theorem as follows.

Theorem 5.1 *Let V be the vector space of row vectors on which $\mathrm{GL}(d,q)$ acts, and let Z be the subgroup of scalar matrices of G. If G is a subgroup of $\mathrm{GL}(d,q)$, then one of the following is true:*

C1. *G acts reducibly.*

C2. *G acts imprimitively: G preserves a decomposition of V as a direct sum $V_1 \oplus V_2 \oplus \cdots \oplus V_r$ of $r > 1$ subspaces of dimension s, which are permuted transitively by G, and so $G \leq \mathrm{GL}(s,q) \wr \mathrm{Sym}(r)$.*

C3. *G acts on V as a group of semilinear automorphisms of a (d/e)-dimensional space over the extension field $\mathrm{GF}(q^e)$ for some $e > 1$, and so G embeds in $\Gamma\mathrm{L}(d/e, q^e)$. (This includes the class of "absolutely reducible" linear groups, where G embeds in $\mathrm{GL}(d/e, q^e)$.)*

C4. *G preserves a decomposition of V as a tensor product $U \otimes W$ of spaces of dimensions $d_1, d_2 > 1$ over $\mathrm{GF}(q)$. Then G is a subgroup of the central product of $\mathrm{GL}(d_1, q)$ and $\mathrm{GL}(d_2, q)$.*

C5. *G is definable modulo scalars over a subfield: for some proper subfield $\mathrm{GF}(q')$ of $\mathrm{GF}(q)$, $G^g \leq \mathrm{GL}(d, q').Z$, for some $g \in \mathrm{GL}(d,q)$.*

C6. *For some prime r, $d = r^n$, and G is contained in the normaliser of an extraspecial group of order r^{2n+1}, or of a group of order 2^{2n+2} and symplectic-type (namely, the central product of an extraspecial group of order 2^{2n+1} with a cyclic group of order 4, amalgamating central involutions).*

C7. *G is tensor-induced: G preserves a decomposition of V as $V_1 \otimes V_2 \otimes \cdots \otimes V_m$, where each V_i has dimension $r > 1$, $d = r^m$, and the set of V_is is permuted transitively by G, and so $G/Z \leq \mathrm{PGL}(r,q) \wr \mathrm{Sym}(m)$.*

C8. *G normalises a classical group in its natural representation.*

C9. *G is almost simple modulo scalars.*

We summarise the outcome: a linear group preserves some natural linear structure in its action on the underlying space and has a normal subgroup related to this structure, or it is almost simple modulo scalars.

In broad outline, it suggests that a first step in investigating a linear group is to determine (at least one of) its categories in the Aschbacher classification. If a category is recognised, then we can investigate the group structure more completely using algorithms designed for this category. Usually, we have reduced the size and nature of the problem. For example, if $G \leq \mathrm{GL}(d,q)$ acts imprimitively, then we obtain a permutation representation of degree dividing d for G; if G preserves a tensor product, we obtain two linear groups of smaller degree. If a proper normal subgroup N exists, we investigate N and G/N recursively, ultimately obtaining a composition series for G.

The *base cases* for the geometric approach are groups in C8 and C9: classical groups in their natural representation, and other groups which are almost simple modulo scalars. Liebeck [74] proved that "most" maximal subgroups of $\mathrm{GL}(d, q)$ have order at most q^{3d}, small by contrast with $|\mathrm{GL}(d, q)|$; the exceptions are known. Further, the absolutely irreducible representations of degree at most 250 of all quasisimple finite groups are now explicitly known: see Hiss & Malle [55] and Lübeck [78].

Landazuri & Seitz [71] and Seitz & Zalesskii [96] provide lower bounds for degrees of non-linear irreducible projective representations of finite Chevalley groups. They show that a faithful projective representation in cross characteristic has degree that is polynomial in the defining characteristic. Hence our principal focus is on matrix representations in *defining characteristic*.

5.1   Deciding membership of an Aschbacher category

In [91] we reported in detail on the algorithms developed to decide if $G = \langle X \rangle \leq \mathrm{GL}(d, q)$, acting on the underlying vector space V, lies in one of the first seven Aschbacher categories. Consequently we only update that report. In Section 6.1 we report on a Monte Carlo algorithm which decides if G is in C8.

5.1.1   Reducible groups

The MEATAXE algorithm of Holt & Rees [57] is Las Vegas and has complexity $O(d^3(d \log d + \log q))$. A key component is a search in the $\mathrm{GF}(q)$-algebra generated by X for an element whose characteristic polynomial has an irreducible factor of multiplicity one. The analysis of [57], completed in [64], shows that the proportion of such elements is at least 0.08.

A matrix A over $\mathrm{GF}(q)$ for which the underlying vector space, considered as a $\mathrm{GF}(q)[A]$-module, has at least one cyclic primary component is *f-cyclic*. Glasby & Praeger [49] present and analyse a test for the irreducibility of G using the set of f-cyclic matrices in G, which contains as a proper subset those considered in [57].

5.1.2   C3 and C5

Holt *et al.* [58] present the SMASH algorithm: effectively an algorithmic realisation of Clifford's theorem [36] about decompositions of V preserved by a non-scalar normal subgroup of G.

If G acts absolutely irreducibly, then we apply SMASH to a normal generating set for its derived group G' to decide if G acts semilinearly. The polynomial-time algorithm of [48] to decide membership in C5 requires that G' acts absolutely irreducibly on V. Implementations of both are available in MAGMA.

Carlson, Neunhöffer & Roney-Dougal [33] present a polynomial-time Las Vegas algorithm to find a non-trivial "reduction" of an irreducible group G that either lies in C3 or C5, or whose derived group does not act absolutely irreducibly on V. In particular, they deduce that G is in one of C2, C3, C4, or C5; or obtain a homomorphism from G to $\mathrm{GF}(q)^{\times}$. An implementation is available in GAP.

5.1.3   Normalisers of p-groups

If G is in C6, then it normalises a group R of order either r^{2n+1} (extraspecial) or 2^{2n+2} (symplectic-type).

Brooksbank, Niemeyer & Seress [25] present an algorithm to produce a non-trivial homomorphism from G to either $\mathrm{GL}(2m, r)$ or $\mathrm{Sym}(r^m)$ where $1 \leqslant m \leqslant n$. They prove that this algorithm runs in polynomial time when G is either the full normaliser in $\mathrm{GL}(d, q)$ of R, or $d = r^2$. The special case where $d = r$ was solved by Niemeyer [89]. Implementations are available in GAP and MAGMA.

5.1.4   Towards polynomial time?

A major theoretical challenge is the following: decide membership of a given group $G \leq \mathrm{GL}(d, q)$ in a *specific* Aschbacher category in polynomial time. This we can always do for C1 and C8, and sometimes for C3, C5 and C6.

Recently Neunhöffer [86] has further developed and analysed variations of the SMASH algorithm, and has also reformulated the Aschbacher categories to facilitate easier membership problems. This work and the "reduction algorithms" of [25] and [33] suggest that, subject to the availability of discrete log and integer factorisation oracles, it may be possible using matrix group algorithms to construct in polynomial time the composition factors of G. We contrast this with the results obtained in the black-box context [10].

6   Naming algorithms

Let b and e be positive integers with $b > 1$. A prime r dividing $b^e - 1$ is a *primitive prime divisor* of $b^e - 1$ if $r | (b^e - 1)$ but $r \nmid (b^i - 1)$ for $1 \leq i < e$. Zsigmondy [107] proved that $b^e - 1$ has a primitive prime divisor unless $(b, e) = (2, 6)$, or $e = 2$ and $b + 1$ is a power of 2. Recall that

$$|\mathrm{GL}(d, q)| = q^{\binom{d}{2}} \prod_{i=1}^{d} (q^i - 1).$$

Hence primitive prime divisors of $q^e - 1$ for various $e \leq d$ divide both the orders of $\mathrm{GL}(d, q)$ and of the other classical groups. We say that $g \in \mathrm{GL}(d, q)$ is a *ppd-element* if its order is divisible by some primitive prime divisor of $q^e - 1$ for some $e \in \{1, \ldots, d\}$.

6.1 Classical groups in natural representation

Much of the recent activity on algorithms for linear groups was stimulated by Neumann & Praeger [84], who presented a Monte Carlo algorithm to decide whether or not a subgroup of $GL(d, q)$ contains $SL(d, q)$.

Niemeyer & Praeger [88] answer the equivalent question for an arbitrary classical group. This they do by refining a classification by Guralnick *et al.* [51] of the subgroups of $GL(d, q)$ which contain ppd-elements for $e > d/2$. The resulting Monte Carlo algorithms have complexity $O(\log \log d(\xi + d^\omega (\log q)^2))$, where ξ is the cost of selecting a random element and d^ω is the cost of matrix multiplication. For an excellent account, see [94]. Their implementation is available in MAGMA.

6.2 Black-box groups of Lie type

Babai *et al.* [8] present a black-box algorithm to name a group G of Lie type in known defining characteristic p. The algorithm selects a sample $\mathcal{L}$ of random elements in G, and determines the three largest integers $v_1 > v_2 > v_3$ such that at least one member of $\mathcal{L}$ has order divisible by a primitive prime divisor of $p^v - 1$ for $v = v_1, v_2$, or v_3. Usually $\{v_1, v_2, v_3\}$ determines $|G|$ and so names G. The algorithm of Altseimer & Borovik [1] distinguishes between $P\Omega(2m + 1, q)$ and $PSp(2m, q)$ for odd q. The central result of [8] is the following.

Theorem 6.1 *Given a black-box group G isomorphic to a simple group of Lie type of known characteristic, the standard name of G can be computed using a polynomial-time Monte Carlo algorithm.*

An implementation developed by Malle and O'Brien is distributed with GAP and MAGMA. It includes naming procedures for the other quasisimple groups: if the non-abelian composition factor is alternating or sporadic, then we identify it by considering the orders of random elements.

6.3 Determining the defining characteristic

Theorem 6.1 assumes that the defining characteristic of the input group of Lie type is *known*.

Problem 6.2 Let G be a group of Lie type in *unknown* defining characteristic r. Determine r.

Liebeck & O'Brien [76] present a Monte Carlo polynomial-time black-box algorithm which proceeds recursively through centralisers of involutions of G to find $SL(2, F)$, where F is a field in characteristic r. It is now easy to read off the value of r.

Kantor & Seress [67] prove that the three largest element orders determine the characteristic of Lie-type simple groups of odd characteristic, and use this result to underpin an alternative algorithm.

The former is distributed in MAGMA, the latter in GAP.

7 Constructing an involution centraliser

Involution centralisers played a key role in the classification of finite simple groups. They were also used extensively in early computations with sporadic groups; see for example [77]. Borovik [15], Parker & Wilson [93] and Yalçınkaya [106] study them in the general context of black-box groups.

The centraliser of an involution in a black-box group having an order oracle can be constructed using an algorithm of Bray [18], who proves the following.

Theorem 7.1 *If x is an involution in a group H, and w is an arbitrary element of H, then $[x, w]$ either has odd order $2k + 1$, in which case $w[x, w]^k$ commutes with x, or has even order $2k$, in which case both $[x, w]^k$ and $[x, w^{-1}]^k$ commute with x. If w is uniformly distributed among the elements of the group for which $[x, w]$ has odd order, then $w[x, w]^k$ is uniformly distributed among the elements of the centraliser of x.*

Thus if the odd order case occurs sufficiently often (with probability at least a positive rational function of the input size), then we can construct random elements of the involution centraliser in Monte Carlo polynomial time. In practice, we also use the output of the even-order case to obtain a generating set for the centraliser.

Parker & Wilson [93] prove the following.

Theorem 7.2 *There is an absolute positive constant c such that if H is a finite simple classical group of Lie rank r defined over a field of odd characteristic, and x is an involution in H, then $[x, h]$ has odd order for at least a proportion c/r of the elements h in H.*

For each class of involutions, they find a dihedral group of twice odd order generated by two involutions of this class, and show that a significant proportion of pairs of involutions in this class generate such a dihedral group.

For exceptional groups, they show that the analogous result is true for at least a positive proportion of elements h in H.

For each sporadic group we can calculate explicitly the proportion of $[x, h]$ which have odd order: for every class of involutions, this proportion is at least 17%.

The work of Liebeck & Shalev [75, Theorem] implies that, with arbitrarily high probability, a constant number of random elements generates the centraliser of an involution in a finite simple group.

Holmes *et al.* [56] establish the cost of constructing an involution centraliser:

Theorem 7.3 *Let H be a simple group of Lie type defined over a field of odd characteristic, having a black-box encoding of length n and equipped with an order oracle. Let ξ and ρ denote the cost of selecting a random element and of an order oracle respectively. The centraliser in H of an involution can be computed in time $O(\sqrt{n}(\xi + \rho) \log(1/\epsilon) + \mu n)$ with probability of success at least $1 - \epsilon$, for positive ϵ.*

8   Constructive recognition

Assume that we wish to construct isomorphisms between the central quotient of a given quasisimple group H and a projective representation $G = \langle X \rangle$ of H. Recall from Section 3 that we do this by defining standard generators for H and constructing the corresponding standard generators of G as SLPs in X.

8.1   Black-box classical groups

Kantor & Seress [65] prove the following.

Theorem 8.1 *There is a Las Vegas algorithm which, when given as input a black-box perfect group G where $G/Z(G)$ is isomorphic to a classical simple group C of known characteristic, produces a constructive isomorphism $G/Z(G) \to C$.*

Recall that $g \in G$ is *p-singular* if its order is divisible by p. As Isaacs, Kantor & Spaltenstein [63] and Guralnick & Lübeck [52] show, a group of Lie type in defining characteristic p has a small proportion of p-singular elements.

Theorem 8.2 *If G is a group of Lie type defined over $\mathrm{GF}(q)$, then $\frac{2}{5q} < \rho(G) < \frac{5}{q}$, where $\rho(G)$ denotes the proportion of p-singular elements in G.*

A necessary first step of the Kantor & Seress algorithm [65] is to find an element of order p: hence its running time has a factor of $q = p^f$ and so it is not polynomial in the size of the input.

Brooksbank & Kantor [22] identify that the obstruction to a polynomial-time algorithm for constructive recognition of the classical groups is $\mathrm{PSL}(2, q)$. Babai & Beals [7] formulate the problem explicitly as follows.

Problem 8.3 Find an element of order p in $\mathrm{PSL}(2, p^f)$ as a word in its defining generators in polynomial time.

Since $\rho(\mathrm{PSL}(2, q)) \leq 2/q$, a random search will involve $O(q)$ selections.

A consequence of the work of [71] is that the degree of a faithful projective representation of $\mathrm{SL}(2, q)$ in cross characteristic is polynomial in q rather than in $\log q$. Hence the critical instances of this problem are matrix representations of $\mathrm{SL}(2, q)$ in defining characteristic.

Conder & Leedham-Green [37] and Conder, Leedham-Green & O'Brien [38] present an algorithm which, subject to the existence of a discrete log oracle, constructively recognises $\mathrm{SL}(2, q)$ as a linear group in defining characteristic in time polynomial in the size of the input. The principal result is the following.

Theorem 8.4 *Let G be a subgroup of $\mathrm{GL}(d, F)$ for $d \geq 2$, where F is a finite field of the same characteristic as $\mathrm{GF}(q)$; assume that G is isomorphic modulo scalars to $\mathrm{PSL}(2, q)$. Subject to a fixed number of calls to a discrete log oracle for $\mathrm{GF}(q)$, there is a Las Vegas algorithm that constructs an epimorphism from G to $\mathrm{PSL}(2, q)$ at a cost of $O(d^5 \tau(d))$ field operations, where $\tau(d)$ denotes the number of divisors of d.*

Brooksbank [21, 24] and Brooksbank & Kantor [22, 26] have exploited this work to produce better constructive recognition algorithms for black-box classical groups. We summarise the outcome.

Theorem 8.5 *There is a Las Vegas algorithm which, when given as input a black-box G such that $C \cong G/Z(G)$ is $\mathrm{PSL}(d,q)$, $\mathrm{PSp}(2m,q)$, $\mathrm{PSU}(d,q)$, or $\mathrm{P\Omega}^\epsilon(d,q)$ for $\epsilon \in \{\pm, 0\}$, and a constructive recognition oracle for $\mathrm{SL}(2,q)$, outputs a constructive isomorphism $G/Z(G) \to C$. Its running time is a polynomial in the input length plus the time of polynomially many calls to the $\mathrm{SL}(2,q)$ oracle.*

A partial implementation of the algorithm of [65], developed by Brooksbank, Seress and others, is available in GAP and MAGMA. The algorithm of [38] is available in MAGMA.

8.2 Classical groups in their natural representation

Leedham-Green & O'Brien [73] developed constructive recognition algorithms for the classical groups in their natural representation, over fields of odd defining characteristic. A key component is the use of involution centralisers, whose structure in such groups is well known; see, for example, [50, Table 4.5.1].

Let ξ denote an upper bound to the number of field operations needed to construct a random element of a group, and let $\chi(q)$ denote an upper bound to the number of field operations equivalent to a call to a discrete logarithm oracle for $\mathrm{GF}(q)$.

Leedham-Green & O'Brien [73] prove the following.

Theorem 8.6 *There is a Las Vegas algorithm that takes as input a subset X of bounded cardinality of $\mathrm{GL}(d,q)$, where X generates a classical group G, and returns standard generators for G as SLPs of length $O(\log^3 d)$ in X. The algorithm has complexity $O(d(\xi + d^3 \log d + d^2 \log d \log \log d \log q + \chi(q)))$ if G is neither of type SO^- or Ω^-. Otherwise the complexity is $O(d(\xi + d^3 \log d + d^2 \log d \log \log d \log q + \chi(q)) + \chi(q^2))$.*

We describe the algorithm for $H = \mathrm{SL}(d,q)$. Let V denote the natural H-module with basis $\{e_1, \ldots, e_d\}$. We first define standard generators $\mathcal{S} = \{s, \delta, u, v\}$ for H. The matrices s, δ, u lie in a copy of $\mathrm{SL}(2,q)$; they fix each of $e_3, \ldots, e_d$ and induce the following action on $\langle e_1, e_2 \rangle$:

$$s \longmapsto \begin{pmatrix} 1 & 1 \\ 0 & 1 \end{pmatrix} \quad \delta \longmapsto \begin{pmatrix} \omega & 0 \\ 0 & \omega^{-1} \end{pmatrix} \quad u \longmapsto \begin{pmatrix} 0 & 1 \\ -1 & 0 \end{pmatrix}.$$

The d-cycle v maps $e_1 \longmapsto e_d \longmapsto -e_{d-1} \longmapsto -e_{d-2} \longmapsto \cdots \longmapsto -e_1$.

The input to the algorithm is $G = \langle X \rangle = \mathrm{SL}(d,q)$. Its task is construct $\mathcal{S}$ as SLPs in X.

A *strong involution* in $\mathrm{SL}(d,q)$ has its -1-eigenspace of dimension in the range $(d/3, 2d/3]$. If $t \in G$ is an involution with 1- and -1-eigenspace E_+ and E_- respectively, then $C_G(t)$ is $(\mathrm{GL}(E_+) \times \mathrm{GL}(E_-)) \cap \mathrm{SL}(d,q)$.

The steps of the recursive algorithm are:

1. Find and construct a strong involution t having its -1-eigenspace of dimension e. Rewrite G with respect to the new basis $E_- \cup E_+$.

2. Now construct $C_G(t)$. Construct the direct summands of its derived group to obtain $\mathrm{SL}(e, q)$ and $\mathrm{SL}(f, q)$ as *subgroups* of G where $f = d - e$.

3. Construct standard generators for $\mathrm{SL}(e, q)$ and $\mathrm{SL}(f, q)$.

4. Construct the centraliser C in G of the involution

$$\begin{pmatrix} I_{e-2} & 0 & 0 \\ 0 & -I_4 & 0 \\ 0 & 0 & I_{f-2} \end{pmatrix}.$$

5. Within C solve constructively for the matrix g

$$\begin{pmatrix} I_{e-2} & 0 & 0 & 0 & 0 & 0 \\ 0 & 0 & 0 & 1 & 0 & 0 \\ 0 & 0 & 0 & 0 & 1 & 0 \\ 0 & -1 & 0 & 0 & 0 & 0 \\ 0 & 0 & -1 & 0 & 0 & 0 \\ 0 & 0 & 0 & 0 & 0 & I_{f-2} \end{pmatrix}.$$

6. Now use g to "glue" the e-cycle $v_e \in \mathrm{SL}(e, q)$ and f-cycle $v_f \in \mathrm{SL}(f, q)$ to obtain the d-cycle $v := v_e g v_f$.

The first step of the algorithm is to search for an element of $\mathrm{SL}(d, q)$ of even order that powers to a strong involution. Lübeck, Niemeyer & Praeger [80] prove the following.

Theorem 8.7 *For some absolute positive constant c, the proportion of $g \in \mathrm{SL}(d, q)$ such that a power of g is a strong involution is at least $c/\log d$.*

Observe that Step 3 is recursive, prompting invocations of the same procedure for $\mathrm{SL}(e, q)$ and $\mathrm{SL}(f, q)$. Since g is a strong involution, each group has degree at most $2d/3$; as shown in [73], the recursive calls do not affect the degree of complexity of the overall algorithm.

Recursion to smaller cases requires additional results about involutions which are not strong. We summarise the relevant results of [73].

Theorem 8.8 *For some absolute positive constant c, the proportion of $g \in \mathrm{SL}(d, q)$ such that a power of g is a "suitable" involution is at least c/d.*

The base cases for the recursion are $\mathrm{SL}(d, q)$ where $d \leq 4$. For $\mathrm{SL}(2, q)$ we use the algorithm of [38] to construct standard generators as SLPs in the input generators; for $\mathrm{SL}(3, q)$ we use the algorithm of [79]; for $\mathrm{SL}(4, q)$ we use the *involution-centraliser algorithm* of [56]. An implementation is available with MAGMA.

Black-box versions of these algorithms are being developed by Damien Burns. We are developing similar algorithms for classical groups in characteristic 2. As Theorem 8.2 indicates, the principal challenge is to construct a strong involution.

Brooksbank [23] also developed constructive recognition algorithms for classical groups in their natural representation: their effective cost is $O(d^5 \log^2 q)$, subject to calls to an $\mathrm{SL}(2, q)$ oracle.

8.3   Small degree matrix representations of $SL(d, q)$

Let $SL(d, q) \leq H \leq GL(d, q)$ with $q = p^f$, where V is the natural H-module and V^* is its dual module. Define the Frobenius map $\delta : GL(d, q) \to GL(d, q)$ by $(a_{i,j})^\delta = (a_{i,j}^p)$ for $(a_{i,j}) \in GL(d, q)$.

Let H act on an irreducible $GF(q)$-module W of dimension at most d^2. Consider $V^* \otimes V$ with basis $\{e_i \otimes e_j \mid 1 \leq i, j \leq d\}$ and let

$$w := \sum_{i=1}^{d} e_i \otimes e_i, \qquad U := \left\{ \sum_{i,j} \alpha_{i,j} e_i \otimes e_j \mid \sum_{i=1}^{d} \alpha_{i,i} = 0 \right\}, \qquad W_1 := U \cap \langle w \rangle.$$

The *adjoint module* of V is $W := U/W_1$. If $d \equiv 0 \bmod p$ then W has dimension $d^2 - 2$, otherwise $d^2 - 1$. The remaining irreducible representations of dimension at most d^2 are $V \otimes V^{\delta^e}$ and $V^* \otimes V^{\delta^e}$ where $0 < e < f$. For a discussion, see [74].

Magaard, O'Brien & Seress [82] describe algorithms which, given as input W, construct a d-dimensional projective representation of H. Their principal result is the following.

Theorem 8.9 *Let $d \geq 2$ and let $q = p^f$ be a prime power. Let $SL(d, q) \leq H \leq GL(d, q)$ where H has natural module V. Let $G = \langle X \rangle$ be a representation of H acting irreducibly on a $GF(q)$-vector space W of dimension $n \leq d^2$.*

Given as input G, the value of d, and error probability $\epsilon > 0$, there is a Las Vegas algorithm that, with probability at least $1 - \epsilon$, constructs the projective action of G on V.

The algorithms are specific to each representation type and in all but one case run in polynomial time. A common feature is to search randomly in G for (a power of) a Singer cycle s, and identify a basis for W consisting of eigenvectors for the action of s on $W \otimes GF(q^d)$. Implementations are available in MAGMA.

Ryba [95] presents a polynomial-time Las Vegas algorithm that, given as input an absolutely irreducible representation in odd defining characteristic of a finite Chevalley group, constructs its action on the adjoint module. A combination of his algorithm and that of [82] can be used to construct the natural projective action of $SL(d, q)$.

8.4   Alternating groups

Beals *et al.* [13] prove the following.

Theorem 8.10 *Black-box groups isomorphic to A_n or S_n with known value of n can be recognised constructively in $O(\xi n + \mu |X| n \log n)$ time, where ξ is the time to construct a random element, μ is the time for a group operation, and X is the input generating set for the group.*

Beals *et al.* [14] present a more efficient algorithm for the deleted permutation module viewed as a linear group. Implementations are available in GAP and MAGMA.

An alternative black-box algorithm, developed by Bratus & Pak [17], was further refined and implemented in MAGMA by Derek Holt.

8.5 Exceptional groups

Algorithms to recognise constructively matrix representations of the Suzuki, large and small Ree groups were developed by Bäärnhielm [11, 12]. Implementations are available in MAGMA.

Kantor & Magaard [68] present black-box Las Vegas algorithms to recognise constructively the exceptional simple groups of Lie type and rank at least 2, other than $^2F_4(q)$, defined over a field of known size.

8.6 Sporadic groups

Wilson [103] introduced the concept of *standard generators* for the sporadic groups. He, Bray and others provide black-box algorithms for their construction. For further details, see the ATLAS web site [104].

9 The constructive membership problem

Recall our definition of the *constructive membership problem* for a quasisimple group $G = \langle X \rangle$: if $g \in G$ then write g as an SLP in X.

Assume we have solved the constructive recognition problem for G: namely, we have constructed standard generators $\bar{S}$ for G as SLPs in X. If we can express $g \in G$ as an SLP in $\bar{S}$, then we rewrite the SLP in $\bar{S}$ for g to obtain one in X. Hence we focus on the task of writing $g \in G$ as an SLP in $\bar{S}$.

9.1 Classical groups

Costi [41] developed algorithms to write an element of a classical group $H \leq GL(d, q)$ in its natural representation as an SLP in the standard generators of H. These algorithms are natural (but quite technical) extensions of row and column operations, and have complexity $O(d^3 \log q)$ field operations.

Consider now the case where G is a defining characteristic (projective) irreducible representation of H. Again Costi [41] developed algorithms to solve the membership problem for G; these have complexity $O(d^4 n^3 \log^3 q + d^2 n^4 \log q)$ where n is the degree of G. Implementations of both are available in MAGMA.

Key components are two polynomial-time algorithms for unipotent groups:

1. SUBSPACE-STABILISER algorithm

   Input: a unipotent matrix group S and a subspace U of its underlying vector space.

   Output: a canonical element $\overline{U}$ of the orbit of U under S; and $s \in S$ such that $U^s = \overline{U}$; and generators for the stabiliser of U in S.

2. An algorithm to solve the constructive membership problem in a unipotent matrix group.

We summarise Costi's algorithm when $H = SL(d, q)$. Let $G \leq GL(n, F)$ be a defining characteristic projective irreducible representation of H. Let G act on the underlying vector space V, and let $\phi : H/Z(H) \to G$ be a constructive isomorphism. Assume that we wish to write $g \in G$ as an SLP in $\bar{S}$.

1. Let K be the maximal parabolic subgroup of H that fixes the space spanned by the first element of the standard basis for the underlying space of H. Namely, elements of K have shape

$$\begin{pmatrix} \det^{-1} & 0 & 0 & 0 \\ \star & & & \\ \vdots & & \mathrm{GL}(d-1,q) & \\ \star & & & \end{pmatrix}$$

where each $\star$ is an arbitrary element of $\mathrm{GF}(q)$. Since $K\phi$ is a p-local subgroup in defining characteristic p, it stabilises a proper $K\phi$-submodule U of V.

2. Consider the elementary abelian subgroup E of H generated by elements

$$\begin{pmatrix} 1 & \star & \cdots & \star \\ 0 & & & \\ \vdots & & I_{d-1} & \\ 0 & & & \end{pmatrix}.$$

Use SUBSPACE-STABILISER to construct $x \in E\phi$ as an SLP that maps U^g to U. Hence $U^{gx} = U$ and so the preimage of gx is in K. Thus we have "killed" the first row of the preimage of gx.

3. Dualise to kill first column, obtaining $g_1 := \begin{pmatrix} \alpha & 0 \\ 0 & A \end{pmatrix}$.

4. Observe that $t\phi := g_1^{-1} \cdot T_{1,j}^{\phi} \cdot g_1 \in E\phi$ where $T_{1,j}$ is a transvection with non-zero entry in $(1, j)$ position. Use the constructive unipotent membership test for $t\phi$ in $E\phi$ to obtain its preimage $t \in E$.

5. Read off from t (a scalar multiple of) the j-th row of the preimage in $\mathrm{SL}(d, q)$ of g_1.

6. We have now reduced the constructive membership problem to the *natural representation* in rank $d - 1$; use the corresponding natural representation algorithm to solve this simpler problem.

The two "unipotent" components of this algorithm depend critically on the assumption that G is a matrix representation of H in defining characteristic.

In ongoing work, Murray, Praeger and Schneider are developing black-box algorithms to solve the problem for classical groups on the standard generators defined in [73]. The basic structure of their algorithms is similar to Costi's, but the problems addressed using the unipotent components must now be solved in a black-box group.

The black-box algorithms of [22, 24, 65] solve the same task, again using a similar approach, on different and significantly larger generating sets. An implementation of [65] is available in MAGMA for $\mathrm{SL}(d, q)$, as are implementations by Brooksbank of some small rank cases from [22, 24].

9.2   Other algorithms

The *centraliser-of-involution* algorithm [56] reduces the problem of testing whether an arbitrary element g of a black-box group G lies in a fixed subgroup H to instances of the same problem for $C_H(t)$ for (at most) three involutions $t \in H$. The reduction occurs in polynomial time. The algorithm is constructive: if $g \in H$ then it returns an SLP for g in the generators of H. Our implementation in MAGMA uses COMPOSITIONTREE, described in Section 11, to solve the problem for each centraliser.

The Schreier-Sims algorithm, and its variations, solves the constructive membership problem for a permutation or matrix group G. First introduced by Sims [99], it constructs a *base* for G which determines a stabiliser chain in G. For a basic outline, see [91]; for an analysis, see [97, p. 64].

9.3 Sporadic groups

For each sporadic group, O'Brien and Wilson developed a black-box algorithm to construct a chain of its subgroups; as described in [91], they exploit the reducibility of members of this chain to obtain a "good" base for the Schreier-Sims algorithm.

With this assistance, either the Schreier-Sims algorithm or the algorithm of [56] solves the constructive membership problem for all ATLAS representations [104] of most sporadic groups; the exceptions are the Baby Monster and the Monster where strategies developed by Wilson and others are employed [105]. Implementations are available in MAGMA.

10 Short presentations

Standard generators may be used to define a surjection from a supplied group $G = \langle X \rangle$ to a simple group H. Is this surjection an isomorphism? If not, what is its kernel? If we have a presentation $\mathcal{P}$ for H on standard generators, then we can evaluate relations of $\mathcal{P}$ in standard generators of G and so obtain normal generators for the kernel of the map from G to H. This motivates our interest in presentations for groups of Lie type on particular generating sets.

Babai & Szemerédi [6] define the *length* of a presentation to be the number of symbols required to write down the presentation. Each generator is a single symbol, and a relator is a string of symbols, where exponents are written in binary. The length of a presentation is the number of generators plus the sum of the lengths of the relators. They also formulated the *Short Presentation Conjecture*: there exists a constant c such that every finite simple group G has a presentation of length $O(\log^c |G|)$.

Perhaps the best known presentations for the finite symmetric groups are those of Moore [83]; see also [42, 6.22]. There, the symmetric group S_n of degree n is presented in terms of the transpositions $t_k = (k, k+1)$ for $1 \le k < n$, which generate S_n and satisfy the defining relations $t_k^2 = 1$ for $1 \le k < n$, and $(t_{k-1} t_k)^3 = 1$ for $1 < k < n$, and $(t_j t_k)^2 = 1$ for $1 \le j < k-1 < n-1$. For $n > 1$ the number of these relations is $n(n-1)/2$, and since each relator has bounded length, the presentation length is $O(n^2)$.

If, for example, S_n acts on the deleted permutation module, then the cost of evaluating these relations is $O(n^5)$: this is *more expensive* than constructive recognition of this representation (which can be performed using the algorithm of [14]).

Hence a goal of both theoretical and practical interest is to obtain "short" presentations for the finite simple groups on particular generating sets.

A key step is to obtain short presentations for A_n and S_n. Independently in 2006, Bray et al. [19] and Guralnick et al. [53] proved the following.

Theorem 10.1 A_n and S_n have presentations with a bounded number of generators and relations, and length $O(\log n)$.

This is best possible since it requires $\log n$ bits to represent n; the previous best result was a modification of the Moore presentation having length $O(n \log n)$.

Guralnick et al. [54] prove that A_n has a presentation on 3 generators, 4 relations, and length $O(\log n)$. Bray et al. [19] prove that S_n has a presentation of length $O(n^2)$ on generators $(1, 2)$ and $(1, 2, \ldots, n)$, and at most 123 relations.

Problem 10.2 Is there a shorter presentation for S_n defined on generators $(1, 2)$ and $(1, 2, \ldots, n)$ with a uniformly bounded number of relations?

In a major extension, Guralnick et al. [53] prove the following.

Theorem 10.3 Every non-abelian finite simple group of rank n over $\mathrm{GF}(q)$, with the possible exception of the Ree groups $^2G_2(q)$, has a presentation with a bounded number of generators and relations and total length $O(\log n + \log q)$.

Again this is best possible. It exploits the following results.

- Campbell, Robertson & Williams [27]: $\mathrm{PSL}(2, q)$ has a presentation on (at most) 3 generators and a bounded number of relations.

- Hulpke & Seress [62]: $\mathrm{PSU}(3, q)$ has a presentation of length $O(\log^2 q)$.

In ongoing work, Leedham-Green and O'Brien are constructing explicit short presentations on our standard generators for the classical groups.

11   The composition tree

In ongoing work, Bäärnhielm, Leedham-Green and O'Brien are developing the concept of a *composition tree*, an integrated framework to realise and exploit the geometric approach. An early design was presented in [72]; our latest is implemented in MAGMA. A variation developed by Neunhöffer & Seress [85] is available in GAP.

A composition series for a group G can be viewed as a labelled rooted binary tree. A node corresponds to a section H of G, the root node to G. If a node is not a leaf, then it has a left child corresponding to a proper normal subgroup K of H and a right child I isomorphic to H/K.

The right child is an image under a homomorphism. Usually these arise naturally from the Aschbacher category of the group, but we exploit additional homomorphisms, including the determinant map and some applicable to unipotent and soluble groups. The left child of a node is the kernel of the chosen homomorphism.

The tree is constructed in *right depth-first order*. Namely, we process the node associated with H: if H is not a leaf, construct recursively the subtree rooted at its right child I, then the subtree rooted at its left child K.

A *leaf* of the composition tree is usually a composition factor of G: however, a non-abelian leaf need only be simple modulo scalars, and cyclic factors are not necessarily of prime order.

Assume $\phi : H \rightarrow I$ where $K = \ker \phi$. It is easy to construct I, since it is the image of H under a homomorphism ϕ. Sometimes it is easy theoretically to construct generating sets for $\ker \phi$, for example if H is in Aschbacher category C3. Otherwise, we first construct a normal generating set for K by evaluating in the generators of H the relators in a presentation for I and then take its normal closure using the algorithm of [97, Chapter 2].

To obtain a presentation for a node, we need only presentations for its associated kernel and image; an algorithm for this task is described in [72]. Hence inductively we require presentations only for the leaves. If we know a presentation on standard generators for the leaf, then this is used; otherwise we use the algorithm of [28] to construct such a presentation.

We solve the constructive membership problem *directly* for a leaf using the techniques of Section 9. If we solve the membership problem for the children of a node, then we readily solve the problem for the node, and so recursively obtain a solution for the root node.

Assume that $G = \langle X \rangle \leq \mathrm{GL}(d, q)$ is input to COMPOSITIONTREE. Some of the algorithms used in constructing a composition tree for G are Monte Carlo. To verify the resulting construction, we write down a presentation for the group defined by the tree and show that G satisfies its relations.

The output of COMPOSITIONTREE is:

- A composition series $1 = G_0 \lhd G_1 \lhd G_2 \lhd \cdots \lhd G_m = G$.
- A representation $S_k = \langle X_k \rangle$ of G_k / G_{k-1}.
- Effective maps $\tau_k : G_k \rightarrow S_k$ and $\phi_k : S_k \rightarrow G_k$. The map τ_k is an epimorphism with kernel G_{k-1}; if $g \in S_k$, then $\phi_k(g)$ is an element of G_k satisfying $\tau_k \phi_k(g) = g$.
- A map to write $g \in G$ as an SLP in X.

12   Applications

Over the past decade, Cannon, Holt and their collaborators have pioneered the development of certain practical algorithms to answer structural questions about finite groups. These exploit the characteristic series $\mathcal{C}$ of a finite group G

$$1 \leq O_\infty(G) \leq S^*(G) \leq P(G) \leq G$$

and a refined series for the soluble radical $O_\infty(G)$

$$1 = N_0 \lhd N_1 \lhd \cdots \lhd N_r = O_\infty(G) \lhd G$$

where $N_i \unlhd G$ and N_i / N_{i-1} is elementary abelian. Since $G/O_\infty(G)$ has a trivial Fitting subgroup, we call it a *TF-group*.

The resulting framework is sometimes called the *Trivial Fitting model of computation*. It suggests the following paradigm to solve a problem.

Solve the problem first in G/N_r, and then, successively, solve it in G/N_i, for $i = r - 1, \ldots, 0$.

Since $H := G/O_\infty(G)$ has the structure outlined in Section 4, the problem may have an "easy" solution in H. In particular, we can usually readily reduce the problem for H to a question about almost simple groups. Increasingly, explicit solutions are available for such groups.

Algorithms which use this paradigm include:

- Determine conjugacy classes of elements (see [29]).

- Determine conjugacy classes of subgroups (see [30]).

- Determine the automorphism group (see [31]).

- Determine maximal subgroups (see [32] and [45]).

While these algorithms are effectively black-box, their current Magma implementations use the Schreier-Sims algorithm for associated computations and so are limited in range. Recently, Holt refined the output of CompositionTree for a group to obtain a chief series exhibiting $\mathcal{C}$. In ongoing work Holt, Leedham-Green, O'Brien and Roney-Dougal are exploring how to exploit CompositionTree and this chief series to provide basic infrastructure for such algorithms.

12.1 Exploiting data for classical groups

We mention two examples where available data for classical groups can be exploited.

In 1963, Wall [102] described theoretically the conjugacy classes and centralisers of elements of classical groups. In ongoing work, Haller and Murray exploit this description and provide algorithms which construct these explicitly in the natural representation of groups contained in the conformal group (the group preserving the corresponding form up to scalars). The constructive isomorphisms obtained from constructive recognition allow us to map the class representatives and centralisers from the natural copy to an arbitrary projective representation.

Kleidman & Liebeck [70] describe the maximal subgroups in the Aschbacher categories C1-C8 of classical groups of degree $d \geq 13$. Holt & Roney-Dougal [60, 61] construct generating sets in the natural representation for these subgroups; in ongoing work with Bray they classify all maximals for $d \leq 12$. Again the constructive isomorphism is used to construct their images in an arbitrary projective representation.

12.2 Constructing the automorphism group of a finite group

As one illustration of the paradigm, we sketch the algorithm of Cannon & Holt [31] to compute the automorphism group of an arbitrary finite group G. (Special purpose algorithms exist for soluble groups.)

Recall that $H := G/O_\infty(G)$ permutes the direct factors of its socle S by conjugation. We embed H in the direct product $D := \prod_i \text{Aut}(T_i) \wr \text{Sym}(d_i)$, where T_i occurs d_i times as socle factor of S. Now $\text{Aut}(H)$ is the normaliser of the image of H in D. Hence we effectively reduce the computation for the TF-group H to the finite simple case.

We now lift results through elementary abelian layers, computing $\mathrm{Aut}(G/N_i)$ successively. Suppose $N \leq M \leq G$, where both M and N are characteristic in G, and M/N is elementary abelian of order p^d.

Assume $\mathrm{Aut}(G/M)$ is known. All automorphisms of G fix both M and N. Observe that $\mathrm{Aut}(G/N)$ has normal subgroups $C \leq B$ where B induces the identity on G/M, and C induces the identity on both G/M and M/N. A key observation is that M/N is a $\mathrm{GF}(p)(G/M)$-module.

- Elements of C correspond to derivations from G/M to M/N and are obtained by solving systems of equations over $\mathrm{GF}(p)$.

- Elements of B/C correspond to module automorphisms of M/N. We can usually choose M and N to ensure that both this and the previous calculation are "easy".

- The remaining – and hardest – task is to determine the subgroup S of $\mathrm{Aut}(G/M)$ which lifts to G/N. Observe that $S \leq T$, the subgroup of $\mathrm{Aut}(G/M)$ whose elements preserve the (module) isomorphism type of M/N. Usually T can be computed readily. If G/N is a split extension of M/N by G/M, then all elements of T lift. Otherwise, we must test each element of T to decide whether it lifts to G/N.

References

[1] Christine Altseimer and Alexandre V. Borovik. Probabilistic recognition of orthogonal and symplectic groups. In *Groups and Computation, III (Columbus, OH, 1999)*, volume 8 of *Ohio State Univ. Math. Res. Inst. Publ.*, pages 1–20. De Gruyter, Berlin, 2001.

[2] Sophie Ambrose. Matrix Groups: Theory, Algorithms and Applications. PhD thesis, University of Western Australia, 2006.

[3] M. Aschbacher. On the maximal subgroups of the finite classical groups. *Invent. Math.* **76**, 469–514, 1984.

[4] László Babai. Local expansion of vertex-transitive graphs and random generation in finite groups. *Theory of Computing*, (Los Angeles, 1991), pp. 164–174. Association for Computing Machinery, New York, 1991.

[5] László Babai. Randomization in group algorithms: conceptual questions. In *Groups and Computation, II (New Brunswick, NJ, 1995)*, 1–17, Amer. Math. Soc., Providence, RI, 1–17, 1997.

[6] László Babai and Endre Szemerédi. On the complexity of matrix group problems, I. In *Proc. 25th IEEE Sympos. Foundations Comp. Sci.*, pages 229–240, 1984.

[7] László Babai and Robert Beals. A polynomial-time theory of black box groups. I. In *Groups St. Andrews 1997 in Bath, I*, volume 260 of *London Math. Soc. Lecture Note Ser.*, pages 30–64, 1999. Cambridge Univ. Press.

[8] László Babai, William M. Kantor, Péter P. Pálfy and Ákos Seress. Black-box recognition of finite simple groups of Lie type by statistics of element orders. *J. Group Theory* **5**, 383–401, 2002.

[9] László Babai, Péter P. Pálfy and Jan Saxl. On the number of p-regular elements in finite simple groups. *LMS J. Comput. Math.* **12**, 82–119, 2009.

[10] László Babai, Robert Beals and Ákos Seress. Polynomial-time Theory of Matrix Groups. In *Proceedings of the 41st Annual ACM Symposium on Theory of Computing, STOC 2009, Bethesda, MD, USA*, pages 55–64, 2009.

[11] Henrik Bäärnhielm. Algorithmic problems in twisted groups of Lie type. PhD thesis, Queen Mary, University of London, 2006.

[12] Henrik Bäärnhielm. Recognising the Suzuki groups in their natural representations. J. Algebra **300**, 171–198, 2006.

[13] Robert Beals, Charles R. Leedham-Green, Alice C. Niemeyer, Cheryl E. Praeger and Ákos Seress. A black-box group algorithm for recognizing finite symmetric and alternating groups. I. *Trans. Amer. Math. Soc.* **355**, 2097–2113, 2003.

[14] Robert Beals, Charles R. Leedham-Green, Alice C. Niemeyer, Cheryl E. Praeger and Ákos Seress. Constructive recognition of finite alternating and symmetric groups acting as matrix groups on their natural permutation modules. *J. Algebra* **292**, 4–46, 2005.

[15] Alexandre V. Borovik. Centralisers of involutions in black box groups. In *Computational and statistical group theory (Las Vegas, NV/Hoboken, NJ, 2001)*, 7–20, *Contemp. Math.*, 298, Amer. Math. Soc., Providence, RI, 2002.

[16] Wieb Bosma, John Cannon and Catherine Playoust. The MAGMA algebra system I: The user language. *J. Symbolic Comput.* **24**, 235–265, 1997.

[17] Sergey Bratus and Igor Pak. Fast constructive recognition of a black box group isomorphic to S_n or A_n using Goldbach's conjecture. *J. Symbolic Comput.* **29**, 33–57, 2000.

[18] John N. Bray. An improved method for generating the centralizer of an involution. *Arch. Math. (Basel)* **74**, 241–245, 2000.

[19] John Bray, M.D.E. Conder, C.R. Leedham-Green and E.A. O'Brien. Short presentations for alternating and symmetric groups. To appear *Trans. Amer. Math. Soc.* 2010.

[20] John Brillhart, D.H. Lehmer, J.L. Selfridge, Bryant Tuckerman, and S.S. Wagstaff, Jr. *Factorizations of $b^n \pm 1$*, volume 22 of *Contemporary Mathematics*. American Mathematical Society, Providence, RI, second edition, 1988. www.cerias.purdue.edu/homes/ssw/cun/index.html.

[21] Peter A. Brooksbank. A constructive recognition algorithm for the matrix group $\Omega(d, q)$. In *Groups and Computation, III (Columbus, OH, 1999)*, volume 8 of *Ohio State Univ. Math. Res. Inst. Publ.*, pages 79–93. De Gruyter, Berlin, 2001.

[22] Peter A. Brooksbank and William M. Kantor. On constructive recognition of a black box PSL(d, q). In *Groups and Computation, III (Columbus, OH, 1999)*, volume 8 of *Ohio State Univ. Math. Res. Inst. Publ.*, pages 95–111. De Gruyter, Berlin, 2001.

[23] Peter A. Brooksbank. Constructive recognition of classical groups in their natural representation. *J. Symbolic Comput.* **35**, 195–239, 2003.

[24] Peter A. Brooksbank. Fast constructive recognition of black-box unitary groups. *LMS J. Comput. Math.* **6**, 162–197, 2003.

[25] Peter Brooksbank, Alice C. Niemeyer and Ákos Seress. A reduction algorithm for matrix groups with an extraspecial normal subgroup. *Finite Geometries, Groups and Computation*, (Colorado), pp. 1–16. De Gruyter, Berlin, 2006.

[26] Peter A. Brooksbank and William M. Kantor. Fast constructive recognition of black box orthogonal groups. *J. Algebra* **300**, 256–288, 2006.

[27] C.M. Campbell, E.F. Robertson and P.D. Williams. On Presentations of PSL$(2, p^n)$. *J. Austral. Math. Soc.* **48**, 333–346, 1990.

[28] John J. Cannon. Construction of defining relators for finite groups. *Discrete Math.* **5**, 105–129, 1973.

[29] John Cannon and Bernd Souvignier. On the computation of conjugacy classes in permutation groups. In *Proceedings of International Symposium on Symbolic and Algebraic Computation, Hawaii, 1997*, pages 392–399. Association for Computing Machinery, 1997.

[30] John J. Cannon, Bruce C. Cox and Derek F. Holt. Computing the subgroups of a permutation group. *J. Symbolic Comput.* **31**, 149–161, 2001.

[31] John J. Cannon and Derek F. Holt. Automorphism group computation and isomorphism testing in finite groups. *J. Symbolic Comput.* **35**, 241–267, 2003.

[32] John J. Cannon and Derek F. Holt. Computing maximal subgroups of finite groups. *J. Symbolic Comput.* **37**, 589–609, 2004.

[33] Jon F. Carlson, Max Neunhöffer and Colva M. Roney-Dougal. A polynomial-time reduction algorithm for groups of semilinear or subfield class. *J. Algebra* **322**, 613–617, 2009.

[34] Frank Celler, Charles R. Leedham-Green, Scott H. Murray, Alice C. Niemeyer and E.A. O'Brien. Generating random elements of a finite group. *Comm. Algebra* **23**, 4931–4948, 1995.

[35] Frank Celler and C.R. Leedham-Green. Calculating the order of an invertible matrix. In *Groups and Computation II*, volume 28 of *Amer. Math. Soc. DIMACS Series*, pages 55–60. (DIMACS, 1995), 1997.

[36] A.H. Clifford. Representations induced in an invariant subgroup. *Ann. of Math.* **38**, 533–550, 1937.

[37] Marston Conder and Charles R. Leedham-Green. Fast recognition of classical groups over large fields. In *Groups and Computation, III (Columbus, OH, 1999)*, volume 8 of *Ohio State Univ. Math. Res. Inst. Publ.*, pages 113–121. De Gruyter, Berlin, 2001.

[38] M.D.E. Conder, C.R. Leedham-Green and E.A. O'Brien. Constructive recognition of PSL(2, q). *Trans. Amer. Math. Soc.* **358**, 1203-1221, 2006.

[39] Gene Cooperman. Towards a practical, theoretically sound algorithm for random generation in finite groups. Posted on arXiv:math, May 2002.

[40] Don Coppersmith and Shmuel Winograd. Matrix multiplication via arithmetic progressions. *J. Symbolic Comput.* **9**, 251–280, 1990.

[41] Elliot Costi. Constructive membership testing in classical groups. PhD thesis, Queen Mary, University of London, 2009.

[42] H.S.M. Coxeter and W.O.J. Moser. *Generators and Relations for Discrete Groups*, 4th ed. Springer-Verlag (Berlin), 1980, ix+169 pp.

[43] A.S. Detinko, B. Eick and D.L. Flannery. Computing with matrix groups over infinite fields. These Proceedings.

[44] John D. Dixon. Generating random elements in finite groups. *Electron. J. Combin.* **15** (2008), no. 1, Research Paper 94, 13 pp.

[45] Bettina Eick and Alexander Hulpke. Computing the maximal subgroups of a permutation group. I. In *Groups and Computation, III (Columbus, OH, 1999)*, volume 8 of *Ohio State Univ. Math. Res. Inst. Publ.*, pages 155–168. De Gruyter, Berlin, 2001.

[46] The GAP Group. GAP – Groups, Algorithms, and Programming, Version 4.4.12; 2008. www.gap-system.org.

[47] Mark Giesbrecht. Nearly optimal algorithms for canonical matrix forms. PhD thesis, University of Toronto, 1993.

[48] S.P. Glasby, C.R. Leedham-Green and E.A. O'Brien. Writing projective representations over subfields. *J. Algebra* **295**, 51–61, 2006.

[49] S.P. Glasby and Cheryl E. Praeger. Towards an efficient MEAT-AXE algorithm using f-cyclic matrices: The density of uncyclic matrices in M(n, q). *J. Algebra* **322**, 766–790, 2009.

[50] Daniel Gorenstein, Richard Lyons and Ronald Solomon. *The classification of the finite simple groups. Number 3.* American Mathematical Society, Providence, RI, 1998.

[51] Robert Guralnick, Tim Penttila, Cheryl E. Praeger and Jan Saxl. Linear groups with orders having certain large prime divisors. *Proc. London Math. Soc.* **78**, 167–214, 1999.

[52] R.M. Guralnick and F. Lübeck. On p-singular elements in Chevalley groups in characteristic p. In *Groups and Computation, III (Columbus, OH, 1999)*, volume 8 of *Ohio State Univ. Math. Res. Inst. Publ.*, pages 169–182, De Gruyter, Berlin, 2001.

[53] R.M. Guralnick, W.M. Kantor, M. Kassabov and A. Lubotzky. Presentations of finite simple groups: a quantitative approach. *J. Amer. Math. Soc.* **21**, 711–774, 2008.

[54] R.M. Guralnick, W.M. Kantor, M. Kassabov and A. Lubotzky. Presentations of finite simple groups: a computational approach. To appear *J. European Math. Soc.*, 2010.

[55] G. Hiss and G. Malle. Low-dimensional representations of quasi-simple groups. *LMS J. Comput. Math.*, 4:22–63, 2001. Also: Corrigenda *LMS J. Comput. Math.* **5**, 95–126, 2002.

[56] P.E. Holmes, S.A. Linton, E.A. O'Brien, A.J.E. Ryba and R.A. Wilson. Constructive membership in black-box groups. *J. Group Theory* **11**, 747–763, 2008.

[57] Derek F. Holt and Sarah Rees. Testing modules for irreducibility. *J. Austral. Math. Soc. Ser. A* **57**, 1–16, 1994.

[58] Derek F. Holt, C.R. Leedham-Green, E.A. O'Brien and Sarah Rees. Computing matrix group decompositions with respect to a normal subgroup. *J. Algebra* **184**, 818–838, 1996.

[59] Derek F. Holt, Bettina Eick and Eamonn A. O'Brien. *Handbook of computational group theory*. Chapman and Hall/CRC, London, 2005.

[60] Derek F. Holt and Colva M. Roney-Dougal. Constructing maximal subgroups of classical groups. *LMS J. Comput. Math.* **8**, 46–79, 2005.

[61] Derek F. Holt and Colva M. Roney-Dougal. Constructing maximal subgroups of orthogonal groups. *LMS J. Comput. Math.* **13**, 164–191, 2010.

[62] Alexander Hulpke and Ákos Seress. Short presentations for three-dimensional unitary groups. *J. Algebra* **245**, 719–729, 2001.

[63] I.M. Isaacs, W.M. Kantor and N. Spaltenstein. On the probability that a group element is p-singular. *J. Algebra* **176**, 139–181, 1995.

[64] Gábor Ivanyos and Klaus Lux. Treating the exceptional cases of the MeatAxe. *Experiment. Math.* **9**, 373–381, 2000.

[65] William M. Kantor and Ákos Seress. Black box classical groups. *Mem. Amer. Math. Soc.*, **149** (708):viii+168, 2001.

[66] William M. Kantor and Ákos Seress. Computing with matrix groups. In *Groups, Combinatorics & Geometry (Durham, 2001)*, 123–137, World Sci. Publishing, River Edge, NJ, 2003.

[67] William M. Kantor and Ákos Seress. Large element orders and the characteristic of Lie-type simple groups. *J. Algebra* **322**, 802–832, 2009.

[68] William M. Kantor and Kay Magaard. Black box exceptional groups of Lie type. Preprint 2009.

[69] W. Keller-Gehrig. Fast algorithms for the characteristic polynomial. *Theoret. Comput. Sci.* **36**, 309–317, 1985.

[70] Peter Kleidman and Martin Liebeck. The subgroup structure of the finite classical groups. London Mathematical Society Lecture Note Series, **129**. Cambridge University Press, Cambridge, 1990.

[71] Vicente Landazuri and Gary M. Seitz. On the minimal degrees of projective representations of the finite Chevalley groups. *J. Algebra* **32**, 418–443, 1974.

[72] C.R. Leedham-Green. The computational matrix group project. In *Groups and Computation, III (Columbus, OH, 1999)*, 229–248. De Gruyter, Berlin, 2001.

[73] C.R. Leedham-Green and E.A. O'Brien. Constructive recognition of classical groups in odd characteristic. *J. Algebra* **322**, 833–881, 2009.

[74] Martin W. Liebeck. On the orders of maximal subgroups of the finite classical groups. *Proc. London Math. Soc.* (3) **50**, 426–446, 1985.

[75] Martin W. Liebeck and Aner Shalev. The probability of generating a finite simple group. *Geom. Ded.* **56**, 103–113, 1995.

[76] Martin W. Liebeck and E.A. O'Brien. Finding the characteristic of a group of Lie type. *J. Lond. Math. Soc.* **75**, 741–754, 2007.

[77] S.A. Linton. The art and science of computing in large groups. *Computational Algebra and Number Theory* (Sydney, 1992), pp. 91–109, 1995. Kluwer Academic Publishers, Dordrecht.

[78] F. Lübeck. Small degree representations of finite Chevalley groups in defining characteristic. *LMS J. Comput. Math.* **4**, 135–169, 2001.

[79] F. Lübeck, K. Magaard and E.A. O'Brien. Constructive recognition of $SL_3(q)$. *J. Algebra* **316**, 619–633, 2007.

[80] Frank Lübeck, Alice C. Niemeyer and Cheryl E. Praeger. Finding involutions in finite Lie type groups of odd characteristic. *J. Algebra* **321**, 3397-3417, 2009.

[81] Eugene M. Luks. Computing in solvable matrix groups. In *Proc. 33rd IEEE Sympos. Foundations Comp. Sci.*, 111–120, 1992.

[82] Kay Magaard, E.A. O'Brien and Ákos Seress. Recognition of small dimensional representations of general linear groups. *J. Aust. Math. Soc.* **85**, 229–250, 2008.

[83] E.H. Moore. Concerning the abstract groups of order $k!$ and $\frac{1}{2}k!$. *Proc. London Math. Soc.* **28**, 357–366, 1897.

[84] Peter M. Neumann and Cheryl E. Praeger. A recognition algorithm for special linear groups. *Proc. London Math. Soc.* (3), **65**, 555–603, 1992.

[85] Max Neunhöffer and Ákos Seress. A data structure for a uniform approach to computations with finite groups. In *Proceedings of ISSAC 2006*, ACM, New York, 2006, pp. 254–261.

[86] Max Neunhöffer. *Constructive Recognition of Finite Groups.* Habilitationsschrift, RWTH Aachen, 2009.

[87] Max Neunhöffer and Cheryl E. Praeger. Computing minimal polynomials of matrices. *LMS J. Comput. Math.* **11**, 252-279, 2008.

[88] Alice C. Niemeyer and Cheryl E. Praeger. A recognition algorithm for classical groups over finite fields. *Proc. London Math. Soc.*, 77:117–169, 1998.

[89] Alice C. Niemeyer. Constructive recognition of normalisers of small extra-special matrix groups. *Internat. J. Algebra Comput.*, **15**, 367–394, 2005.

[90] E.A. O'Brien and M.R. Vaughan-Lee. The 2-generator restricted Burnside group of exponent 7. *Internat. J. Algebra Comput.*, **12**, 575–592, 2002.

[91] E.A. O'Brien. Towards effective algorithms for linear groups. *Finite Geometries, Groups and Computation, (Colorado)*, pp. 163-190. De Gruyter, Berlin, 2006.

[92] Igor Pak. The product replacement algorithm is polynomial. In *41st Annual Symposium on Foundations of Computer Science (Redondo Beach, CA, 2000)*, 476–485, IEEE Comput. Soc. Press, Los Alamitos, CA, 2000.

[93] Christopher W. Parker and Robert A. Wilson. Recognising simplicity of black-box groups by constructing involutions and their centralisers. To appear *J. Algebra*, 2010.

[94] Cheryl E. Praeger. Primitive prime divisor elements in finite classical groups. In *Groups St. Andrews 1997 in Bath, II*, 605–623, Cambridge University Press, 1999.

[95] Alexander J.E. Ryba. Identification of matrix generators of a Chevalley group. *J. Algebra* **309**, 484–496, 2007.

[96] Gary M. Seitz and Alexander E. Zalesskii. On the minimal degrees of projective representations of the finite Chevalley groups. II. J. Algebra **158**, 233–243, 1993.

[97] Ákos Seress. *Permutation group algorithms*, volume 152 of *Cambridge Tracts in Mathematics*. Cambridge University Press, Cambridge, 2003.

[98] Igor E. Shparlinski. *Finite fields: theory and computation. The meeting point of number theory, computer science, coding theory and cryptography.* Mathematics and

its Applications, 477. Kluwer Academic Publishers, Dordrecht, 1999.

[99] Charles C. Sims. Computational methods in the study of permutation groups. In *Computational problems in abstract algebra (Proc. Conf., Oxford, 1967)*, pages 169–183, Pergamon Press, Oxford, 1970.

[100] V. Strassen. Gaussian elimination is not optimal. *Numer. Math.* **13**, 354–356, 1969.

[101] Joachim von zur Gathen and Jürgen Gerhard. *Modern Computer Algebra*, Cambridge University Press, 2002.

[102] G.E. Wall. On the conjugacy classes in the unitary, symplectic and orthogonal groups. *J. Austral. Math. Soc.* **3**, 1–62, 1963.

[103] Robert A. Wilson. Standard generators for sporadic simple groups. *J. Algebra* **184**, 505–515, 1996.

[104] R.A. Wilson et al. ATLAS of Finite Group Representations. `brauer.maths.qmul.ac.uk/Atlas`.

[105] R.A. Wilson. Computing in the Monster. In *Groups, Combinatorics & Geometry (Durham, 2001)*, 327–335, World Sci. Publishing, River Edge, NJ, 2003.

[106] Şükrü Yalçınkaya. Black box groups. *Turkish J. Math.* **31**, 171–210, 2007.

[107] K. Zsigmondy. Zur Theorie der Potenzreste. *Monatsh. für Math. u. Phys.* **3**, 265–284, 1892.

RESIDUAL PROPERTIES OF 1-RELATOR GROUPS

MARK SAPIR

Department of Mathematics, Vanderbilt University, USA
Email: m.sapir@vanderbilt.edu

Abstract

This text is based on my lectures given at the "Groups St Andrews 2009" conference in Bath (August, 2009). I am going to survey the proof of the fact that almost all 1-related groups with at least 3 generators are residually finite (and satisfy some other nice properties) [BS1, BS2, SS], plus I will sketch proofs of some related facts. There are several unsolved problems at the end of the paper.

1 Residually finite groups

Definition 1.1 A group G is called *residually finite* if for every $g \in G$, $g \neq 1$, there exists a homomorphism ϕ from G onto a finite group H such that $\phi(g) \neq 1$. If H can be always chosen a p-group for some fixed prime p, then G is called *residually (finite p-group)*.

Example 1.2 Free groups F_k, cyclic extensions of free groups, finitely generated nilpotent groups, etc. are residually finite. The additive group $\mathbb{Q}$, infinite simple groups, free Burnside groups of sufficiently large exponents are not residually finite. The latter result follows from the combination of Novikov-Adyan's solution of the Bounded Burnside Problem [Ad] and Zelmanov's solution of the Restricted Burnside Problem [Zel].

Theorem 1.3 *Groups acting faithfully on rooted locally finite trees are residually finite.*

A rooted binary tree T.

Proof Indeed, every automorphism f of the tree T must fix the root, and so fixes the levels of the tree. If $f \neq 1$ on level number n, we consider the homomorphism

This work was supported in part by the NSF grant DMS-0700811.

from $\mathrm{Aut}(T)$ to the (finite) group of automorphisms of the finite tree consisting of the first n levels of T: the homomorphisms restricting automorphisms of T to vertices of levels at most n. The automorphism f survives this homomorphism. Thus $\mathrm{Aut}(T)$ and all its subgroups are residually finite. $\square$

Conversely (Kaluzhnin, see [GNS]) every finitely generated residually finite group acts faithfully on a locally finite rooted tree.

1.1 Linear groups

Theorem 1.4 (A. Malcev, 1940, [Mal]) *Every finitely generated linear group is residually finite. Moreover, in characteristic 0, it is a virtually residually (finite p-group) for all but finitely many primes p.*

Note that a linear group itself may not be residually (finite p-group) for any p. Example: $\mathrm{SL}_3(\mathbb{Z})$ by the Margulis' normal subgroup theorem [Mar].

1.2 Problems

Residual finiteness of a finitely presented group $G = \langle\, X \mid R \,\rangle$ is in fact a property of finite groups: we need to find out if there are "enough" finite groups with $|X|$ generators satisfying the relations from R.

We shall consider two outstanding problems in group theory.

Problem 1.5 (Gromov, [Gr1]) Is every hyperbolic group residually finite?

This problem is hard because every non-elementary hyperbolic group has (by Gromov [Gr1] and Olshanskii [Ol91, Ol93, Ol95]) very many homomorphic images satisfying very strong finiteness properties (some of them are torsion and even of bounded torsion). This is in sharp contrast with CAT(0)-groups and even groups acting properly by isometries on finite dimensional CAT(0) cubical complexes, which by a result of Wise [Wise1] may be not residually finite and by a result of Burger and Mozes [BM] can be infinite and simple. By a result of Olshanskii [Ol00], Problem 1.5 has a positive solution if and only if any two non-elementary hyperbolic groups have infinitely many common finite simple non-abelian factors.

Problem 1.6 When is a one-relator group $\langle\, X \mid R = 1 \,\rangle$ residually finite?

This problem is related to Problem 1.5 because it is easy to see that "most" 1-relator groups satisfy the small cancellation condition $C'(\lambda)$ (with arbitrary small λ) and so are hyperbolic. Note, though, that residual finiteness of small cancellation groups is as far from being proved (or disproved) as residual finiteness of all hyperbolic groups. Probably the strongest non-probabilistic result concerning Problem 1.6 was obtained by D. Wise [Wise2]: every 1-relator group with a positive relator satisfying a small cancellation condition $C'(\lambda)$ with $\lambda \leq 1/6$ is residually finite.

Example 1.7 $BS(2,3) = \langle a,t \mid ta^2t^{-1} = a^3 \rangle$ is not residually finite ($a \mapsto a^2, t \mapsto t$ is a non-injective surjective endomophism, showing that the group is not even Hopfian).

Example 1.8 $BS(1,2) = \langle a,t \mid tat^{-1} = a^2 \rangle$ is metabelian, and representable by matrices over a commutative ring (as are all finitely generated metabelian groups), so it is residually finite by Malcev's result [Mal].

1.3 The result

The main goal of this article is to explain the proof that the probabilistic version of Problem 1.6 has a positive solution. Namely we are going to discuss the following.

Theorem 1.9 (Borisov, Sapir, Špakulová [BS1], [BS2], [SS]) *Almost surely as $n \to \infty$, every 1-relator group with 3 or more generators and relator of length n, is*

- *residually finite,*
- *a virtually residually (finite p-group) for all but finitely many primes p,*
- *coherent (that is, all finitely generated subgroups are finitely presented).*

1.4 Three probabilistic models

The words "almost surely" in Theorem 1.9 need clarification. There are three natural probabilistic models to consider.

- **Model 1.** Uniform distribution on words of length $\leq n$.
- **Model 2.** Uniform distribution on cyclically reduced words of length $\leq n$.
- **Model 3.** Uniform distribution on 1-relator groups given by cyclically reduced relators of length $\leq n$ (up to isomorphism)

These models turn out to be equivalent [SS]. We prove that if the statement of the theorem is true for Model 2, then it is true for the other two models. Note that the equivalence Model 3 ≡ Model 2 uses a strong result of Kapovich-Schupp-Shpilrain [KSS].

2 Some properties of 1-relator groups with 3 or more generators

The following property has been mentioned above.

Fact 1. (Gromov [Gr1]) Almost every 1-relator group is hyperbolic.

Fact 2. (Sacerdote and Schupp [SSch]) Every 1-relator group with 3 or more generators is SQ-universal (that is every countable group embeds into a quotient of that group).

Fact 3. (B. Baumslag–Pride [BP]) Every group with the number of generators minus the number of relators at least 2 is *large*, that is it has a subgroup of finite index that maps onto F_2.

Fact 3 and a result of P. Neumann [N] imply Fact 2.

Here is a close to complete proof of the result of Baumslag and Pride.

Let $G = \langle x_1, \ldots, x_g \mid u_1, \ldots, u_r \rangle$, $g - r \geq 2$. First note that if $g - r \geq 1$ then we can assume that $t = x_1$ occurs in each relation with total exponent 0, i.e. G maps onto $(\mathbb{Z}, +)$, $t \to 1$.

Rewrite the relators u_k in terms of $s_{j,i} = t^i x_j t^{-i}$. Assume that for each $j \geq 2$, m generators are involved, so $0 \leq i \leq m - 1$.

Consider the subgroup H that is a normal closure of $s = t^n$ and all $x_i, i \geq 2$, the kernel of the map $G \to \mathbb{Z}/n\mathbb{Z}$. It is of index n.

Consider the homomorphic image $\bar{H}$ of H obtained by killing $s_{j,i}$, $2 \leq j \leq g, 0 \leq i \leq m - 1$.

The standard Reidemeister–Schreier shows that $\bar{H}$ has $(g - 1)(n - m) + 1$ generators $s, s_{j,i}$, $2 \leq j \leq g$, $m \leq i \leq n$, and nr relators not involving s.

So $\bar{H}$ is $\langle s \rangle * K$ where K has $(g - 1)(n - m)$ generators and nr relators. For large enough n, then #generators - #relators of K is ≥ 1. So K maps onto $\mathbb{Z}$, and $\bar{H}$ maps onto F_2. Q.E.D.

2.1 Lackenby's result

Theorem 2.1 (Lackenby [La]) *For every large group G and every $g \in G$ there exists n such that $G/\langle\langle g^n \rangle\rangle$ is large.*

The (very easy) proof given by Olshanskii and Osin [OO] is essentially the same as the proof of the result of Baumslag and Pride above.

Application. There exists an infinite finitely generated group that is:

- residually finite
- torsion
- all sections are residually finite
- every finite section is solvable; every nilpotent finite section is Abelian.

3 The Magnus procedure

In order to deal with 1-relator groups, the main tool is the procedure invented by Magnus in the 30s. Here is an example.

Example 3.1 (Magnus procedure) Consider the group

$$\langle a, b \mid aba^{-1}b^{-1}aba^{-1}b^{-1}a^{-1}b^{-1}a = 1 \rangle$$

For simplicity, we chose a relator with total exponent of a equal 0 (as in the proof of Baumslag-Pride above, the general case reduces to this). We can write the relator as

$$(aba^{-1})(b^{-1})(aba^{-1})(b^{-1})(a^{-1}b^{-1}a).$$

Replace $a^i b a^{-i}$ by b_i. The index i is called *the Magnus a-index* of that letter. So we have a new presentation of the same group.

$$\langle\, b_{-1}, b_0, b_1, a \mid b_1 b_0^{-1} b_1 b_0^{-1} b_{-1}^{-1} = 1, ab_{-1}a^{-1} = b_0, ab_0 a^{-1} = b_1 \,\rangle.$$

Note that b_{-1} appears only once in $b_1 b_0^{-1} b_1 b_0^{-1} b_{-1}^{-1}$. So we can replace b_{-1} by $b_1 b_0^{-1} b_1 b_0^{-1}$, remove this generator, and get a new presentation of the same group.

$$\langle\, b_0, b_1, a \mid a^{-1}b_0 a = b_1 b_0^{-1} b_1 b_0^{-1}, a^{-1}b_1 a = b_0 \,\rangle.$$

This is clearly an ascending HNN extension of the free group $\langle b_0, b_1 \rangle$. Thus the initial group is an ascending HNN extension of a free group. We shall see below that this happens quite often.

3.1 Ascending HNN extensions

Definition 3.2 Let G be a group, $\phi \colon G \to G$ be an injective endomorphism. The group

$$\mathrm{HNN}_\phi(G) = \langle\, G, t \mid tat^{-1} = \phi(a), a \in G \,\rangle$$

is called an *ascending HNN extension* of G or the *mapping torus* of ϕ.

Example 3.3 Here is the main motivational example for us: $H_T = \langle\, x, y, t \mid txt^{-1} = xy, tyt^{-1} = yx \,\rangle$, corresponding to the Thue endomorphism of the free group. We shall return to this example several times later.

Ascending HNN extensions have nice geometric interpretations. They are fundamental groups of the mapping tori of the endomorphisms. In the picture below, the handle is attached on one side to the curves x and y, and on the other side to xy and yx. The fundamental group of this object is H_T.

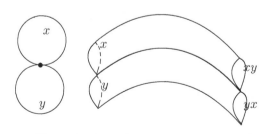

3.2 Facts about ascending HNN extensions

- Every element in an ascending HNN extension of G can be represented in the form $t^{-k}gt^{\ell}$ for some $k, \ell \in \mathbb{Z}$ and $g \in G$. The number $\ell - k$ is an invariant, the representation is unique for a given k.

- (Feighn–Handel [FH]) If G is free then $\mathrm{HNN}_\phi(G)$ is *coherent* i.e. every finitely-generated subgroup is finitely presented.

- (Geoghegan–Mihalik–Sapir–Wise [GMSW]) If G is free then $\mathrm{HNN}_\phi(G)$ is *Hopfian* i.e. every surjective endomorphism is injective.

- (Sapir–Wise [SW]) An ascending HNN extension of a residually finite group can be non-residually finite (an example is Grigorchuk's group and its Lysenok extension).

3.3 Walks in $\mathbb{Z}^2$

Consider a 1-relator group $\langle a, b \mid R \rangle$ with 2 generators. We can consider the 2-dimensional grid $\mathbb{Z}^2$, label horizontal edges by a, vertical edges by b. Then the word R corresponds to a walk in $\mathbb{Z}^2$ starting at $(0,0)$. Suppose that the total sum of exponents of a in R is 0. Then the projections of the vertical steps of the walk onto the horizontal axis give the Magnus a-indices of the corresponding b's. Hence the group is an ascending HNN extension of a free group if one of the two support vertical lines intersects the walk in exactly one step.

A similar fact holds when the total sum of exponents of a is not 0.

3.4 Results of Ken Brown

Theorem 3.4 (Ken Brown, [Br]) *Let $G = \langle x_1, \ldots, x_k \mid R = 1 \rangle$ be a 1-relator group. Let w be the corresponding walk in $\mathbb{Z}^k$, connecting point O with point M.*

- *If $k = 2$ then G is an ascending HNN extension of a free group if and only if one of the two support lines of w that is parallel to $\vec{OM}$ (i.e. one of the two lines that are parallel to $\vec{OM}$, intersect w and such that the whole of w is on one side of that line) intersects w in a single vertex or a single edge.*

- *If $k > 2$ then G is never an ascending HNN extension of a free group.*

3.5 The Dunfield-Thurston result

Let $k = 2$. What is the probability that Brown's condition holds? Consider a random word R in a and b and the corresponding random walk w in $\mathbb{Z}^2$. The projection of w onto the line R perpendicular to $\vec{OM}$ is a random *bridge* (i.e. a walk that starts and ends at the same point). Support lines map to the two extreme points of the bridge.

A bridge is called *good* if it visits one of its extreme points only once, otherwise it is *bad*. Good bridges correspond to words which satisfy Brown's condition, bad bridges correspond to words that do not satisfy that condition. The number of good (bad) walks of length n is denoted by $\#good(n)$ (resp. $\#bad(n)$). The probability for the walk to be good (bad) is denoted by p_{good} (resp. $p_{bad} = 1 - p_{good}$).

Note that we can turn any good bridge into a bad one by inserting one or two subwords of length 8 in the corresponding word. Hence

$$\#good(n) \leq \#bad(n+16).$$

The number of words of length $n+16$ is at most 4^{16} times the number of words of length n. Hence

$$p_{good} \leq 4^{16} p_{bad}$$

Hence $p_{good} < 1$. Similarly $p_{bad} < 1$. In fact the Monte Carlo experiments (pick a large n, and a large number of group words R in two generators; for each word R check if Magnus rewriting of R produces a word with the maximal or minimal index occurring only once) show that p_{good} is about .96.

3.6   The Congruence Extension Property

Theorem 3.5 (Olshanskii [Ol95]) *Let K be a collection of (cyclic) words in $\{a,b\}$ that satisfy $C'(1/12)$. Then the subgroup N of F_2 generated by K satisfies the congruence extension property: that is, for every normal subgroup $L \lhd N$, $\langle\langle L \rangle\rangle_F \cap N = L$. Hence $H = N/L$ embeds into $G = F_2/\langle\langle L \rangle\rangle$.*

Remark 3.6 In fact [Ol95] contains a much stronger result for arbitrary hyperbolic groups.

Proof   Consider L as the set of relations of G. We need to show that the kernel of the natural map $N \to G$ is L, that is if $w(K) = 1$ modulo L (here $w(K)$ is obtained from w by plugging elements of K for its letters), then $w \in L$. Consider a van Kampen diagram Δ for the equality $w(K) = 1$ with minimal possible number of cells. The boundary of every cell is a product of words from K, called *blocks*. If two cells touch by a subpath of their boundaries that includes a "large" portion (say, $\frac{1}{12}$) of a block, then by the small cancellation condition, these boundaries have a common block. Hence we can remove the common part of the boundary consisting of several blocks, and form a new cell with boundary label still in L. That contradicts the assumption that Δ is minimal. This implies that cells in Δ do not have common large parts of their boundaries. Hence Δ is a small cancellation map. By the standard Greendlinger lemma, one of the cells π in Δ has a large part (say, $> \frac{1}{12}$) of a block in common with the boundary of Δ. Then we can cut π out of Δ and obtain a smaller diagram with the same property. The proofs concludes by induction. The proof is illustrated by the following picture. The left part of it illustrates "cancellation" of cells with large common parts of the boundaries. The right part illustrates the induction step.   □

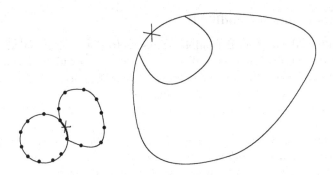

3.7 Embedding into 2-generated groups

Theorem 3.7 (Sapir, Špakulová [SS]) *Consider a group*

$$G = \langle x_1, x_2, \ldots, x_k | R = 1 \rangle,$$

where R is a word in the free group on $\{x_1, x_2, \ldots, x_k\}$, $k \geq 2$. Assume the sum of exponents of x_k in R is zero and that the maximal Magnus x_k-index of x_1 is unique. Then G can be embedded into a 2-generated 1-relator group which is an ascending HNN extension of a finitely generated free group.

The embedding is given by the map $x_i \mapsto w_i$, $i = 1, \ldots, k$ where

$$w_1 = aba^2b \ldots a^n ba^{n+1} ba^{-n-1} ba^{-n} b \ldots a^{-2} ba^{-1} b$$
$$w_i = ab^i a^2 b^i \ldots a^n b^i a^{-n} b^i \ldots a^{-2} b^i a^{-1} b^i, \quad \text{for} \quad 1 < i < k$$
$$w_k = ab^k a^2 b^k \ldots a^n b^k a^{-n} b^k \ldots a^{-2} b^k$$

The injectivity of that map follows from Theorem 3.5.

4   Probability theory and brownian motions

Let C be the space of all continuous functions $f \colon [0, +\infty] \to \mathbb{R}^k$ with $f(0) = 0$. We can define a σ-algebra structure on that space generated by the sets of functions of the form $U(t_1, x_1, t_2, x_2, \ldots, t_n, x_n)$ where $t_i \in [0, +\infty]$, $x_i \in \mathbb{R}^k$. This set consists of all functions $f \in C$ such that $f(t_i) = x_i$. A measure μ on C is called the *Wiener's measure* if for every Borel set A in $\mathbb{R}^k$ and every $t < s \in [0, +\infty]$ the probability that $f(t) - f(s)$ is in A is

$$\frac{1}{\sqrt{2\pi(t-s)}} \int_A e^{-|x|^2/2(t-s)} \, dx.$$

This means that (by definition) Brownian motion is a continuous Markov stationary process with normally distributed increments.

4.1 Donsker's theorem (modified)

A standard tool in dealing with random walks is to consider rescaled limits of them. The rescaled limit of random walks is "usually" a Brownian motion. This is the case in our situation too. Cyclically reduced relators correspond to cyclically reduced walks, i.e. walks without backtracking such that the labels of the first and the last steps are not mutually inverse.

Theorem 4.1 (Sapir, Špakulová [SS]) *Let P_n^{CR} be the uniform distribution on the set of cyclically reduced random walks of length n in $\mathbb{R}^k$. Consider a piecewise linear function $Y_n(t) : [0,1] \to \mathbb{R}^k$, where the line segments connect points $Y_n(t) = S_{nt}/\sqrt{n}$ for $t = 0, 1/n, 2/n, \ldots, n/n = 1$, where (S_n) has a distribution according to P_n^{CR}. Then $Y_n(t)$ converges in distribution to a Brownian motion, as $n \to \infty$.*

The main ingredient in the proof is Rivin's Central Limit Theorem for cyclically reduced walks [Riv].

4.2 Convex hull of Brownian motion and maximal Magnus indices.

Once again, let w be the walk in $\mathbb{Z}^k$ corresponding to the relator R in k generators. Suppose that it connects O and M. Consider the hyperplane P that is orthogonal to $\vec{OM}$, the projection w' of w onto P, and the convex hull Δ of that projection. From Theorem 3.7, it follows that the 1-relator group G is inside an ascending HNN extension of a free group if there exists a vertex of Δ that is visited only once by w'. The idea to prove that this happens with probability tending to 1 is the following.

Step 1. We prove that the number of vertices of Δ is growing (a.s.) with the length of w (here we use that $k \geq 3$; if $k = 2$, then Δ has just two vertices). Indeed, if the number of vertices is bounded with positive probability, then with positive probability the limit of random walks w' (which is a Brownian bridge) would have non-smooth convex hull which is impossible by a theorem about Brownian motions (Cranston-Hsu-March, [CHM]).

Step 2. For every vertex of Δ, and for any "bad" walk w' of length r we construct (in a bijective manner) a "good" walk w' of length $r + 4$. This implies that the number of vertices of "bad" walks is bounded almost surely if the probability of a "bad" walk is > 0.

Here is the walk in $\mathbb{Z}^3$ corresponding to the word

$$cb^{-1}acac^{-1}b^{-1}caca^{-1}b^{-1}aab^{-1}c.$$

and its projection onto the plane perpendicular to $\vec{OM}$. The second (bottom left) picture shows an approximal projection. One can see that the walk is bad: every vertex of the convex hull of the projection is visited twice.

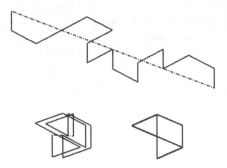

The walk corresponds to 5 different good walks (5 is the number of vertices in the convex hull) by inserting squares after the second visit of the vertex of the convex hull. Here is the walk and its projection corresponding to the word

$$cb^{-1}acac^{-1}b^{-1}caca^{-1}b^{-1}((b^{-1}cbc^{-1}))aab^{-1}c.$$

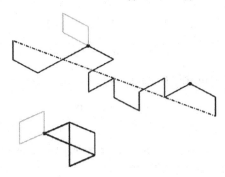

Thus almost surely every 1-relator group with at least 3 generators is inside a 2-generated 1-relator group which is an ascending HNN extension of a free group. This implies, by Feighn and Handel [FH], that almost surely every 1-relator group with at least 3 generators is coherent. Residual properties need new ideas.

5   Algebraic geometry

5.1   Periodic points of a word map

Consider the group

$$G = \langle x_1, \ldots, x_k, t \mid x_1^t = \phi(x_1), \ldots, x_k^t = \phi(x_k) \rangle$$

for some injective endomorphism ϕ of F_k.

For example,

$$H_T = \langle x, y, t \mid txt^{-1} = xy, tyt^{-1} = yx \rangle,$$

so the endomorphism ϕ is given by

$$\phi \colon x \mapsto xy, y \mapsto yx.$$

It is easy to see that every element in G has the form $t^k w(x_1, \ldots, x_k)t^l$. If $k+l \neq 0$, then the element survives the homomorphism $x_i \mapsto 0$, $t \mapsto 1$, $G \to \mathbb{Z}$. Hence we can assume that $k + l = 0$, and the element is conjugate to $w(x_1, \ldots, x_k)$. Thus it is enough to consider elements from F_k only. Consider any $w = w(x, y) \neq 1$. We want to find $\psi: G \to V$ with $\psi(w) \neq 1$, $|V| < \infty$. Suppose that ψ exists.

Let us denote $\psi(x), \psi(y), \psi(t)$ by $\bar{x}, \bar{y}, \bar{t}$. So we want

$$w(\bar{x}, \bar{y}) \neq 1.$$

Note:

$$\bar{t}(\bar{x}, \bar{y})\bar{t}^{-1} = (\bar{x}\bar{y}, \bar{y}\bar{x}) = (\phi(\bar{x}), \phi(\bar{y}))$$

We can continue:

$$\bar{t}^2(\bar{x}, \bar{y})\bar{t}^{-2} = (\bar{x}\bar{y}\bar{y}\bar{x}, \bar{y}\bar{x}\bar{x}\bar{y}) = (\phi^2(\bar{x}), \phi^2(\bar{y})).$$

$$\cdots$$

$$\bar{t}^k(\bar{x}, \bar{y})\bar{t}^{-k} = (\phi^k(\bar{x}), \phi^k(\bar{y})).$$

Since $\bar{t}$ has finite order in V, for some k, we must have

$$(\phi^k(\bar{x}), \phi^k(\bar{y})) = (\bar{x}, \bar{y}).$$

So $(\bar{x}, \bar{y})$ is a periodic point of the map

$$\tilde{\phi}: (a, b) \mapsto (ab, ba).$$

on the "space" $V \times V$.

So if G is residually finite, then for every $w(x, y) \neq 1$, we find a finite group V and a periodic point $(\bar{x}, \bar{y})$ of the map

$$\tilde{\phi}: (a, b) \mapsto (\phi(a), \phi(b))$$

on $V \times V$ such that

$$w(\bar{x}, \bar{y}) \neq 1.$$

So the periodic point should be outside the "subvariety" given by $w = 1$.

The key observation. If $(\bar{x}, \bar{y})$ is periodic with period of length ℓ and $w(\bar{x}, \bar{y}) \neq 1$, then there exists a homomorphism from G into the wreath product $V' = V \wr \mathbb{Z}/\ell\mathbb{Z} = (V \times V \times \cdots \times V) \rtimes \mathbb{Z}/\ell\mathbb{Z}$ which separates w from 1.

Indeed, the map

$$x \mapsto ((\bar{x}, \phi(\bar{x}), \phi^2(\bar{x}), \ldots, \phi^{\ell-1}(\bar{x})), 0),$$

$$y \mapsto ((\bar{y}, \phi(\bar{y}), \phi^2(\bar{y}), \ldots, \phi^{\ell-1}(\bar{y})), 0),$$

$$t \mapsto ((1, 1, \ldots, 1), 1)$$

extends to a homomorphism $\gamma: G \to V'$ and

$$\gamma(w) = ((w(\bar{x}, \bar{y}), \ldots), 0) \neq 1.$$

The idea.   Thus in order to prove that the group $\mathrm{HNN}_\phi(F_k)$ is residually finite, we need, for every word $w \neq 1$ in F_k, to find a finite group G and a periodic point of the map $\tilde{\phi}\colon G^k \to G^k$ outside the "subvariety" given by the equation $w = 1$.

Consider again the group $H_T = \langle a, b, t \mid tat^{-1} = ab, tbt^{-1} = ba \rangle$. Consider two matrices

$$U = \begin{bmatrix} 1 & 2 \\ 0 & 1 \end{bmatrix}, \qquad V = \begin{bmatrix} 1 & 0 \\ 2 & 1 \end{bmatrix}.$$

They generate a free subgroup in $\mathrm{SL}_2(\mathbb{Z})$ (Sanov [San]). Then the matrices

$$A = UV = \begin{bmatrix} 5 & 2 \\ 2 & 1 \end{bmatrix}, \qquad B = VU = \begin{bmatrix} 1 & 2 \\ 2 & 5 \end{bmatrix}$$

also generate a free subgroup. Now let us iterate the map $\psi\colon (x, y) \to (xy, yx)$ starting with (A, B) mod 5 (that is we are considering the finite group $\mathrm{SL}_2(\mathbb{Z}/5\mathbb{Z})$):

$$\left(\begin{bmatrix} 5 & 2 \\ 2 & 1 \end{bmatrix}, \begin{bmatrix} 1 & 2 \\ 2 & 5 \end{bmatrix} \right) \to \left(\begin{bmatrix} 4 & 0 \\ 4 & 4 \end{bmatrix}, \begin{bmatrix} 4 & 4 \\ 0 & 4 \end{bmatrix} \right) \to$$

$$\left(\begin{bmatrix} 1 & 1 \\ 1 & 2 \end{bmatrix}, \begin{bmatrix} 2 & 1 \\ 1 & 1 \end{bmatrix} \right) \to \left(\begin{bmatrix} 3 & 2 \\ 4 & 3 \end{bmatrix}, \begin{bmatrix} 3 & 4 \\ 2 & 3 \end{bmatrix} \right) \to$$

$$\left(\begin{bmatrix} 3 & 3 \\ 3 & 0 \end{bmatrix}, \begin{bmatrix} 0 & 3 \\ 3 & 3 \end{bmatrix} \right) \to \left(\begin{bmatrix} 4 & 3 \\ 0 & 4 \end{bmatrix}, \begin{bmatrix} 4 & 0 \\ 3 & 4 \end{bmatrix} \right) \to$$

$$\left(\begin{bmatrix} 5 & 2 \\ 2 & 1 \end{bmatrix}, \begin{bmatrix} 1 & 2 \\ 2 & 5 \end{bmatrix} \right).$$

Thus the point (A, B) is periodic in $\mathrm{SL}_2(\mathbb{Z}/5\mathbb{Z})$ with period 6.

5.2   Dynamics of polynomial maps over local fields I

Let us replace 5 by 25, 125, etc. It turns out that (A, B) is periodic in $\mathrm{SL}_2(\mathbb{Z}/25\mathbb{Z})$ with period 30, in $\mathrm{SL}_2(\mathbb{Z}/125\mathbb{Z})$ with period 150, etc.

Moreover there exists the following general Hensel-like statement:

Theorem 5.1 *Let $P\colon \mathbb{Z}^n \to \mathbb{Z}^n$ be a polynomial map with integer coefficients. Suppose that a point $\vec{x}$ is periodic with period d modulo some prime p, and the Jacobian $J_P(x)$ is not zero. Then $\vec{x}$ is periodic modulo p^k with period $p^{k-1}d$ for every k.*

This theorem does not apply to our situation straight away, because it is easy to see that the Jacobian of our map is 0, but slightly modifying the map (decreasing the dimension), we obtain the result.

Now take any word $w \neq 1$ in x, y. Since $\langle A, B \rangle$ is free in $\mathrm{SL}_2(\mathbb{Z})$, the matrix $w(A, B)$ is not 1, and there exists $k \geq 1$ such that $w(A, B) \neq 1 \mod 5^k$.

Therefore our group H_T is residually finite.

5.3 Reduction to polynomial maps over finite fields

Let us try to generalize the example of H_T.

Consider an arbitrary ascending HNN extension of a free group

$$G = \langle a_1, \dots, a_k, t \mid ta_it^{-1} = w_i, i = 1, \dots, k \rangle.$$

Consider the ring of matrices $M_2(\mathbb{Z})$.

The map $\psi \colon M_2(\mathbb{Z})^k \to M_2(\mathbb{Z})^k$ is given by

$$\vec{x} \mapsto (w_1(\vec{x}), \dots, w_k(\vec{x})).$$

It can be considered as a polynomial map $A^{4k} \to A^{4k}$ (where A^l is the affine space of dimension l). In fact there is a slight problem here with the possibility that w_i may contain inverses of elements. To resolve it, we replace inverses by the adjoint operation applicable to all matrices. For invertible matrices the inverse and the adjoint differ by a scalar multiple. Thus the group we consider will be $\mathrm{PGL}_2(\cdot)$ instead of $\mathrm{SL}_2(\cdot)$. Here "$\cdot$" is any finite field of, say, characteristic p.

Thus our problem is reduced to the following:

Problem 5.2 Let P be a polynomial map $A^n \to A^n$ with integer coefficients. Show that the set of periodic points of P is Zariski dense.

5.4 Deligne problem

Fixed points are not enough, for example $x \mapsto x + 1$ does not have fixed points.

Deligne suggested considering quasi-fixed points,

$$P(x) = x^{p^n} (= Fr^n(x))$$

where Fr is the Frobenius map (raising all coordinates to the power p).

Note that all quasi-fixed points are periodic because Fr is an automorphism of finite order, and commutes with P since all coefficients of P are integers.

Deligne conjecture proved by Fujiwara and Pink: If P is dominant and quasi-finite then the set of quasi-fixed points is Zariski dense. Unfortunately in our case the map is rarely dominant or quasi-finite. For example, in the case of H_T, the matrices xy, yx have the same trace, so they satisfy a polynomial equation that does not hold for the pair (x, y). Hence the map is not dominant.

5.5 The main results

Theorem 5.3 (Borisov, Sapir [BS1]) *Let* $P^n \colon A^n(\mathbb{F}_q) \to A^n(\mathbb{F}_q)$ *be the n-th iteration of* P. *Let* V *be the Zariski closure of* $P^n(A^n)$. *The set of its geometric points is* $V(\overline{\mathbb{F}_q})$, *where* $\overline{\mathbb{F}_q}$ *is the algebraic closure of* $\mathbb{F}_q$. *Then the following hold.*

1. *All quasi-fixed points of* P *belong to* $V(\overline{\mathbb{F}_q})$.

2. *Quasi-fixed points of* P *are Zariski dense in* V. *In other words, suppose* $W \subset V$ *is a proper Zariski closed subvariety of* V. *Then for some* $Q = q^m$ *there is a point*

$$(a_1, \dots, a_n) \in V(\overline{\mathbb{F}_q}) \setminus W(\overline{\mathbb{F}_q})$$

such that

$$\begin{cases} f_1(a_1, \ldots, a_n) = a_1^Q \\ f_2(a_1, \ldots, a_n) = a_2^Q \\ \vdots \\ f_n(a_1, \ldots, a_n) = a_n^Q \end{cases} \tag{1}$$

This, as before, implies

Theorem 5.4 (Borisov, Sapir [BS1]) *Every ascending HNN extension of a free group is residually finite.*

The proof of Theorem 5.3 from [BS1] is non-trivial but relatively short. Unfortunately the naive approach based on the Bézout theorem fails. The Bézout theorem (that the number of solutions of the system of equations (1) is Q^n for large enough Q, see [F]) only gives the number of solutions of our system of equations, but does not tell us that there are solutions outside a given subvariety (in fact all solutions can, in principle, coincide, as for the equation $x^Q = 0$). Our proof gives a very large Q and uses manipulations with ideals in the ring of polynomials, and its localizations.

Here is a more detailed description of the proof of Theorem 5.4. We denote by I_Q the ideal in $\overline{\mathbb{F}_q}[x_1, \ldots, x_n]$ generated by the polynomials $f_i(x_1, \ldots, x_n) - x_i^Q$, for $i = 1, 2, \ldots, n$.

Step 1. For a big enough Q the ideal I_Q has finite codimension in the ring $\overline{\mathbb{F}_q}[x_1, \ldots, x_n]$.

Step 2. For all $1 \leq i \leq n$ and $j \geq 1$

$$f_i^{(j)}(x_1, \ldots, x_n) - x_i^{Q^j} \in I_Q.$$

Step 3. There exists a number k such that for every quasi-fixed point $(a_1, \ldots, a_n)$ with big enough Q and for every $1 \leq i \leq n$ the polynomial

$$(f_i^{(n)}(x_1, \ldots, x_n) - f_i^{(n)}(a_1, \ldots, a_n))^k$$

is contained in the localization of I_Q at $(a_1, \ldots, a_n)$.

Let us fix some polynomial D with the coefficients in a finite extension of $\mathbb{F}_q$ such that it vanishes on W but not on V.

Step 4. There exists a positive integer K such that for all quasi-fixed points $(a_1, \ldots, a_n) \in W$ with big enough Q we get

$$R = (D(f_1^{(n)}(x_1, \ldots, x_n), \ldots, f_n^{(n)}(x_1, \ldots, x_n)))^K \equiv 0 \mod I_Q^{(a_1, \ldots, a_n)}$$

We know that all points with $P(x) = x^Q$ belong to V. We want to prove that some of them do not belong to W. We suppose that they all do, and we are going to derive a contradiction.

Step 5. First of all, we claim that in this case R lies in the localizations of I_Q with respect to all maximal ideals of the ring of polynomials.

This implies that $R \in I_Q$. Therefore there exist polynomials $u_1, \ldots, u_n$ such that

$$R = \sum_{i=1}^{n} u_i \cdot (f_i - x_i^Q) \tag{2}$$

Step 6. We get a set of u_i's with the following property:
- For every $i < j$ the degree of x_i in every monomial in u_j is smaller than Q.

Step 7. We look how the monomials cancel in the equation (1) and get a contradiction.

5.6   Hrushovsky's result

Hrushovsky [Hr] managed to replace A^n in our statement by an arbitrary variety V. His proof is very non-trivial and uses model theory as well as deep algebraic geometry. Using his result we prove

Theorem 5.5 (Borisov, Sapir, [BS1]) *The ascending HNN extension of any finitely generated linear group is residually finite.*

This theorem is stronger than Theorem 5.4 because free groups are linear. It also applies to, say, right-angled Artin groups (which usually have many injective endomorphisms).

For non-linear residually finite groups the statement of Theorem 5.5 is not true: the Lysenok HNN extension of Grigorchuk's group [L] is not residually finite (see Sapir and Wise [SW]).

5.7   Dynamics of polynomial maps over local fields II

It remains to show that the ascending HNN extensions of free groups are virtually residually (finite p-groups) for almost all primes p.

Example 5.6 Consider again the group $H_T = \langle a, b, t \mid tat^{-1} = ab, tbt^{-1} = ba \rangle$. Let us prove that G is virtually residually (5-group). We know that the pair of matrices

$$A = \begin{bmatrix} 5 & 2 \\ 2 & 1 \end{bmatrix}, B = \begin{bmatrix} 1 & 2 \\ 2 & 5 \end{bmatrix}$$

generates a free subgroup and is a periodic point of the map $\psi \colon (x, y) \to (xy, yx)$ modulo 5^d. The period is $\ell_d = 6 \cdot 5^{d-1}$ by Theorem 5.1.

Then the group H_T is approximated by subgroups of the finite groups

$$\mathrm{SL}_2(\mathbb{Z}/5^d\mathbb{Z})^{\ell_d} \rtimes \mathbb{Z}/\ell_d\mathbb{Z}.$$

Let ν_d be the corresponding homomorphisms.

Let $G_d = \mathrm{SL}_2(\mathbb{Z}/5^d\mathbb{Z})$. There exists a natural homomorphism $\mu_d \colon G_d^{\ell_d} \to G_1^{\ell_d}$. The kernel is a 5-group.

The image $\langle x_d, y_d \rangle$ of $F_2 = \langle x, y \rangle$ under $\nu_d\mu_d$ is inside the direct power $G_1^{\ell_d}$, hence $\nu_d\mu_d(F_2)$ is a 2-generated group in the variety of groups generated by G_1, hence it is of bounded size, say, M.

Then there exists a characteristic subgroup of $\mu_d(F_2)$ of index $\leq M_1$ which is residually 5-group.

The image of H_T under μ_d is an extension of a subgroup that has a 5-subgroup of index $\leq M_1$ by a cyclic group which is an image of $\mathbb{Z}/6 * 5^{d-1}\mathbb{Z}$.

Since $\mathbb{Z}/6*5^{d-1}\mathbb{Z}$ has a 5-subgroup of index 6, $\mu_d\nu_d(H)$ has a 5-subgroup of index at most some constant M_2 (independent of d). Hence H_T has a subgroup of index at most M_2 which is residually (finite 5-group). Thus H_d is virtually residually (finite 5-group).

Consider now the general situation.

To show that $G = \text{HNN}_\phi(F_k)$ is virtually residually (finite p-group) for almost all p we do the following:

- Instead of a tuple of matrices $(A_1, A_2, \ldots, A_k)$ from $\text{SL}_2(\mathbb{Z})$ such that $(A_1 \bmod p, \ldots, A_k \bmod p)$ is periodic for the map ϕ, we find a periodic tuple of matrices in $\text{SL}_2(\mathcal{O}/p\mathcal{O})$ where $\mathcal{O}$ is the ring of integers of some finite extension of $\mathbb{Q}$ unramified at p (this is possible to do using Theorem 5.3).

- Then using a version of Theorem 5.1, and a result of Breuillard and Gelander [BG] we lift the matrices $A_1, \ldots, A_n$ to the p-adic completion of $SL_2(\mathcal{O})$ in such a way that the lifts generate a free subgroup.

This completes the proof of Theorem 1.9.

6 Applications

6.1 An application to pro-finite groups

Theorem 5.3 is equivalent to the following statement about pro-finite completions of free groups.

Theorem 6.1 (Borisov, Sapir [BS1]) *For every injective homomorphism $\phi: F_k \to F_k$ there exists a pro-finite completion F of F_k, and an automorphism ϕ' of F such that $\phi'_{F_k} = \phi$.*

Proof Consider the homomorphisms μ_n from $\text{HNN}_\phi(F_k)$ to finite groups. The images $\mu_n(t)$ (t is the free generator of $\text{HNN}_\phi(F_k)$) induce automorphisms of the images of F_k. The intersections of kernels of μ_n with F_k give a sequence of subgroups of finite index. The corresponding pro-finite completion of F_k is what we need. □

6.2 Some strange linear groups

Let $\text{HNN}_\phi(F_k) = \langle F_k, t \mid F_k^t = \phi(F_k) \rangle$. Consider the normal closure N of F_k in $\text{HNN}_\phi(F_k)$. Since $F_k^t \subseteq F_k$, we have $F_k \subseteq F_k^{t^{-1}}$:

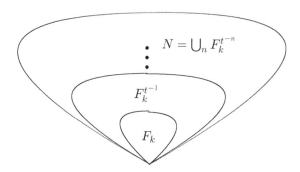

Hence the group N is a locally free group where every finitely generated subgroup is inside a free group of rank k. If $\phi(F_k) \subset [F_k, F_k]$, then $[N, N] = N$. Nevertheless, N is linear and in fact is inside $SL_2(\mathbb{C})$. This immediately follows from the proof of Theorem 5.4.

7 Two possible approaches to Problem 1.5 and some other open problems

It was proved by Minasyan (unpublished) that the group H_T is hyperbolic. Hence by [Gr1, Ol91], it would remain hyperbolic if we impose an additional relation of the form $t^p = 1$, where t is the free generator of H_T, and $p \gg 1$ is a prime.

Problem 7.1 Is the group $\langle\, a, b, t \mid a^t = ab, b^t = ba, t^p = 1 \,\rangle$ residually finite?

If the answer is "yes", then there are arbitrary large finite 2-generated groups $\langle a, b \rangle$ for which the pair (a, b) is periodic with respect to the map $(x, y) \mapsto (xy, yx)$ with the fixed period p (note that in the proof above the periods of points rapidly increase with p). We can add more torsion by imposing $a^p = 1, b^p = 1$ as well (but keeping the factor-group hyperbolic).

Let us also recall a problem by Olshanskii from Kourovka Notebook [KN, Problem 12.64] which can be slightly modified as follows.

Problem 7.2 Let $d \gg p \gg 1$ (p a prime). Are there infinitely many 2-generated finite simple groups G with generators a, b such that for every word $w(x, y)$ of length at most d, the group G satisfies $w(a, b)^p = 1$? If the answer is "no", then there exists a non-residually finite hyperbolic group.

Note that since (assuming the Classification of Finite Simple Groups) there are only 18 types of non-abelian finite simple group, one can ask this question for each type separately. In particular, the specification to A_n gives a nice, but still unsolved, combinatorial problem about permutations.

Problem 7.3 Are there two permutations a and b of $\{1, \ldots, n\}$, which together act transitively on the set, such that $w(a, b)^p = 1$ for every word w of length $\leq d$?

Here p and d are sufficiently large (say, $d = \exp(\exp(p)), p \gg 1$), n runs over an infinite set of numbers (the set depends on p and d). To exclude the case when $\langle a, b \rangle$ is a p-group (in which case we run into the restricted Burnside problem) we can assume that p does not divide n.

Another approach to Problem 1.5 is to consider multiple HNN extensions of free groups. Consider, for example, the group $H_T(u,v) = \langle a, b, t, s \mid a^t = ab, b^t = ba, a^s = u, b^s = v \rangle$ where u, v are words in a, b "sufficiently independent" from ab, ba. This group is again hyperbolic by, say, [Ol95]. Suppose that $H_T(u, v)$ has a homomorphism onto a finite group G, and $\bar{a}, \bar{b}$ are images of a, b, with $\bar{a}, \bar{b} \neq 1$. Then the pair $(\bar{a}, \bar{b})$ is periodic for both maps $\phi : (x, y) \mapsto (xy, yx)$ and $\psi : (x, y) \mapsto (u(x, y), v(x, y))$. Moreover for every word $w(x, y)$, the pair $w(\phi, \psi)(a, b)$ is periodic both for ϕ and for ψ. If ϕ and ψ are sufficiently independent, then it seems unlikely that these two maps have so many common periodic points. MAGMA refuses to give any finite simple homomorphic image with $\bar{a} \neq 1, \bar{b} \neq 1$ for almost any choice of u, v.

Here is an "easier" problem.

Problem 7.4 Is the group $H_T = \langle a, b, t \mid tat^{-1} = ab, tbt^{-1} = ba \rangle$ linear?

Recall that the normal closure N of $\{a, b\}$ is linear (see Section 6.2). So H_T is an extension of a linear group by a cyclic group. Still we conjecture that the answer to Problem 7.4 is "No". One way to prove it would be to consider the minimal indices of subgroups of H_T that are residually (finite p-groups). In linear groups these indices grow polynomially in terms of p. Our proof gives a much faster growing function (double exponential). Non-linear hyperbolic groups are known [M.Kap] but their presentations are much more complicated.

In [DS], we asked whether an ascending HNN extension $\text{HNN}_\phi(F_k)$, where ϕ is a proper endomorphism and $k \geq 2$, can be inside $SL_2(\mathbb{C})$. Calegary and Dunfield constructed such an example [CD]. But in their example, ϕ is reducible. Its components are an automorphism and a Baumslag-Solitar endomorphism of F_1 $(x \mapsto x^\ell)$. Both Baumslag-Solitar groups and many finitely generated free-by-cyclic groups are inside $SL_2(\mathbb{C})$, so the result of [CD] is not very surprising. Thus we still have the following

Problem 7.5 Is there an irreducible proper injective endomorphism of F_k, $k \geq 2$, such that $\text{HNN}_\phi(F_k)$ is embeddable into $SL_2(\mathbb{C})$?

References

[Ad]   S. I. Adian, *The Burnside problem and identities in groups*, Ergebnisse der Mathematik und ihrer Grenzgebiete [Results in Mathematics and Related Areas], **95**, (Springer-Verlag, Berlin-New York, 1979).

[BP]   Benjamin Baumslag and Stephen J. Pride, Groups with two more generators than relators, *J. London Math. Soc. (2)* **17** (1978), 425–426.

[BS1]   Alexander Borisov and Mark Sapir, Polynomial maps over finite fields and residual finiteness of mapping tori of group endomorphisms, *Invent. Math.* **160** (2005), 341–356.

[BS2]   Alexander Borisov and Mark Sapir, Polynomial maps over finite and p-adic fields and residual properties of ascending HNN extensions of free groups, *Int. Math. Res. Not. IMRN* **16** (2009), 3002–3015.

[BG]   Emanuel Breuillard and Tsachik Gelander, A topological Tits alternative, *Ann. of Math. (2)* **166** (2007), 427–474.

[Br]   Kenneth S. Brown, Trees, valuations, and the Bieri-Neumann-Strebel invariant, *Invent. Math.* **90** (1987), 479–504.

[DS]   Cornelia Druţu and Mark Sapir, Non-linear residually finite groups, *J. Algebra* **284** (2005), 174–178.

[BM]   Marc Burger and Shahar Mozes, Lattices in product of trees, *Inst. Hautes Études Sci. Publ. Math.* **92** (2000), 151–194 (2001).

[CD]   Danny Calegari and Nathan Dunfield, An ascending HNN extension of a free group inside SL(2,C), *Proc. Amer. Math. Soc.* **134** (2006), 3131-3136.

[CHM]   M. Cranston, P. Hsu, and P. March, Smoothness of the convex hull of planar Brownian motion, *Ann. Probab.* **17** (1989) 144–150.

[DT]   Nathan M. Dunfield and Dylan P. Thurston, A random tunnel number one 3-manifold does not fiber over the circle, *Geom. Topol.* **10** (2006), 2431–2499.

[FH]   M. Feighn and M. Handel, Mapping tori of free group automorphisms are coherent, *Ann. Math. (2)* **149** (1999), 1061–1077.

[F]   William Fulton, *Intersection theory, Second edition*, Ergebnisse der Mathematik und ihrer Grenzgebiete. 3. Folge. A series of modern surveys in Mathematics [Results in Mathematics and Related Areas. 3rd Series. A Series of Modern Surveys in Mathematics], **2** (Springer-Verlag, Berlin, 1998.)

[GMSW] Ross Geoghegan, Michael L. Mihalik, Mark Sapir and Daniel T. Wise, Ascending HNN extensions of finitely generated free groups are Hopfian, *Bull. London Math. Soc.* **33** (2001), 292–298.

[GNS]   R. I. Grigorchuk, V. V. Nekrashevich and V. I. Sushchanskii, Automata, dynamical systems, and groups (Russian), *Tr. Mat. Inst. Steklova* **231** Din. Sist., Avtom. i Beskon. Gruppy (2000), 134–214.

[Gr1]   M. Gromov, Hyperbolic groups, in *Essays in Group Theory*, (S. M. Gersten, ed) M.S.R.I. Pub. **8** (Springer-Verlag, Berlin-New York, 1987), 75–263.

[Hr]   E. Hrushovski, The Elementary Theory of the Frobenius Automorphisms, arXiv:math.LO/0406514.

[KSS]   I. Kapovich, P. Schupp and V. Shpilrain, Generic properties of Whitehead's algorithm and isomorphism rigidity of random one-relator groups, *Pacific J. Math.* **223** (2006), 113–140.

[M.Kap] Michael Kapovich, Representations of polygons of finite groups, *Geom. Topol.* **9** (2005), 1915–1951.

[KN]   V. D. Mazurov and E. I. Khukhro, eds, *The Kourovka notebook. Unsolved problems in group theory. Sixteenth edition. Including archive of solved problems* (Russian Academy of Sciences Siberian Division, Institute of Mathematics, Novosibirsk, 2006).

[La]   Marc Lackenby, Adding high powered relations to large groups, *Math. Res. Lett.* **14** (2007), 983–993.

[L]   I. G. Lysenok, A set of defining relations for the Grigorchuk group, *Mat. Zametki* **38** (1985), 503–516.

[Mal]   A. I. Malcev, On isomorphic matrix representations of infinite groups, *Rec. Math. [Mat. Sbornik] N.S.* **8** (1940), 405–422.

[Mar]   Gregory A. Margulis, *Discrete subgroups of semisimple Lie groups*, Ergebnisse der Mathematik und ihrer Grenzgebiete **17** (Springer-Verlag, Berlin, 1991).

[N]   P. M. Neumann, The SQ-universality of some finitely presented groups, *J. Austral.*

Math. Soc. **16** (1973), 1–6.

[Ol91]   A. Yu. Olshanskii, Periodic quotient groups of hyperbolic groups, *Mat. Sb.* **182** (1991), 543–567.

[Ol93]   A. Yu. Olshanskii, On residualing homomorphisms and *G*-subgroups of hyperbolic groups, *Internat. J. Algebra Comput.* **3** (1993), 365–409.

[Ol95]   A. Yu. Olshanskii, SQ-universality of hyperbolic groups, *Mat. Sb.* **186** (1995), 119–132.

[Ol00]   A. Yu. Olshanskii, On the Bass-Lubotzky question about quotients of hyperbolic groups, *J. Algebra* **226** (2000), 807–817.

[OO]   A. Yu. Olshanskii and D. V. Osin, Large groups and their periodic quotients, *Proc. Amer. Math. Soc.* **136** (2008), 753–759.

[Pink]   Richard Pink, On the calculation of local terms in the Lefschetz-Verdier trace formula and its application to a conjecture of Deligne, *Ann. of Math. (2)* **135** (1992), 483–525.

[Riv]   Igor Rivin, Growth in free groups (and other stories), arXiv:math/9911076v2 (1999).

[SSch]   George S. Sacerdote and Paul E. Schupp, SQ-universality in HNN groups and one relator groups, *J. London Math. Soc. (2)* **7** (1974), 733–740.

[San]   I. N. Sanov, A property of a representation of a free group, *Doklady Akad. Nauk SSSR (N. S.)* **57** (1947), 657–659.

[SS]   Mark Sapir and Iva Špakulová, Almost all one-relator groups with at least three generators are residually finite, arXiv:math0809.4693 (2008).

[SW]   Mark Sapir and Daniel T. Wise, Ascending HNN extensions of residually finite groups can be non-Hopfian and can have very few finite quotients, *J. Pure Appl. Algebra* **166** (2002), 191–202.

[Wise1] Daniel T. Wise, A non-Hopfian automatic group, *J. Algebra* **180** (1996), 845–847.

[Wise2] Daniel T. Wise, The residual finiteness of positive one-relator groups, *Comment. Math. Helv.* **76** (2001), 314–338.

[Zel]   E. I. Zelmanov, Solution of the restricted Burnside problem for groups of odd exponent, *Izv. Akad. Nauk SSSR Ser. Mat.* **54** (1990), 42–59, 221.

WORDS AND GROUPS

DAN SEGAL

All Souls College, Oxford, OX1 4AL, U.K.
Email: dan.segal@all-souls.ox.ac.uk

The most basic question you can ask about a group is: what happens when you multiply elements together? To make this sound more like mathematics, we can rephrase it as a series of questions about *verbal mappings*. A group *word* is an expression

$$w = w(\mathbf{x}) = \prod_{j=1}^{s} x_{i(j)}^{\varepsilon(j)}$$

where $i(1), \ldots, i(s) \in \{1, \ldots, k\}$ and $\varepsilon(j) = \pm 1$ for each j. It's convenient to think of k as a fixed, large number. The corresponding *verbal mapping* on a group G is $f_w : G^{(k)} \to G$ defined by evaluating w, so

$$\mathbf{g} f_w = w(\mathbf{g}) = \prod_{j=1}^{s} g_{i(j)}^{\varepsilon(j)}$$

where $\mathbf{g} = (g_1, \ldots, g_k)$. Obviously f_w only depends on the equivalence class of w, i.e. the element represented by w in the free group F_k on $x_1, \ldots, x_k$: if $\pi_{\mathbf{g}} : F_k \to G$ sends x_i to g_i $(i = 1, \ldots, k)$ then $\mathbf{g} f_w = w \pi_{\mathbf{g}}$.

An algebraic geometer will want to describe the fibres and the image of a map like $f = f_w$; a number theorist will want to know how big they are. Both kinds of question are easily answered when f happens to be a *homomorphism* (the group theorists' comfort zone). Of course, verbal mappings are not usually homomorphisms, unless G is an abelian group. But we can take this case as a paradigm; the extent to which the 'geometry' and 'arithmetic' of f_w deviate from the paradigm will then give some sort of measure of how far the group G is from being abelian, and/or how far the word w is from being trivial.

These musings don't amount to a theory; they are a common background to several families of results and problems. I will discuss a small selection of these, under the following headings.

1. Fibres over finite groups.

2. Ellipticity in profinite groups.

3. Ellipticity in finite groups.

4. Algebraic groups.

5. Finite simple groups.

My book [S] covers some of this material in more detail, as well as other things. I have tried to avoid repeating too much stuff from the book; only Sections 2 and 3 overlap significantly with [S].

Notation and general remarks

For a subset S of a group G and $m \in \mathbb{N}$,

$$S^{*m} = \{s_1 s_2 \ldots s_m \mid s_i \in S\}.$$

For a word w in k variables,

$$G_w = \left\{w(\mathbf{g})^{\pm 1} \mid \mathbf{g} \in G^{(k)}\right\},$$
$$w(G) = \langle G_w \rangle.$$

$w(G)$ is the *verbal subgroup* of G corresponding to w. While the 'symmetrized' set G_w is often more convenient, we also need a notation for the actual set of w-values, which I will denote by G_{+w}, so

$$G_{+w} = G^{(k)} f_w = \left\{w(\mathbf{g}) \mid \mathbf{g} \in G^{(k)}\right\}.$$

Warning: the papers by Shalev *et al* cited below use a different notation (writing $w(G)$ to mean G_{+w}).

The word w has *width* m in G if $w(G) = G_w^{*m}$, and *positive width* m if $w(G) = G_{+w}^{*m}$.

It is sometimes necessary to consider 'generalized words'. Given a group G, a *generalized word* over G is an expression

$$w = w(\mathbf{x}) = \prod_{l=1}^{t} x_{i_l}^{\varepsilon(l)\alpha(l)} \tag{1}$$

where $i_1, \ldots, i_t \in \{1, \ldots, k\}$, $\varepsilon(l) = \pm 1$ and each $\alpha(l)$ is some automorphism of G. It is *positive* if $\varepsilon(l) = 1$ for each l. It defines a verbal mapping $f_w : G^{(k)} \to G$ where

$$\mathbf{g} f_w = w(\mathbf{g}) = \prod_{l=1}^{t} g_{i_l}^{\varepsilon(l)\alpha(l)}.$$

Given a *finite* group G and a word w, there is a *positive* word w^* such that $f_w = f_{w^*}$ on $G^{(k)}$: supposing G has order m, we obtain w^* from w by replacing each occurrence of x^{-1} in w by x^{m-1}, for each variable x. So we can often restrict attention to positive words without loss of generality.

The minimal size of a generating set for G is denoted $\mathrm{d}(G)$; when G is a topological group, this means *topological* generating set, i.e. a subset X such that the subgroup $\langle X \rangle$ generated by X is dense in G.

"CFSG" means "the classification of the finite simple groups".

1   Fibres over finite groups

If G is a finite group, it makes sense to compare the sizes of the fibres $f_w^{-1}(g)$, $g \in G$. If G is *abelian*, f_w is a homomorphism and as long as $g \in G_{+w}$ we have

$$\left| f_w^{-1}(g) \right| = \left| f_w^{-1}(1) \right| = \frac{|G|^k}{|w(G)|} \geq |G|^{k-1}.$$

This can be expressed in a more invariant way — i.e. independently of k, as long as k is large enough — by setting

$$P(G, w = g) = \frac{\left| f_w^{-1}(g) \right|}{|G|^k};$$

this is the *probability that* $w(\mathbf{a}) = g$ as $\mathbf{a}$ ranges over $G^{(k)}$. We also write

$$P(G, w) = P(G, w = 1)$$

for the probability that $w(\mathbf{a}) = 1$. Thus

$$G \text{ abelian} \implies$$
$$P(G, w = g) = P(G, w) \geq |G|^{-1} \quad (\forall w, \; \forall g \in G_{+w}). \tag{2}$$

Should we expect something similar when G is not abelian? Hardly! Let's see to what extent such numerical conditions on the fibres of verbal mappings characterize commutativity properties of a group.

Suppose that G is *not nilpotent*. Let w_n be the left-normed repeated commutator $[x_1, \ldots, x_n]$. Then $w_n(G) \neq 1$ for each n, so for each n there exists $h_n \in G$ with $1 \neq h_n \in G_{+w_n}$. Since $w_n(\mathbf{g}) = 1$ if $g_i = 1$ for any i, we have

$$\frac{1}{|G|^n} \leq P(G, w_n = h_n) \leq \frac{(|G| - 1)^n}{|G|^n} \longrightarrow 0 \text{ as } n \to \infty.$$

Thus $P(G, w = h)$ takes arbitrarily small positive values as w varies over all words, and the outer inequality in (2) is violated.

Now suppose that G is not *soluble*. Then G has a just-non-soluble quotient Q; that is, Q contains a unique minimal normal subgroup $M = S_1 \times \cdots \times S_r$ where $S_1, \ldots, S_r$ are isomorphic non-abelian simple groups, and $M = \delta_l(Q)$, the lth term of the derived series of Q, for some l.

Lemma 1.1 *For each $n \in \mathbb{N}$ there is a word w_n in n variables such that for* $\mathbf{g} = (g_1, \ldots, g_n) \in Q^{(n)}$,

$$w_n(\mathbf{g}) = 1 \implies \langle g_1, \ldots, g_n \rangle \neq Q. \tag{3}$$

I'll prove this below, but first let's see what can be deduced. Let $P_Q(n)$ denote *the probability that a random n-tuple in Q generates Q*. This quantity has been much studied, notably by Avinoam Mann (see [LS], Chapter 11), but here we only need the trivial estimate

$$P_Q(n) \geq 1 - m2^{-n} \tag{4}$$

where m is the number of maximal subgroups of Q. To see this, put

$$Y = \left\{ \mathbf{g} \in Q^{(n)} \mid \langle g_1, \ldots, g_n \rangle \neq Q \right\}$$

and observe that

$$|Y| = \left| \bigcup_{M <_{\max} Q} M^{(n)} \right| \leq m \left(\frac{|Q|}{2} \right)^n.$$

Recall now that Q is a quotient of our group G; combining (3) and (4) we see that for $w = w_n$,

$$P(G, w) \leq P(Q, w) \leq 1 - P_Q(n) \leq m2^{-n}.$$

Thus if G is not soluble, $P(G, w)$ takes arbitrarily small values, and the right-hand inequality in (2) is violated.

Proof of Lemma 1.1   We may assume that $n \geq \mathrm{d}(G)$. Let F be the free group on $\{x_1, \ldots, x_n\}$, let K be the intersection of the kernels of all epimorphisms from F onto Q, and set $E = K\delta_l(F)$. If $\pi : F \to Q$ is an epimorphism then $E\pi = \delta_l(Q) = M$; it follows that E/K is a subdirect product of copies of $M = S_1 \times \cdots \times S_r$.

Now any subdirect product H in a direct product $\prod_{i \in X} T_i$ of non-abelian simple groups takes the form $\Delta_1 \times \cdots \times \Delta_r$, where Δ_j is a diagonal subgroup in $\prod_{i \in X(j)} T_i$ and $\bigcup_{j=1}^r X(j) = X$ ([C], Exercise 4.3). It follows that H contains an element whose projection to each factor T_i is non-trivial, and hence lies in no proper normal subgroup of H. Applying this to $H = E/K$, we can find $w \in E$ such that $\langle w^E \rangle K = E$.

Now suppose that $Q = \langle g_1, \ldots, g_n \rangle$. Define $\pi : F \to Q$ by $x_i \pi = g_i$ $(i = 1, \ldots, n)$. Then $w(\mathbf{g}) = w\pi$, and so

$$\left\langle w(\mathbf{g})^M \right\rangle = \left\langle w^E \right\rangle \pi = E\pi = M.$$

Hence $w(\mathbf{g}) \neq 1$ and the lemma follows.

None of this is particularly surprising: if G is far from commutative there is no reason to expect (2) to be true. The nice thing is that it works both ways:

Theorem 1.2 ([A], [NS]) *Let G be a finite group, and put $\varepsilon(G) = p^{-|G|}$ where p is the largest prime divisor of $|G|$.*
  (i) *G is soluble if and only if*

$$\inf_w P(G, w) > 0,$$

where w ranges over all words, and this holds if and only if $\inf_w P(G, w) > \varepsilon(G)$.

(ii) *G is nilpotent if and only if*

$$\inf_{w,g} P(G, w = g) > 0,$$

where w ranges over all words and g ranges over G_{+w}, and this holds if and only if $\inf_{w,g} P(G, w) > \varepsilon(G)$.

The 'if' statements have been proved above. To prove the other direction, we begin with the case where G is a p-group. Say G has order $p^h = m$, and fix a 'basis' $\mathbf{b} = (b_1, \ldots, b_h)$ for G, so that

$$1 < \langle b_1 \rangle < \langle b_1, b_2 \rangle < \ldots < \langle b_1, b_2, \ldots, b_h \rangle = G$$

is a central series with cyclic factors of order p. Then each element of G is uniquely of the form

$$g = b_1^{x_1} \cdots b_h^{x_h} = \mathbf{b}^{\mathbf{x}}$$

with $x_1, \ldots, x_h \in \mathbf{P} = \{0, 1, 2, \ldots, p-1\}$.

Here $\mathbf{P}$ is a set of of integers; I would like to 'pretend' that $\mathbf{P}$ is the same as $\mathbb{F}_p$, but it is not obvious how to do this, since the elements b_i need not have order p. The trick is to identify G with a subgroup of $\mathrm{GL}_m(\mathbb{F}_p)$ by taking the regular representation. For each $s \geq 1$ let V_s denote the linear span of the set $(G-1)^s$ in $\mathrm{M}_m(\mathbb{F}_p)$, and set $V_0 = \{1\}$. As G is a unipotent group we have $V_n = 0$ for all $n \geq m$. For a tuple $\mathbf{j} = (j_1, j_2, \ldots)$ we set $|\mathbf{j}| = j_1 + j_2 + \ldots$.

Lemma 1.3 *There exist matrices $B_\mathbf{j} = B_\mathbf{j}(\mathbf{b}) \in V_{|\mathbf{j}|}$ and polynomials $F_\mathbf{j} \in \mathbb{F}_p[X_1, \ldots, X_h]$ for $\mathbf{j} \in \mathbf{P}^{(h)}$ such that*

$$\mathbf{b}^{\mathbf{x}} = \sum_{\mathbf{j} \in \mathbf{P}^{(h)}} F_\mathbf{j}(x_1, \ldots, x_h) B_\mathbf{j} \qquad (\mathbf{x} \in \mathbf{P}^{(h)});$$

each $F_\mathbf{j}$ has total degree at most $|\mathbf{j}|$.

Proof For each j with $1 \leq j \leq p-1$ there is an integer $c(j)$ with $c(j) \cdot j! \equiv 1 \pmod{p}$. Now put $a_i = b_i - 1$ for each i. Then for $0 \leq x \leq p-1$ we have

$$b_i^x = \sum_{j=0}^{x} \binom{x}{j} a_i^j = 1 + \sum_{j=1}^{x} c(j) x(x-1) \ldots (x-j+1) a_i^j$$

$$= \sum_{j=0}^{p-1} F_j(x) a_i^j$$

where $F_0(X) = 1$ and $F_j(X) = c(j) X(X-1) \ldots (X-j+1)$ for $j \geq 1$. The lemma follows on setting

$$F_\mathbf{j}(X_1, \ldots, X_h) = F_{j_1}(X_1) \ldots F_{j_h}(X_h),$$
$$B_\mathbf{j} = a_1^{j_1} a_2^{j_2} \ldots a_h^{j_h}.$$

□

Next, let w be a positive generalized word over G, as in (1).

Lemma 1.4 *There exist matrices $B(w)_{\mathbf{j}} \in V_{|\mathbf{j}|}$ and polynomials $F(w)_{\mathbf{j}} \in \mathbb{F}_p[X_{11}, \dots, X_{kh}]$ for $\mathbf{j} \in \mathbf{P}^{(ht)}$ such that*

$$w(\mathbf{b}^{\mathbf{x}_1}, \dots, \mathbf{b}^{\mathbf{x}_k}) = \sum_{\mathbf{j} \in \mathbf{P}^{(ht)}} F(w)_{\mathbf{j}}(\mathbf{x}_1, \dots, \mathbf{x}_k) B(w)_{\mathbf{j}} \qquad (\mathbf{x}_1, \dots, \mathbf{x}_k \in \mathbf{P}^{(h)});$$

each $F(w)_{\mathbf{j}}$ has total degree at most $|\mathbf{j}|$.

Proof For each l, the tuple $\mathbf{b}^{\alpha(l)} = (b_1^{\alpha(l)}, \dots, b_h^{a(l)})$ is again a basis for G, and for $\mathbf{j} \in \mathbf{P}^{(h)}$ we put $B(l)_{\mathbf{j}} = B_{\mathbf{j}}(\mathbf{b}^{\alpha(l)})$. Then for $\mathbf{x}_1, \dots, \mathbf{x}_k \in \mathbf{P}^{(h)}$ we have

$$w(\mathbf{b}^{\mathbf{x}_1}, \dots, \mathbf{b}^{\mathbf{x}_k}) = \prod_{l=1}^{t} \sum_{\mathbf{j} \in \mathbf{P}^{(h)}} F_{\mathbf{j}}(\mathbf{x}_{i_l}) B(l)_{\mathbf{j}}$$

$$= \sum_{\mathbf{j}_1, \dots, \mathbf{j}_t} F(w)_{\mathbf{j}}(\mathbf{x}_1, \dots, \mathbf{x}_k) B(w)_{\mathbf{j}}$$

where for $\mathbf{j} = (\mathbf{j}_1, \dots, \mathbf{j}_t)$

$$F(w)_{\mathbf{j}}(\mathbf{X}_1, \dots, \mathbf{X}_k) = F_{\mathbf{j}_1}(\mathbf{X}_{i_1}) \dots F_{\mathbf{j}_t}(\mathbf{X}_{i_t}),$$
$$B(w)_{\mathbf{j}} = B(1)_{\mathbf{j}_1} \dots B(t)_{\mathbf{j}_t}.$$

$\square$

We can now deduce

Proposition 1.5 *Let $c \in G$ and suppose that $c = w(\mathbf{h})$ for some $\mathbf{h} \in G^{(k)}$. Then*

$$\left| f_w^{-1}(c) \right| \geq p |G|^k \varepsilon(G).$$

Proof Let's take the elements of G as basis for the regular representation. Then for $g \in G$ we have

$$g = c \iff g_{1c} = 1,$$

where g_{1c} denotes the $(1, c)$-entry of the matrix g.

Say $\mathbf{h} = (\mathbf{b}^{\mathbf{z}_1}, \dots, \mathbf{b}^{\mathbf{z}_k})$. Define a map $\psi : \mathbf{P}^{hk} \to \mathbb{F}_p$ by

$$\psi(\mathbf{x}_1, \dots, \mathbf{x}_k) = 1 - w(\mathbf{b}^{\mathbf{x}_1}, \dots, \mathbf{b}^{\mathbf{x}_k})_{1c}.$$

Lemma 1.4 shows that ψ is equal to a polynomial of total degree at most $m - 1$, since for $|\mathbf{j}| \geq m$ we have

$$B(w)_{\mathbf{j}} \in V_{|\mathbf{j}|} = 0.$$

Also $\psi(\mathbf{z}_1, \dots, \mathbf{z}_k) = 0$. Identifying $\mathbf{P}$ with $\mathbb{F}_p$, we can now apply the *Chevalley–Warning Theorem* [W] (see [LN], Theorem 6.11) to infer that ψ has at least p^{hk-m+1} zeros in $\mathbf{P}^{hk}$. Each one corresponds to a solution of $w(\mathbf{b}^{\mathbf{x}_1}, \dots, \mathbf{b}^{\mathbf{x}_k}) = c$, giving the result since $\varepsilon(G) = p^{-m}$. $\square$

This completes the proof of Theorem 1.2 for p-groups.

If G is any finite group and $G = AB$ where A and B are proper subgroups then

$$\varepsilon(G) \leq \varepsilon(A)\varepsilon(B); \qquad (5)$$

for if p is the largest prime factor of $|G|$ and of $|A|$ and q is the largest prime factor of $|B|$ then $q \leq p$ so

$$p^{|G|} \geq p^{|A|+|B|} \geq p^{|A|}q^{|B|}.$$

Suppose now that $G = P_1 \times \cdots \times P_r$ is nilpotent, where P_i is a p_i-group and $p_1, \ldots, p_r$ are distinct primes. Let w be a positive generalized word over G. As the P_i are characteristic, w may be considered a generalized word over each of them. If $c_i \in P_i$ and $c = c_1 \ldots c_r \in Gf_w$ then $c_i \in P_i f_w$ for each i, and we deduce that

$$|f_w^{-1}(c)| = \prod |f_w^{-1}(c_i)| \geq \prod p_i |P_i|^k \varepsilon(P_i) \geq \prod p_i \cdot |G|^k \varepsilon(G). \qquad (6)$$

In particular, taking w to be an ordinary word (which we may assume to be positive), we see that

$$P(G, w = c) = \frac{|f_w^{-1}(c)|}{|G|^k} > \varepsilon(G),$$

which completes the proof of Theorem 1.2(ii).

Now we prove the 'if' claim in Theorem 1.2(i). Fix a positive word w. Suppose that G is soluble, but not nilpotent, let N be the Fitting subgroup of G and K/N a minimal normal subgroup of G/N. Then K/N is a p-group for some prime p; we choose a Sylow p-subgroup P of K and set $H = N_G(P)$. Then $K = NP$ and $G = KH = NH$ by the Frattini argument; also $H < G$ because K is not nilpotent.

Arguing by induction on the group order, we may suppose that $w(\mathbf{h}) = 1$ for more than $|H|^k \varepsilon(H)$ elements $\mathbf{h} \in H^{(k)}$.

Now fix $\mathbf{h} \in H^{(k)}$ such that $w(\mathbf{h}) = 1$. There is a generalized word $w'_{\mathbf{h}}$ over N such that

$$w(\mathbf{a} \cdot \mathbf{h}) = w'_{\mathbf{h}}(\mathbf{a})w(\mathbf{h})$$

for all $\mathbf{a} \in N^{(k)}$, where

$$\mathbf{a} \cdot \mathbf{h} = (a_1 h_1, \ldots, a_k h_k);$$

here $w'_{\mathbf{h}}$ is obtained from w by twisting with (the restriction to N of) certain inner automorphisms of G, induced by suitable products of $h_1^{-1}, \ldots, h_k^{-1}$. Applying (6) to the group N we find that

$$w'_{\mathbf{h}}(\mathbf{a}) = 1$$

for more than $|N|^k \varepsilon(N)$ elements $\mathbf{a} \in N^{(k)}$, and then for each of these we have $w(\mathbf{a} \cdot \mathbf{h}) = 1$. Thus altogether we have more than

$$|H|^k \varepsilon(H) \cdot |N|^k \varepsilon(N) = |H \cap N|^k |G|^k \varepsilon(H)\varepsilon(N) \geq |H \cap N|^k |G|^k \varepsilon(G)$$

pairs $(\mathbf{a}, \mathbf{h}) \in N^{(k)} \times H^{(k)}$ for which $w(\mathbf{a} \cdot \mathbf{h}) = 1$. As the fibres of the map $(\mathbf{a}, \mathbf{h}) \mapsto \mathbf{a} \cdot \mathbf{h}$ each have size $|H \cap N|^k$ it follows that $w(\mathbf{g}) = 1$ for more than $|G|^k \varepsilon(G)$ elements $\mathbf{g} \in G^{(k)}$, so $P(G, w) > \varepsilon(G)$ as claimed.

Here is a curious consequence:

Corollary 1.6 *Let G be a finite group. Then G is soluble if and only if for every sufficiently large n, every n-generator one-relator group maps onto G.*

Proof Suppose that G is not soluble, let Q be a just-non-soluble quotient of G, as above, and let $n \in \mathbb{N}$. According to Lemma 1.1, there exists a word w in n variables such that

$$w(\mathbf{g}) = 1 \implies \langle g_1, \ldots, g_n \rangle \neq Q.$$

The one-relator group $\langle x_1, \ldots, x_n; w \rangle$ then does *not* map onto Q, and a fortiori it doesn't map onto G.

Now suppose that G is soluble. Recall that $P_G(n)$ denotes the probability that a random n-tuple in G generates G. Suppose that w is a word in n variables. Then the probability that $\mathbf{g} \in G^{(n)}$ satisfies both $w(\mathbf{g}) = 1$ and $\langle g_1, \ldots, g_n \rangle = G$ is

$$\pi_n(w) := P(G, w) + P_G(n) - \kappa,$$

where κ is the probability that at least one of the two conditions holds.

We saw above (4) that $P_G(n) \geq 1 - m2^{-n}$ where m denotes the number of maximal subgroups of G. On the other hand, $P(G, w) > \varepsilon(G)$. So as long as $m2^{-n} \leq \varepsilon(G)$ we have

$$\pi_n(w) > \varepsilon(G) + 1 - m2^{-n} - \kappa \geq 0.$$

Thus $w(\mathbf{g}) = 1$ for at least one generating set $\{g_1, \ldots, g_n\}$ for G, and $\langle x_1, \ldots, x_n; w \rangle$ maps onto G by $x_i \mapsto g_i$ $(i = 1, \ldots, n)$. $\qquad\square$

There is a huge gap between the estimate (2), which holds in abelian groups, and the lower bound $\varepsilon(G)$ for $P(G, w)$ given in Theorem 1.2(ii). It seems unlikely that the latter is close to best-possible; indeed, A. Amit (who first proved the existence of a lower bound in the nilpotent case) conjectures that $P(G, w) \geq |G|^{-1}$ for every w if G is nilpotent. This may be too strong, but there is scope for experimentation here. Nothing is known beyond the results stated above.

2   Ellipticity in profinite groups

It's also interesting to look at the *image* of a verbal mapping. If f_w is a homomorphism (e.g. when G is an abelian group) then G_w is a subgroup, i.e.

$$w(G) = G_w = G_w^{*1}.$$

In general, $w(G)$ is an ascending union

$$w(G) = \bigcup_{n=1}^{\infty} G_w^{*n}, \tag{7}$$

which may or may not be a finite union. We say that w has *width* m in G if $w(G) = G_w^{*m}$; so w having finite width is another way for f_w to resemble a homomorphism. I will write

$$m_G(w)$$

for the least finite m such that w has width m in G (and $m_G(w) = \infty$ if there is no such m).

If $m_G(w)$ is finite, the group G is said to be w-*elliptic*. G is *verbally elliptic* if it is w-*elliptic* for every word w. This holds if G is abelian. In the 1960s P. Hall and his students looked for generalizations of this; some results along these lines are

Theorem 2.1 (P. Stroud) *Every finitely generated abelian-by-nilpotent group is verbally elliptic.*

Theorem 2.2 (D. Segal) *Every virtually soluble minimax group is verbally elliptic.*

Theorem 2.3 (P. Stroud) *The free two-generator $\mathfrak{N}_2$-by-abelian group is* not *verbally elliptic.*

Theorem 2.4 (A. Rhemtulla) *Non-abelian free groups are* not *w-elliptic for any non-trivial, non-universal word w.*

Important earlier contributions were made by V. A. Romankov and K. George. I will not discuss these results here; they are expounded at length in the book [S]. They exemplify the general principle that verbal ellipticity is some sort of weak commutativity condition.

Of course, it is almost trivial that every *finite* group is verbally elliptic; but it might be interesting to seek non-trivial bounds for the width of words in finite groups. A good way of discussing such questions is to use the language of *profinite* groups.

A profinite group G is the inverse limit of some inverse system of finite groups; in fact

$$G = \varprojlim \mathcal{F}(G)$$

where

$$\mathcal{F}(G) = \{G/N \mid N \vartriangleleft_o G\}$$

is the family of all (continuous) finite quotients of G (here, $N \vartriangleleft_o G$ means 'N is an open normal subgroup of G'). As a general principle, one can say that properties of the topological group G reflect *uniform* properties of the family $\mathcal{F}(G)$ of finite groups. In particular:

1. G is (topologically) finitely generated iff $d(G)$ is finite, and

$$d(G) = \sup\{d(Q) \mid Q \in \mathcal{F}(G)\}.$$

Here $d(G)$ denotes the minimal size of a *topological* generating set for G (while $d(Q)$ is the minimal size of a generating set for Q in the ordinary sense — finite groups are always given the discrete topology).

2. Let w be a word. Then G is w-elliptic iff $m_G(w)$ is finite, and

$$m_G(w) = \sup\{m_Q(w) \mid Q \in \mathcal{F}(G)\}.$$

All the 'verbal' concepts are understood *algebraically*: in applying the definition of $m_G(w)$ we forget the topology. The same applies to the definition of $w(G)$: it is the subgroup generated *algebraically* by the w-values.

In the profinite situation, there is a striking alternative interpretation:

Proposition 2.5 (B. Hartley) *Let G be a profinite group and w a word. Then G is w-elliptic if and only if $w(G)$ is closed.*

Proof Suppose $m_G(w) = m$ is finite. Then

$$w(G) = G_w^{*m}$$
$$= \bigcup G_{+w}^{\varepsilon(1)} \ldots G_{+w}^{\varepsilon(m)}$$

where $(\varepsilon(1), \ldots, \varepsilon(m))$ ranges over $(\pm 1)^{(m)}$. For each such tuple $(\varepsilon(1), \ldots, \varepsilon(m))$, the set $G_{+w}^{\varepsilon(1)} \ldots G_{+w}^{\varepsilon(m)}$ is the image of the map

$$f_w^\varepsilon : \left(G^{(k)}\right)^{(m)} \to G$$
$$(\mathbf{g}_1, \ldots, \mathbf{g}_m) \mapsto w(\mathbf{g}_1)^{\varepsilon(1)} \ldots w(\mathbf{g}_m)^{\varepsilon(m)}.$$

This map is continuous, so its image is compact and hence closed in G. Therefore $w(G)$ is closed.

The converse is a little more subtle: it is an application of the Baire category theorem to the expression (7). $\square$

This result serves as a bridge between algebra and topology. Let's say that a word w is *elliptic* in a *family* of groups $\mathcal{C}$ if

$$\exists m \in \mathbb{N} \text{ such that } m_G(w) \le m \ \ \forall G \in \mathcal{C}.$$

Recall that a profinite group is a *pro-$\mathcal{C}$ group* if $\mathcal{F}(G) \subseteq \mathcal{C}$. Now combining **1.**, **2.** and Proposition 2.5 we get

Proposition 2.6 *A word w is elliptic in a family of finite groups $\mathcal{C}$ if and only if $w(G)$ is closed in G for every pro-$\mathcal{C}$ group G.*

Why am I bothering you with this general nonsense? Let's have a little history. The following is well known:

Lemma 2.7 *If $G = \langle g_1, \ldots, g_d \rangle$ is nilpotent then the derived group G' satisfies*

$$G' = [G, g_1] \ldots [G, g_d].$$

This shows that the word $[x, y]$ is elliptic in the class $\mathfrak{N} \cap \mathfrak{G}_d$ of d-generator nilpotent groups. It also implies that the word

$$v(x, y, z) = [x, y]z^p$$

(here p is any prime) is elliptic in $\mathfrak{N} \cap \mathfrak{G}_d$ since

$$v(G) = G'G^p = G' \cdot \{g^p \mid g \in G\}.$$

So Proposition 2.6 gives

Corollary 2.8 *If G is a finitely generated pro-p group then the subgroups G' and $G'G^p$ are closed.*

Now the word v has another special property: it is *locally finite*; indeed any d-generator group H with $v(H) = 1$ satisfies $|H| \leq p^d$. In general, let's call a word w d-*locally finite* if

$$\beta(w) := |F_d/w(F_d)| < \infty.$$

Lemma 2.9 *Let G be a d-generator profinite group and w a d-locally finite word. If $w(G)$ is closed then $w(G)$ is open.*

Proof If $w(G) \leq N \lhd_o G$ then G/N is a finite d-generator group killed by w, so $|G/N| \leq \beta(w)$. Hence there is a smallest such N, call it M, and as $w(G)$ is closed we have

$$w(G) = \bigcap_{w(G) \leq N \lhd_o G} N = M.$$

$\square$

Remark (for later use): it's enough to assume that there is a finite upper bound $\overline{\beta}(d, w)$ for the size of each *finite* quotient of $F_d/w(F_d)$.

Anyway, we can now deduce:

- *if G is a finitely generated pro-p group then $G'G^p$ is open in G.*

This implies that $G'G^p$ has finite index in G, hence $G'G^p$ is again a finitely generated pro-p group.

Now let K be any normal subgroup of finite index in G. It is an exercise to show that G/K is necessarily a p-group (note: this is *not* just the definition of 'pro-p group', since we are not assuming that K is *open*). Writing

$$\Phi(G) = G'G^p,$$

we therefore have

$$\Phi^n(G) \leq K$$

for some n. As $\Phi^n(G)$ is open in G, K is also open. As every subgroup of finite index contains a normal subgroup of finite index, we have proved

Theorem 2.10 (Serre) *In a finitely generated pro-p group every subgroup of finite index is open.*

When Serre proved this, around 1975, he asked: is the same true for finitely generated profinite groups in general? It took nearly 30 years to answer this question:

Theorem 2.11 ([NS2]) *In a finitely generated profinite group every subgroup of finite index is open.*

Why is this interesting? Since every open subgroup has finite index, what it says is that open subgroups are the same thing as finite-index subgroups. But the open subgroups are a base for the neighbourhoods of 1: this is one definition of 'profinite group'. As a consequence, we have

Corollary 2.12 *In a finitely generated profinite group the topology is uniquely determined by the group structure.*

More generally, one can say:

- Every group homomorphism from a finitely generated profinite group to any profinite group is continuous.

Let's get back to words. First, we need a substitute for that little argument with p-groups:

Lemma 2.13 *Let $d \in \mathbb{N}$ and let H be a finite group. Then there exists a d-locally finite word w such that $w(H) = 1$.*

Proof Exercise! □

Now the algebraic theorem at the heart of our topological result is:

Theorem 2.14 ([NS2]) *Every d-locally finite word is elliptic in the class of d-generator finite groups.*

Proof of Theorem 2.11 Let G be a d-generator profinite group and K a normal subgroup of finite index in G. There exists a d-locally finite word w such that $w(G/K) = 1$. Theorem 2.14 implies that $w(G)$ is closed, and then Lemma 2.9 shows that $w(G)$ is open. As $K \geq w(G)$ it follows that K is open.

In the same work we established a companion result:

Theorem 2.15 ([NS2]) *Let G be a profinite group and H a closed normal subgroup of G. Then the subgroup*

$$[G, H] = \langle [g, h] \mid g \in G, \ h \in H \rangle$$

is closed in G.

Starting with $H = G$ and arguing by induction we infer that each term $\gamma_n(G)$ of the lower central series is closed, and hence get

Corollary 2.16 *For each d and n the word $\gamma_n = [x_1, \ldots, x_n]$ is elliptic in the class of d-generator finite groups.*

These results suggest

Definition The word w is *uniformly elliptic* in a class $\mathcal{C}$ if for each $d \in \mathbb{N}$ there exists $f_w(d) \in \mathbb{N}$ such that

$$m_G(w) \le f_w(\mathrm{d}(G))$$

for every finite group $G \in \mathcal{C}$.

As we have seen,

- *w is uniformly elliptic in $\mathcal{C}$ iff $w(G)$ is closed in G for every finitely generated pro-$\mathcal{C}$ group G.*

Which words are uniformly elliptic? This seems to be a tricky question; V. A. Romankov showed that the second derived group G'' is *not closed* in the free 3-generator pro-p group; thus the word $\delta_2 = [[x_1, x_2], [x_3, x_4]]$ is *not* uniformly elliptic in p-groups.

In the context of pro-p groups, the complete answer was found by Andrei Jaikin. Here we identify a word with an element of the free group

$$F = F(x_1, x_2, \ldots)$$

on countably many generators:

Theorem 2.17 ([JZ]) *Let p be a prime and let $1 \ne w \in F$. Then $w(G)$ is closed in G for every finitely generated pro-p group G if and only if*

$$w \notin F''(F')^p.$$

Moreover, this holds if and only if $G/\overline{w(G)}$ is virtually nilpotent for every finitely generated pro-p group G.

In honour of Andrei, let's make the

Definition $w \in F$ is a *J-word* if $w \notin F''(F')^p$ for every prime p, or $w = 1$.

This is therefore a *necessary* condition for w to be uniformly elliptic in all finite groups. Whether it is *sufficient* I don't know; to explain some partial answers to this question, I need another

Definition $\mathcal{Y}$ denotes the *smallest locally-closed and residually-closed class of groups that contains all virtually soluble groups.*

Now combining Theorems 2.15 and 2.14, I can prove

Theorem 2.18 ([S], Chapter 4)

(i) *Let G be a finitely generated profinite group and w a J-word. If $G/w(G) \in \mathcal{Y}$ then $w(G)$ is closed in G.*

(ii) *Let E be the free profinite group on $d \geq 2$ generators. If $w(E)$ is closed in E then w is a J-word, and $G/w(G)$ is virtually nilpotent for every d-generator profinite group G.*

(iii) *Let w be a word such that $F/w(F) \in \mathcal{Y}$. Then w is uniformly elliptic in all finite groups if and only if w is a J-word.*

Note that $F/w(F) \in \mathcal{Y}$ if and only if the variety defined by w is generated by $\mathcal{Y}$-groups, for example by finite groups. The proof also uses some results of Burns, Macedońska and Medvedev about varieties of groups.

Problem (Well known!) Let $w = [x, y, \ldots, y]$ be an Engel word. Is $F_\infty/w(F_\infty)$ residually finite?

If the answer is 'yes', then w is uniformly elliptic.

Another obvious family of J-words are the *Burnside words* x^q. If q is large, the negative solution of the Burnside Problem shows that x^q is not locally finite. However, the positive solution of the Restricted Burnside Problem (RBP) says that this fact is invisible in the universe of finite groups; the arguments used to deal with locally finite words might therefore be expected to work also for Burnside words.

They don't, however. For there is a subtle way in which the infinitude of a Burnside group might manifest itself within finite group theory. This will be discussed in the next lecture, along with the proof of the following result:

Theorem 2.19 ([NS4]) *For each $q \in \mathbb{N}$ the Burnside word x^q is uniformly elliptic in all finite groups.*

This is the same as saying that the power subgroup G^q is closed in every finitely generated profinite group G. With the positive solution of RBP and a remark above, this in turn is equivalent to

Corollary 2.20 *In a finitely generated profinite group G the power subgroups G^q are open.*

In turn, this now implies

Theorem 2.21 *Every non-commutator word is uniformly elliptic in all finite groups.*

Proof A non-commutator word takes the form

$$w = x_1^{e_1} \ldots x_k^{e_k} u$$

where $u \in F'$ and at least one of the e_i is non-zero. Say $|e_i| = q > 0$. Let G be a finitely generated profinite group. Then $w(G) \geq G^q$ which is open, so $w(G)$ is open, hence closed. $\square$

3 Ellipticity in finite groups

Given a word w and a finite d-generator group G, we want to prove that

$$m_G(w) \leq f = f(w,d).$$

How does this work for $w = [x,y]$ when $G = \langle g_1, \ldots, g_d \rangle$ is nilpotent? Recall Lemma 2.7:

$$G' = [G, g_1] \ldots [G, g_d].$$

We argue by induction on the nilpotency class c. Put $K = \gamma_c(G)$. Let $h \in G'$. Inductively, we assume that

$$h = b[g_1, x_1] \ldots [g_d, x_d]$$

for some $x_1, \ldots, x_d \in G$ and some 'error term' $b \in K$. To kill off the error term, it suffices to find $y_i \in G$ such that

$$[g_1, x_1 y_1] \ldots [g_d, x_d y_d] = b[g_1, x_1] \ldots [g_d, x_d]. \tag{8}$$

To expand out the left-hand side and cancel the factors $[g_i, x_i]$ is quite messy; but if we restrict the y_i to lie in $\gamma_{c-1}(G) = N$ the equation simply reduces to

$$[g_1, y_1] \ldots [g_d, y_d] = b. \tag{9}$$

Now the mapping from $N^{(d)} \to K$ given by

$$\mathbf{y} \mapsto [\mathbf{g}, \mathbf{y}] = [g_1, y_1] \ldots [g_d, y_d]$$

is *surjective*, so we can solve (9) to complete the proof.

Note that $[\mathbf{g}, \mathbf{x}] = v(\mathbf{x})$, say, is actually a generalized word over G. We have shown that the equation $v(\mathbf{x}) = h$ is solvable by successive approximation. Just as in Hensel's Lemma, the key step is to show that a 'derived mapping' — which in this special case happens to be $f_v : N^{(d)} \to K$ — is surjective. The hypothesis that ensures this is that $\{g_1, \ldots, g_d\}$ generates G.

We can try to apply this method in a more general situation, using induction on the group order. However, three issues arise:

1. How does the induction start?

2. If G is not nilpotent, what should K and N be?

3. If w is some complicated word, the derived mapping may be arbitrarily horrible.

Let's discuss **Issue 3** first. Suppose

$$w(G) = \langle r_1, \ldots, r_\delta \rangle \quad \text{with } r_1, \ldots, r_\delta \in G_w. \tag{10}$$

Let H be a normal subgroup of $w(G)$ such that $w(G)/H$ is nilpotent. Then

$$w(G)' = [w(G), r_1] \ldots [w(G), r_\delta] H$$
$$\subseteq G_w^{*2\delta} H,$$

since any commutator $[g, r]$ with $r \in G_w$ is a product of 2 conjugates of r, each of which is again in G_w.

Suppose also that w *is not a commutator word.* Then

$$q = q(w) := |\mathbb{Z}/w(\mathbb{Z})| < \infty, \qquad (11)$$

which implies that

$$g^q \in G_w \quad \forall g \in G.$$

Then for each $r \in G_w$ we have $\langle r \rangle \subseteq G_w^{*q}$, whence

$$w(G) = \langle r_1 \rangle \cdots \langle r_\delta \rangle \, w(G)'$$
$$\subseteq G_w^{*(\delta q + 2\delta)} H. \qquad (12)$$

So we're off to a start, anyway. This motivates the

Definition The word w is *d-restricted* if both (11) and (10) hold for every finite d-generator group G, where δ depends only on w and d.

Parenthesis Let me now elucidate a remark made at the end of the previous lecture about the Burnside word $w = x^q$. Let $F = F_d = \langle x_1, \ldots, x_d \rangle$ and suppose that the Burnside group $B(d, q) = F/F^q$ is finite. Then F^q is finitely generated, and each generator is a product of qth powers; say

$$F^q = \langle u_1(\mathbf{x})^q, \ldots, u_\delta(\mathbf{x})^q \rangle.$$

Now if $G = \langle g_1, \ldots, g_d \rangle$ is any finite group and $\pi : F \to G$ sends x_i to g_i then

$$G^q = F^q \pi = \langle u_1(\mathbf{g})^q, \ldots, u_\delta(\mathbf{g})^q \rangle.$$

Conclusion: *If for each $\delta \in \mathbb{N}$ there exists a finite d-generator group G such that G^q is not generated by δ qth powers then $B(d, q)$ is infinite.*

In other words, if x^q is not d-restricted — a property entirely detected in *finite* groups — then $B(d, q)$ is infinite! As it turns out, this *won't* lead to an alternative proof of the Novikov–Adian theorem, because x^q *is* in fact d-restricted, for every d; but this only emerges as a *consequence* of this whole project.

Now our main result is

Theorem 3.1 ([NS2], [NS4]) *Every d-restricted word is elliptic in d-generator finite groups.*

As every d-locally finite word is easily seen to be d-restricted, this implies Theorem 2.14 of the preceding lecture. It is *not* good enough to show that the Burnside words are uniformly elliptic; this depends on another theorem that I will come to later.

Now fix a d-restricted word w and let G be a finite d-generator group. Then

$$W = w(G) = \langle r_1, \ldots, r_\delta \rangle$$

where $r_1, \ldots, r_\delta \in G_w$. Let

$$H = \gamma_c(W) = [H, W]$$

be the nilpotent residual of W.

In view of (12), we can now *forget* about G and w, and concentrate on W and H: it will suffice to show that every element of H is a product of boundedly many commutators $[h, r_j]$ with $h \in H$, together with boundedly many qth powers. This leads to relatively straightforward equations like (8).

Issue 1: *Starting the induction.* What if G happens to be a simple group? We quote

Theorem 3.2 ([MZ], [SW]) *There exists $m = m(q)$ such that x^q has width m in every finite simple group.*

This and related results will be discussed in lecture 5.

Using Theorem 3.2 and the ideas discussed above, we can reduce Theorem 3.1 to the following key result:

Theorem 3.3 *Let $W = \langle r_1, \ldots, r_\delta \rangle$ be a finite group and $H = [H, W]$ an acceptable normal subgroup of W. Let $q \in \mathbb{N}$. Then*

$$H = \left(\prod_{i=1}^{\delta} [H, r_i] \right)^{*f_1(\delta, q)} \cdot H_q^{*f_2(q)} \tag{13}$$

Acceptable is a small technical condition we'll ignore, and $H_q = \{ h^q \mid h \in H \}$.

Both for the reduction argument mention above, and for the special application to Burnside words, we also need a variant:

Theorem 3.4 *Let W be a finite group and $H = [H, W]$ an acceptable normal subgroup of W. Suppose that $\mathrm{d}(W) = \delta$ and that $\mathrm{Alt}(n)$ is not involved in W. If $W = H \langle s_1, \ldots, s_t \rangle$ then*

$$H = \left(\prod_{i=1}^{t} [H, s_i] \right)^{*f_3(n, \delta)}.$$

The difference in the hypothesis is that we are not given generators for W, only generators for W modulo H. In applications, the s_i will be w-values where w is a word that isn't known to be d-restricted, so that we can't a priori find a bounded set of w-values that generates W.

The difference this makes in the proof is technical, but highlights one of the key ideas. We consider the action of W on some minimal normal (non-central) subgroup A. When $W = \langle r_1, \ldots, r_\delta \rangle$ it follows that at least one of the sets $[A, r_j]$ must be quite big. In the other case, we need to know that one of the sets $[A, s_j]$ is quite big; this is proved using special representation-theoretic properties of groups that don't involve $\text{Alt}(n)$.

In particular, we can use this for a Burnside word, and deduce

Proposition 3.5 ([NS2]) *For each* $n \in \mathbb{N}$, *the word* x^q *is uniformly elliptic in the class of finite groups that do not involve* $\text{Alt}(n)$.

Issue 2. The proof of Theorem 3.3 is by induction on $|H|$. Since $H = [H, W]$, we can choose a normal subgroup N of W minimal subject to

$$H \geq N = [N, W].$$

There are two possibilities:

Case 1: N is nilpotent of class at most 2 and exponent prime or 4;

Case 2: N is quasi-semisimple: $N = N'$ and $N/Z(N)$ is a direct product of non-abelian simple groups.

I'll assume for simplicity that $Z(W) = 1$.
We set $K = N$ unless $N > N' > 1$ in which case set $K = N'$.

To prove (13), let $h \in H$. We assume inductively that

$$h = b \cdot \prod_{i=1}^{f_1} \prod_{j=1}^{\delta} [g_{ij}, r_j] \cdot \prod_{j=1}^{f_2} h_j^q \tag{14}$$

where $h_j, g_{ij} \in H$, $b \in K$ is the 'error term', and the g_{ij} satisfy an *extra* condition

$$\left\langle r_1^{\tau_1(\mathbf{r}, \mathbf{g})}, \ldots, r_\delta^{\tau_\delta(\mathbf{r}, \mathbf{g})} \right\rangle K = W; \tag{15}$$

here the τ_j are certain words.

The inductive step: we have to find $\mathbf{a}_1, \ldots, \mathbf{a}_\delta \in N^{(f_1)}$ and $\mathbf{b} \in N^{(f_2)}$ such that

$$h = \prod_{i=1}^{f_1} \prod_{j=1}^{\delta} [a_{ij} g_{ij}, r_j] \cdot \prod_{j=1}^{f_2} (b_j h_j)^q \tag{16}$$

and

$$\left\langle r_1^{\tau_1(\mathbf{r}, \mathbf{a}.\mathbf{g})}, \ldots, r_\delta^{\tau_\delta(\mathbf{r}, \mathbf{a}.\mathbf{g})} \right\rangle = W. \tag{17}$$

This can be done provided the error term b lies in the image of a certain mapping

$$\Phi : N^{(f_1 \delta + f_2)} \to K;$$

here Φ is a generalized word over N, got by expanding the commutators and powers in (16) and then cancelling out (14). The hypothesis (15) is designed precisely to ensure that Φ is actually surjective.

In fact we have to show that (16) has *very many* solutions, so many that some of them are sure to satisfy (17). In other words, we need to show that the fibres of Φ are sufficiently large.

The analysis of Φ needs different methods in Case 1 and Case 2.

Case 1: N *nilpotent*. This comes down to the study of certain quadratic maps over finite fields. The machinery for this was developed in [S1].

Case 2: N *quasi-semisimple*. Here we study certain equations over a direct product of quasisimple groups. These are eventually reduced to an equation over *one* quasisimple group:

Theorem 3.6 ([NS3]) *Let* $q \in \mathbb{N}$. *Suppose that* $m \in \mathbb{N}$ *and the quasisimple finite group* S *are sufficiently large. Let* $q_1, \ldots, q_m$ *be divisors of* q *and* $\beta_1, \ldots, \beta_m$ *automorphisms of* S. *Then there exist inner automorphisms* α_i *of* S *such that*

$$S = [S, (\alpha_1 \beta_1)^{q_1}] \ldots [S, (\alpha_m \beta_m)^{q_m}].$$

The proof depends on CFSG. When S is of Lie type, it again comes down ultimately to solving equations over finite fields.

A typical application of this is to the part of Φ coming from the qth powers.

Proposition 3.7 *Let* $q \in \mathbb{N}$. *Let* N *be a quasi-semisimple normal subgroup of a finite group* H, *and suppose that the simple factors of* N *are sufficiently large. Then*

$$N \cdot H_q^{*f} = H_q^{*f}$$

where f *depends only on* q.

As well as being an essential part of the main proof, this can be combined with Proposition 3.5 to establish

Theorem 3.8 ([NS4]) *The Burnside word* x^q *is* d-*restricted for every* d.

As explained earlier, this implies

Corollary 3.9 *Every non-commutator word is uniformly elliptic in finite groups.*

4 Algebraic groups

Fix an algebraically closed field K; by a *variety* I will mean a Zariski-closed subset of $K^{(n)}$ for some n (more properly, this is an 'affine algebraic set'). Topological terms will refer to the Zariski topology, and $\overline{S}$ will denote the closure of a set S. A *morphism* is a map $f : X \to Y$ between varieties defined by polynomials in the coordinates. If X is irreducible and Xf is dense in Y one says that f is *dominant*. In this case, Y is also irreducible and Xf contains a dense open subset of Y. For precise statements see [H], §4.

A *linear algebraic group is a* Zariski-closed subgroup of $\mathrm{SL}_n(K)$, for some n; i.e. a closed subset of the affine space $M_n(K)$ that is also a subgroup of $\mathrm{SL}_n(K)$. If w is a word, the entries of the matrix $w(g_1, \ldots, g_k)$ are given by polynomials in the entries of the matrices $g_1, \ldots, g_k$; so $f_w : G^{(k)} \to G$ is a morphism. Using this observation, Merzlyakov established

Theorem 4.1 ([M]) *Let G be a linear algebraic group over an algebraically closed field. Then w has finite width in G, and $w(G)$ is a Zariski-closed subgroup of G. If G is connected then so is $w(G)$.*

Proof I'll assume that G is *connected*; the general case needs a bit more argument but is not essentially harder (see [S], §5.1). This means that G is irreducible as a variety.

For $j \in \mathbb{N}$ set

$$P_j = (G_{+w} \cdot G_{+w}^{-1})^{*j} = \mathrm{Im}\, f(j)$$

where $f(j) : G^{(2kj)} \to G$ is the morphism given by

$$(g_1, \ldots, g_{2j}) \mapsto \prod_{i=1}^{j} w(g_{2i-1}) w(g_{2i})^{-1}.$$

Now $f(j)$ is dominant as a morphism from $G^{(2kj)}$ into $\overline{P_j}$, so $\overline{P_j}$ is irreducible. Since

$$\overline{P_1} \subseteq \overline{P_2} \subseteq \ldots \subseteq \overline{P_j} \subseteq \overline{P_{j+1}} \subseteq \ldots \tag{18}$$

and $\dim \overline{P_j} \leq \dim G$ for each j, there exists m such that

$$\dim \overline{P_j} = \dim \overline{P_m} \quad \forall j \geq m.$$

But an irreducible variety can't contain a proper subvariety of the same dimension, so the chain (18) becomes stationary at $\overline{P_m} = T$, say.

As $f(m)$ is dominant as a map into T, we have $P_m \supseteq U$ for some dense open subset U of T. Now

$$U \subseteq P_m \subseteq w(G) = \bigcup_{j=1}^{\infty} P_j \subseteq T = \overline{U}.$$

It follows that $T = \overline{w(G)}$ is a closed subgroup of G (the closure of a subgroup is always a subgroup).

Let $y \in T$. Then yU is non-empty and open in T, so $yU \cap U \neq \varnothing$ and hence

$$y \in U \cdot U^{-1} \subseteq P_{2m} \subseteq w(G).$$

Thus $T \subseteq P_{2m} \subseteq w(G) \subseteq T$ and so

$$w(G) = \overline{w(G)} = P_{2m} \subseteq G_w^{*4m}.$$

$\square$

Now I would like to sketch the proof of a sharper result that applies when G is *semisimple*.

Theorem 4.2 ([B2], [L]) *Let G be a connected semisimple algebraic group over an algebraically closed field, and w a non-trivial word. Then $f_w : G^{(k)} \to G$ is dominant. Hence G_{+w} contains a dense open subset of G.*

Corollary 4.3 *Let u, w be non-trivial words. Then $G = G_{+u} \cdot G_{+w}$. Every word has positive width 2 in G.*

Proof Let $U \subseteq G_{+u}$ and $W \subseteq G_{+w^{-1}}$ be dense open subsets of G. Let $g \in G$. Then gW is dense so $gW \cap U \neq \varnothing$. Thus

$$g \in U \cdot W^{-1} \subseteq G_{+u} \cdot G_{+w}.$$

$\square$

The key case for the theorem is where $G = \mathrm{SL}_n(K)$; this is interesting enough if the reader is not very familiar with the theory of algebraic groups. Until further notice, we fix $n \geq 2$ and take $G = \mathrm{SL}_n(K)$.

We can take K to be as big as we like: if an algebraic geometry statement is true over some extension field of K it is true over K, provided K is algebraically closed (this is essentially the meaning of the Nullstellensatz).

I will write χ_g to denote the characteristic polynomial of a matrix g. If $g \in G$ then the first and last coefficients of χ_g are 1 and $(-1)^n$, and we get a morphism $\chi : G \to K^{n-1}$ by setting $g\chi = (a_1, \ldots, a_{n-1})$ where

$$\chi_g(T) = T^n - a_1 T^{n-1} + \cdots + (-1)^{n-1} a_{n-1} T + (-1)^n.$$

Let G_{reg} denote the set of matrices in G having n distinct eigenvalues (the regular semisimple elements); thus $G_{reg} = \chi^{-1}(W)$ where W is the dense open subset of K^{n-1} corresponding to polynomials with non-zero discriminant. The Jordan normal form shows that for each $v \in W$, the fibre $\chi^{-1}(v)$ is exactly one conjugacy class in G.

The following is easy to see:

Lemma 4.4 *Put $G_1 = \mathrm{diag}(\mathrm{SL}_{n-1}(K), 1) < G$. Then $G_1\chi$ is a vector space of codimension 1 in K^{n-1}.*

Below we will prove

Lemma 4.5 *G has a dense subgroup H such that $\chi_h(1) \neq 0$ for every element $h \neq 1$ of H.*

On the other hand, we have

Lemma 4.6 *If H is any dense subgroup of G then $H_{+w} \neq \{1\}$.*

Proof W.l.o.g. K contains two algebraically independent elements x and y. Then $G = \mathrm{SL}_n(K) \geq \mathrm{SL}_2(K) \geq F$ where

$$F = \left\langle \begin{pmatrix} y & x \\ 0 & y^{-1} \end{pmatrix}, \begin{pmatrix} y & 0 \\ x & y^{-1} \end{pmatrix} \right\rangle;$$

this is a non-abelian free group by [W], Ex. 2.2. It follows that $G_{+w} \supseteq F_{+w} \neq \{1\}$. But $f_w^{-1}(\{1\})$ is a closed subset of $G^{(k)}$ so it can't contain the dense subset $H^{(k)}$. (If $\mathrm{char}(K) = 0$ we can instead take $y = 1$, $x = 2$.) $\qquad\square$

We can now prove the theorem for $G = \mathrm{SL}_n(K)$ by induction on n. Let $X = \overline{G_{+w}}$ be the closure of G_{+w} in G. We have to show that $X = G$.

If $n = 2$ then $G_1 = \{1\} \subseteq X$. If $n > 2$, we may suppose inductively that $G_{1,+w}$ is dense in G_1 so again $G_1 \subseteq \overline{G_{1,+w}} \subseteq \overline{G_{+w}} = X$.

Set $Y = \overline{X\chi} \subseteq K^{n-1}$, let H be the subgroup given in Lemma 4.5, and take $h \in H_{+w} \setminus \{1\}$. Then $h \in X$ but $h\chi \notin G_1\chi$ since every element of G_1 has 1 as an eigenvalue. Hence Y properly contains the $(n-2)$-dimensional space $G_1\chi$. But Y is an irreducible variety, because it is the closure of $G^{(k)} f_w \chi$ and $G^{(k)}$ is irreducible (it is a connected algebraic group, having no proper subgroups of finite index: see [W], Chapter 5). As $Y \subseteq K^{n-1}$ and K^{n-1} is irreducible of dimension $n - 1$ it follows that $Y = K^{n-1}$.

Thus $\chi_{|X} : X \to K^{n-1}$ is dominant, and so $X\chi$ contains a dense open subset U of K^{n-1}. I claim that $\chi^{-1}(U \cap W) \subseteq X$. To see this, suppose that $g \in G$ and $g\chi \in U \cap W$. Then $g \in G_{reg}$ and $g\chi = x\chi$ for some $x \in X$, which implies that g is conjugate to x. As X is clearly invariant under conjugation it follows that $g \in X$. Finally, since U and W are both open and dense in K^{n-1} it follows that $\chi^{-1}(U \cap W)$ is non-empty and open in G; as G is irreducible this means that $\chi^{-1}(U \cap W)$ is dense in G, and it follows that $X = G$ as required.

To deduce the result for an arbitrary semisimple group G one uses the fact that a large chunk of G consists of subgroups like $\mathrm{SL}_n(K)$. Recall that the *rank* of G is the dimension $\mathrm{rk}(G)$ of a maximal torus in G (connected diagonalizable subgroup).

Theorem 4.7 ([B2]) *Let G be a connected simple algebraic group over K. If G is not isogenous to $\mathrm{SL}_n(K)$ for any n then G has a closed connected semisimple subgroup L such that $\dim(L) < \dim(G)$ and $\mathrm{rk}(L) = \mathrm{rk}(G)$.*

(Recall that an *isogeny* is an (algebraic group) epimorphism with finite kernel, such as that from SL_n to PSL_n.)

This is proved by an analysis of the possible root systems, explained in [B2]. Larsen [L] gives an explicit list of inclusions $\mathrm{SL}_2^{(n)} < \mathrm{Sp}_{2n}$, $\mathrm{SL}_2^{(2n)} = \mathrm{Spin}_4^{(n)} < \mathrm{Spin}_{4n} < \mathrm{Spin}_{4n+1}, \ldots, A_2^{(3)} < E_6, \ldots, A_2 < G_2$.

If Theorem 4.2 holds for groups G_i then it holds for the direct product $G = \prod_{i=1}^m G_i$, and for any image $G\pi$ of G under an isogeny π. Every connected semisimple algebraic group is of the form $G\pi$ with each G_i being a connected, simply connected simple algebraic group. So to prove the theorem, we may assume that

G is simple. If G is isogenous to $\mathrm{SL}_n(K)$ for some n we are done. If not, let L be a semisimple subgroup as given in Theorem 4.7. Arguing by induction on $\dim(G)$ we may suppose that L_{+w} is dense in L. Now let T be a maximal torus in L. Then T is also a maximal torus of G, and

$$T \subseteq L = \overline{L_{+w}} \subseteq \overline{G_{+w}}.$$

Now the union of all conjugates of T is a dense subset of G (it contains the regular semisimple elements: see [H], §§22.2, 26.2); as G_{+w} is invariant under conjugation it follows that $\overline{G_{+w}} = G$, and the proof is complete, modulo Lemma 4.5.

Proof of Lemma 4.5   W.l.o.g. K is big enough to contain a local field, say K contains either $P = \mathbb{Q}_p$ for some prime p (when $\mathrm{char}(K) = 0$) or $P = \mathbb{F}_p((t))$ (when $\mathrm{char}(K) = p$). Now for any local field P and each $n \geq 2$ there exists a central division algebra Δ over P of index n: this means that $\dim_P(\Delta) = n^2$. As K is algebraically closed, we have $\Delta \otimes_P K \cong \mathrm{M}_n(K)$. We identify Δ with a P-subalgebra of $\mathrm{M}_n(K)$, and put $H = \Delta \cap \mathrm{SL}_n(K)$. If $1 \neq h \in H$ then $h - 1$ is a non-zero element of Δ, hence invertible; therefore 1 is not an eigenvalue of h.

That H is dense in $\mathrm{SL}_n(K)$ follows from two facts, (a) $H = \mathcal{H}(P)$ for a certain algebraic group $\mathcal{H}$, a 'P-form of SL_n': this means that $\mathcal{H}(K) = \mathrm{SL}_n(K)$ (see [PR], Proposition 2.17); and (b) that $\mathcal{H}(P)$ is dense in $\mathcal{H}(K)$, which holds whenever $\mathcal{H}$ is a connected reductive algebraic group and P is an infinite field ([B1], AG 13.7, 18.2).

For the existence of Δ, recall that the Brauer group of P is $\mathbb{Q}/\mathbb{Z}$ (see [CF], Chapter VI §1). Let Δ be a central division algebra over P with invariant $\frac{1}{n} \bmod \mathbb{Z}$; it is clear from the discussion in *loc. cit.* that the index of Δ is n. An explicit construction of Δ as a 'cyclic algebra' — a twisted group algebra of C_n over P — is given in [Co], §7.7.

This concludes our discussion of algebraic groups over an algebraically closed field K. We have seen that if G is simple then every word has width 2 in $G(K)$. What about other fields? The question of verbal width in groups such as $\mathrm{SL}_n(F)$ when F is an algebraic number field seems not to have been explored, and should be interesting. The following is deduced from Theorem 4.2 in [B2] and [L]:

Corollary 4.8 *If G is a simple algebraic group defined over $\mathbb{R}$ and w is a non-trivial word then $G(\mathbb{R})_{+w}$ contains a non-empty set that is open in the real topology of $G(\mathbb{R})$.*

I think the same will hold if $\mathbb{R}$ is replaced by $\mathbb{Q}_p$ (and the real topology by the p-adic topology).

On the other hand, the case of finite fields has received a lot of attention recently; I will discuss some of the results and conjectures in the next lecture.

5   Finite simple groups

The classification of finite simple groups (CFSG) shows that these (the non-abelian ones) are of three kinds: 26 sporadic groups, alternating groups, and groups of Lie type over finite fields. To establish uniform bounds that hold over all finite simple groups, one usually breaks the problem into parts:

1) The *sporadic groups*, and maybe finitely many more small groups: these can be ignored.

2) *Groups of Lie type and small Lie rank.* As there are only a few Lie types of each given rank, this comes down to studying groups of a fixed Lie type; these are 'essentially' the $\mathbb{F}_q$-points of simple algebraic groups defined over $\mathbb{Z}$, and Theorem 4.2 suggests that the image of a verbal mapping in such groups should be 'generically' quite large. To make this precise requires some quite sophisticated algebraic geometry; slightly different-flavoured proofs of the following result are given in [L] and in [LaSh], §2:

Proposition 5.1 ([L]) *Let w be a non-trivial word. Then for each r there exists $c = c(w, r) > 0$ such that for every finite simple group G of Lie type of Lie rank r we have*
$$|G_{+w}| > c\,|G|\,.$$
Moreover, there is an absolute constant $c_0 > 0$ such that
$$|G_{+w}| > c_0 r^{-1}\,|G|$$
provided $|G|$ is sufficiently large and G is not of type A_r or ^2A_r.

A different approach to this case is provided by model theory: I will sketch this later.

3a, 3b) *Groups of Lie type and large Lie rank; large alternating groups.* Groups of large Lie rank are classical groups PSL_n, PSp_{2n} etc. These, and the large alternating groups, can often be dealt with by finding matrices, or permutations, of a nice form inside them.

As a companion reult to Proposition 5.1, Larsen [L] also proves:

Proposition 5.2 *Let w be a non-trivial word and let $\varepsilon > 0$. Then there exists N such that*
$$|G_{+w}| > |G|^{1-\varepsilon}$$
whenever $n > N$ and G is either $\mathrm{Alt}(n)$ or a simple group of Lie type of Lie rank n.

With CFSG, the two propositions together imply

Theorem 5.3 ([L]) *Let w be a non-trivial word and let $\varepsilon > 0$. Then $|G_{+w}| > |G|^{1-\varepsilon}$ for all sufficiently large finite simple groups G.*

A useful portmanteau result can sometimes be used to further reduce the work required:

Theorem 5.4 ([N]) *Let k be a perfect field and let $G = G(k)$ be a classical quasisimple group over k. Then G has a quasisimple subgroup H isomorphic to $\mathrm{SL}_n(k_1)$ or $\mathrm{PSL}_n(k_1)$, for some n and a subfield k_1 of k, such that G is the product of 200 conjugates of H.*

It follows that if a word w has width m in $\mathrm{SL}_n(k_1)$, then it has width $200m$ in G. Obviously this method won't give sharp bounds, but it provides a quicker route to some general existence theorems.

The most general theorem about verbal width in finite simple groups is due to Aner Shalev:

Theorem 5.5 ([Sh]) *Every word has positive width 3 in every sufficiently large finite simple group.*

I must also mention another outstanding recent result, the proof of *Ore's conjecture*:

Theorem 5.6 ([LOST]) *The commutator word $[x, y]$ has width one in every finite simple group.*

The proof of Theorem 5.6 involves many techniques from character theory, algebraic geometry and number theory as well as heavy computation; I will say no more about it here, and concentrate now on various approaches to results like Theorem 5.5.

5.1 A model-theoretic method

Let's consider simple groups of a fixed Lie type X. We've seen that if K is an algebraically closed field, then verbal mappings on the simple group $X(K)$ behave nicely; then algebraic geometry in the hands of experts like Larsen can be used to deduce uniform bounds for the width of a word in all the finite groups $X(\mathbb{F}_q)$. Results of a similar flavour can be obtained more easily by 'magic': the method is in principle ineffective, i.e. it can't yield explicit bounds. But it is powerful in its simplicity, and was used to good effect by John Wilson and Jan Saxl [Wi], [SW].

The key to this method is the following theorem of Françoise Point:

Theorem 5.7 ([P]) *Let $(F_n \mid n \in \mathbb{N})$ be a family of finite fields, let $\mathcal{U}$ be a non-principal ultrafilter on $\mathbb{N}$ and let $E = \prod_n F_n / \mathcal{U}$ be the corresponding ultraproduct. Then E is an infinite field and the ultraproduct of groups*

$$G = \prod_n X(F_n) / \mathcal{U}$$

is isomorphic to $X(E)$.

(The statement in [P] doesn't require the F_n to be finite but stipulates some extra conditions, needed when X is a twisted type; these conditions are automatic in the case of finite fields.)

Now let w be a non-trivial word, and suppose that w does not have bounded width in $X(F)$ as F ranges over all finite fields. Then there is an infinite sequence of finite fields (F_n) and for each $n \in \mathbb{N}$ an element

$$g_n \in w(X(F_n)) \setminus X(F_n)_w^{*n}.$$

Let $\tilde{g}$ be the image of $(g_n)_{n \in \mathbb{N}}$ in G.

Suppose $\tilde{g} \in w(G)$. Then $\tilde{g} \in G_w^{*m}$ for some finite m; this implies that some subset of $\{1, \ldots, m-1\}$ is a member of $\mathcal{U}$, a contradiction as a non-principal ultrafilter can't contain finite sets.

On the other hand, $G \cong X(E)$ is a simple group (cf. [SW]), so if $w(G) \neq G$ then $w(G) = 1$. Thus the first-order statement '$w(x_1, \ldots, x_k) = 1 \ \forall x_1, \ldots, x_k$' holds in $\prod_n X(F_n)/\mathcal{U}$, and so holds in $X(F_n)$ for each n in some member of $\mathcal{U}$ ('Łoś's theorem'). So $g_n = 1$ for infinitely many n, another contradiction.

Hence w *does* have bounded width in $X(F)$ as F ranges over all finite fields. As there are only finitely many different Lie types of each Lie rank, we may conclude:

Theorem 5.8 *Let w be a non-trivial word. Then for each r there exists $m = m(w, r)$ such that w has width m in every finite simple group of Lie type and Lie rank at most r.*

5.2   A combinatorial method

This is based on an interesting discovery due to Tim Gowers, generalized by Babai, Nikolov and Pyber as follows. In this result, $k(G)$ denotes *the minimal dimension of a non-trivial $\mathbb{R}$-linear representation of G.*

Theorem 5.9 ([BNP]) *Let $S_1, \ldots, S_t$ be subsets of a finite group G, where $t \geq 3$. If*

$$\prod_{i=1}^{t} |S_i| \geq \frac{|G|^t}{k(G)^{t-2}}$$

then $S_1 \cdot S_2 \cdot \ldots \cdot S_t = G$.

Note that this applies to *any* finite group; in fact, the proof is purely combinatorial, and is quite accessible in [BNP] (in particular, it makes no reference to CFSG, or indeed to any structural aspects of group theory).

The group theory comes into play when we want to apply the result. In particular one needs a lower bound for $k(G)$; such lower bounds are known for the simple groups. If G is simple of Lie type over $\mathbb{F}_q$, of Lie rank r and dimension d, then

$$k(G) \geq cq^r,$$

where c is an absolute constant, while $|G| \sim q^d$.

If G is not of type A_r or ${}^2A_r$, Proposition 5.1 gives a good lower bound for the size of sets G_{+w}. Nikolov and Pyber prove a similar weaker result for these excluded types. Combining these they deduce:

Proposition 5.10 ([NP]) *Let w be a non-trivial word. Then*

$$|G_{+w}| \geq |G| / k(G)^{1/3}$$

for every simple group G of Lie type and sufficiently large order.

Taking $S_i = G_{+w_i}$ in theorem 5.9 now gives

Theorem 5.11 ([Sh]) *Let w_1, w_2 and w_3 be non trivial words. Then*

$$G_{+w_1} G_{+w_2} G_{+w_3} = G$$

for every sufficiently large finite simple group G of Lie type.

This was originally proved by Aner Shalev using a different method, which I'll come to.

5.3 Character theory

The sharpest results rely on character theory. This approach was pioneered by Frobenius, who discovered a closed formula for the number of solutions of an equation in a finite group. In what follows, G denotes a finite group and χ is supposed to range over all irreducible (complex) characters of G. Given conjugacy classes $C_1, \ldots, C_s$ of G, let

$$N(\mathbf{C}; g)$$

denote the number of solutions to the equation

$$x_1 \ldots x_s = g, \ x_1 \in C_1, \ldots, x_s \in C_s.$$

Then we have

$$N(\mathbf{C}; g) = \frac{\prod |C_i|}{|G|} \sum_\chi \frac{\chi(C_1) \ldots \chi(C_s) \overline{\chi(g)}}{\chi(1)^{s-1}}$$

(see [I], Ex. 3.9). To show that there is at least one solution, it therefore suffices to show that this sum is not zero. In general, sums like this are hard to estimate; a favourable situation is when all non-trivial character-values are very small, so that the term corresponding to the principal character dominates the sum. This method is used by Liebeck and Shalev to prove

Theorem 5.12 ([LiSh]) *There is an absolute constant c such that if G is any finite simple group and S is a normal subset of G with $|S|^t \geq |G|$ then*

$$m \geq ct \implies S^{*m} = G.$$

Now let w be a non-trivial word, and let N be the number provided by Theorem 5.3 such that $|G_{+w}| > |G|^{1/2}$ for all finite simple groups G with $|G| > N$. Suppose that G is a finite simple group with $w(G) \neq 1$, and set $S = G_{+w}$. Then $|S|^t \geq |G|$ where $t = \max\{2, \log_2 N\}$; so taking $m(w) = \lceil ct \rceil$ we immediately obtain a weaker version of Theorem 5.5:

Theorem 5.13 ([LiSh]) *For each word w there exists $m(w) \in \mathbb{N}$ such that w has positive width $m(w)$ in every finite simple group.*

The proof in [LiSh], which preceded [L] and [LaSh] by several years, proceeds by showing that if G is sufficiently large then G_{+w} contains a relatively large conjugacy class of G. I would like to sketch some of the ideas, which don't need the sophisticated algebraic geometry of Larsen's approach. There are three cases to consider:

Case 1. Where G is of Lie type and bounded Lie rank r. In this case, we have

$$|C|^{8r} \geq |G|$$

for *every* non-trivial conjugacy class C, and we are done provided $G_{+w} \neq \{1\}$; this holds for all but finitely many simple groups G [J].

Case 2. $G = \text{Alt}(n)$, where n is large. There exists s, depending on w, such that $w(\text{Alt}(s)) \neq 1$. Write $n = ds + r$ where $0 \leq r < s$, and let $1 \neq \sigma \in \text{Alt}(s)_{+w}$. Then G_{+w} contains the permutation $\tau = \sigma \times \sigma \times \cdots \times \sigma \times 1$ (with d copies of σ) which has support of size at least $3d$. Now we apply

Lemma 5.14 ([LiSh]) *Let $\delta > 0$. Then for all sufficiently large n, if $\tau \in \text{Alt}(n)$ has support of size m, the conjugacy class C of τ satisfies*

$$|C| \geq n^{(1/3-\delta)m}.$$

Taking $\delta = \frac{1}{12}$ and n sufficiently large we find that G_{+w} contains a conjugacy class C with

$$|C| \geq n^{n/2s} > |G|^{1/2s}.$$

Case 3. Groups of Lie type and large Lie rank. These are the classical groups $\text{PSL}_n(\mathbb{F}_q)$, $\text{PSp}_{2n}(\mathbb{F}_q)$ etc., and are all dealt with in a similar way. We consider instead their covering groups. Suppose for example that $G = \text{SL}_n(\mathbb{F}_q)$. There exists s such that $w(\text{SL}_s(\mathbb{F}_q)) \neq 1$; again write $n = ds + r$ where $0 \leq r < s$, and let $1 \neq \sigma \in \text{SL}_s(\mathbb{F}_q)_{+w}$. Then G_{+w} contains a block-diagonal matrix τ having d identical blocks σ; let C be the conjugacy class of τ, let ρ be a power of τ with prime order, and denote the conjugacy class of ρ by C_1. Obviously $|C| \geq |C_1|$, and a result from [LiSh2] shows that

$$|C_1| \geq c|G|^{1/6s}$$

where $c > 0$ is an absolute constant.

The same technique is applied to the other classical groups. However, all of these can be directly reduced to the previous case by using Theorem 5.4.

The sharpest results are obtained by Larsen and Shalev in [Sh] and [LaSh]. To begin with, let $G = G_r(q)$ be a finite simple group of Lie type, of Lie rank r over $\mathbb{F}_q$, and let C_1, C_2 and C_3 be conjugacy classes in G.

Proposition 5.15 ([Sh])   (i) *If $|G|$ is sufficiently large and C_1, C_2 and C_3 consist of regular semisimple elements, or*

(ii) *if r is sufficiently large and*

$$|C_1|\,|C_2|\,|C_3| \geq q^{-15/4}\,|G|^3\,,$$

then $C_1 C_2 C_3 = G$.

Proposition 5.16 ([Sh]) *Let w be a non-trivial word. If r is sufficiently large then G_{+w} contains a conjugacy class C with $|C| > q^{-5r/4}\,|G|$.*

Proposition 5.17 ([GL]) *The number of regular semisimple elements in G is at least $(1 - aq^{-1})\,|G|$, where a is an absolute constant.*

Now let w_1, w_2 and w_3 be non trivial words, and put $S_i = G_{+w_i}$ for each i.

If r is large and G is sufficiently large, Proposition 5.16 together with Proposition 5.15(ii) show that $S_1 S_2 S_3 = G$.

If r is small and G is sufficiently large, Proposition 5.17 and Proposition 5.1 together imply that each S_i contains a regular semisimple element, and then Proposition 5.15(i) shows again that $S_1 S_2 S_3 = G$.

Thus in any case

$$S_1 S_2 S_3 = G,$$

which is Shalev's Theorem 5.11. This was his original proof.

Next, we turn to the alternating groups. For a permutation $\sigma \in \text{Alt}(n)$ we denote by $\text{cyc}(\sigma)$ the number of orbits of $\langle \sigma \rangle$ in $\{1, \ldots, n\}$, i.e. the number of cycles (including 1-cycles) in σ. Again, the first key result is based on character theory:

Proposition 5.18 ([LaSh]) *Let $k \in \mathbb{N}$. Then for all sufficiently large n, if $\sigma \in \text{Alt}(n)$ and $\text{cyc}(\sigma) \leq k$ then the conjugacy class C of σ satisfies $C^{*2} = \text{Alt}(n)$.*

The application to verbal mappings is made via

Proposition 5.19 ([LaSh]) *There exists a sequence (σ_n) of permutations with $\sigma_n \in \text{Alt}(n)$ such that*

(i) *$\text{cyc}(\sigma_n) \leq 23$ for each n, and*

(ii) *if w is a non-trivial word then $\sigma_n \in \text{Alt}(n)_{+w}$ for all sufficiently large n.*

Let C_n denote the conjugacy class of σ_n in $\text{Alt}(n)$, let w_1 and w_2 be non trivial words and set $S_i = G_{+w_i}$ for each i. The two last propositions together imply that for all sufficiently large n we have

$$S_1 S_2 \supseteq C_n^{*2} = \text{Alt}(n).$$

This gives

Theorem 5.20 ([LaSh]) *Let u and w be non trivial words. Then for all sufficiently large n,*

$$\mathrm{Alt}(n)_{+u}\mathrm{Alt}(n)_{+w} = \mathrm{Alt}(n).$$

Theorem 5.5 follows from Theorems 5.11 and 5.20, in view of CFSG.

Is this the end of the story? Maybe not: Larsen and Shalev conjecture that $G_{+u}G_{+w} = G$ for all sufficiently large finite simple groups G ('sufficiently large' depending on the non-trivial words u and w). Moreover, they prove this for the case of Lie-type groups of bounded Lie rank, so only the case of classical groups of large rank remains open.

Added in proof (June 2010): the proof of this conjecture has recently been announced by Larsen, Shalev and Tiep.

References

[A]   M. Abért, On the probability of satisfying a word in a group, *J. Group Theory* **9** (2006), 685–694.

[BNP]   L. Babai, N. Nikolov and L. Pyber, Expansion and product decompositions of finite groups: variations on a theme of Gowers, to appear.

[B1]   A. Borel, *Linear algebraic groups*, 2nd ed., Springer-Verlag, New York, 1991.

[B2]   A. Borel, On free subgroups of semi-simple groups, *L'Enseignement Math.* **29** (1983), 151–164.

[C]   P. J. Cameron, *Permutation groups*, London Math. Soc. Student Texts **45**, Cambridge Univ. Press, Cambridge, 1999.

[CF]   J. W. S. Cassels and A. Fröhlich, *Algebraic Number Theory*, Academic Press, London, 1967.

[Co]   P. M. Cohn, *Algebra 2nd ed., Vol. 3*, John Wiley, Chichester, 1991.

[GL]   R. M. Guralnick and F. Lübeck, On *p*-singular elements in Chevalley groups in characteristic *p*, in *Groups and computation III*, 169-182, Ohio State Univ. Math. Res. Inst. Publ. **8**, de Gruyter, Berlin, 2001.

[H]   J. E. Humpreys, *Linear algebraic groups*, Springer-Verlag, New York, 1975.

[I]   I. M. Isaacs, *Character theory of finite groups*, Academic Press, New York, 1976.

[JZ]   A. Jaikin-Zapirain, On the verbal width of finitely generated pro-*p* groups, *Revista Mat. Iberoamericana* **24** (2008), 617–630.

[J]   G. A. Jones, Varieties and simple groups, *J. Austral. Math. Soc.* **17** (1974), 163–173.

[L]   M. Larsen, Word maps have large image, *Israel J. Math.* **139** (2004), 149–156.

[LaSh] M. Larsen and A. Shalev, Word maps and Waring type problems, *J. Amer. Math. Soc.* **22** (2009), 437–466.

[LN]   R. Lidl and H. Niederreiter, *Finite Fields* (2nd Edn), Encyclopedia of Mathematics and its Applications 20, Cambridge Univ. Press, Cambridge, 1996.

[LiSh] M. Liebeck and A. Shalev, Diameter of simple groups: sharp bounds and applications, *Annals of Math.* **154** (2001), 383–406.

[LiSh2] M. Liebeck and A. Shalev, Simple groups, permutation groups and probability, *J. Amer. Math. Soc.* **12** (1999), 497–520.

[LOST] M. Liebeck, E. O'Brien, A. Shalev and P. Tiep, The Ore conjecture, *J. European Math. Soc.* **12** (2010), 939–1008.

[LS]   A. Lubotzky and D. Segal, *Subgroup growth*, Birkhäuser, Basel, 2003.

[M]   Ju. I. Merzljakov, Algebraic linear groups as full groups of automorphisms and the closure of their verbal subgroups (Russian; English summary), *Algebra i Logika Sem.* **6** (1967) no. 1, 83–94.

[MZ]   C. Martinez and E. Zelmanov, Products of powers in finite simple groups, *Israel J. Math.* **96** (1996), 469–479.

[N]   N. Nikolov, A product decomposition for the classical quasisimple groups, *J. Group Theory* **10** (2007), 43–53.

[NP]   N. Nikolov and L. Pyber, Product decompositions of quasi-random groups and a Jordan-type theorem, to appear. See arXiv:math/0703343.

[NS]   N. Nikolov and D. Segal, A characterization of finite soluble groups, *Bull. London Math. Soc.* **39** (2007), 209–213.

[NS2] N. Nikolov and D. Segal, On finitely generated profinite groups, I: strong completeness and uniform bounds, *Annals of Math.* **165** (2007), 171–238.

[NS3] N. Nikolov and D. Segal, On finitely generated profinite groups, II: products in quasisimple groups, *Annals of Math.* **165** (2007), 239–273.

[NS4] N. Nikolov and D. Segal, Powers in finite groups, *Groups, Geometry and Dynamics*, to appear. See arXiv:0909.6439

[PR]   V. P. Platonov and A. S. Rapinchuk, *Algebraic groups and number theory*, Algebraic groups and number theory, Academic Press, New York, 1994.

[P]   F. Point, Ultraproducts and Chevalley groups, *Arch. Math. Logic* **38** (1999), 355–372.

[SW]   J. Saxl and J. S. Wilson, A note on powers in simple groups, *Math. Proc. Cambridge Philos. Soc.* **122** (1997), 91–94.

[S]   D. Segal, *Words: notes on verbal width in groups*, London Math. Soc. Lecture Notes Series **361**, Cambridge Univ. Press, Cambridge, 2009.

[S1]   D. Segal, Closed subgroups of profinite groups, *Proc. London Math. Soc.* **81** (2000), 29–54.

[Sh]   A. Shalev, Word maps, conjugacy classes and a non-commutative Waring-type theorem, *Annals of Math.* **170** (2009), 1383–1416.

[Wa]   E. Warning, Bemerkung zur vorstehenden Arbeit von Herr Chevalley, *Abh. Math. Sem. Univ. Hamburg* **11** (1936), 76–83.

[W]   B. A. F. Wehrfritz, *Infinite linear groups*, Springer-Verlag, Berlin, 1973.

[Wi]   J. S. Wilson, On simple pseudofinite groups, *J. London Math. Soc.* **51** (1995), 471–490.

THE MODULAR ISOMORPHISM PROBLEM FOR THE GROUPS OF ORDER 512

BETTINA EICK* and ALEXANDER KONOVALOV†

*Institut Computational Mathematics, Technical University Braunschweig, Pockelsstrasse 14, Braunschweig, 38106 Germany
Email: beick@tu-bs.de
†School of Computer Science & Centre for Interdisciplinary Research in Computational Algebra, University of St Andrews, St Andrews, Fife, KY16 9SX, Scotland
Email: alexk@mcs.st-and.ac.uk

Abstract

For a prime p, let G be a finite p-group and K a field of characteristic p. The Modular Isomorphism Problem (MIP) asks whether the modular group algebra KG determines the isomorphism type of G. We briefly survey the history of this problem and report on our computer-aided verification of the Modular Isomorphism Problem for the groups of order 512 and the field K with 2 elements.

1 Introduction

The Modular Isomorphism Problem has been known for more than 50 years. Despite various attempts to prove it or to find a counterexample to it, it is still open and remains one of the challenging problems in the theory of finite p-groups bordering on the theory of associative algebras.

Solutions for the modular isomorphism problem are available for various special types of p-groups. For example, the MIP holds for

- abelian p-groups (Deskins [14]; an alternative proof was given by Coleman [12]);
- p-groups G of class 2 with G' elementary abelian (Sandling [34, Theorem 6.25]);
- metacyclic p-groups (Bagiński [1] for $p > 3$; completed by Sandling [36]);
- 2-groups of maximal class (Carlson [11]; alternative proof by Bagiński [3]);
- p-groups G of maximal class, $p \neq 2$, where $|G| \leq p^{p+1}$ and G contains an abelian maximal subgroup (Caranti and Bagiński [2]);
- elementary abelian-by-cyclic groups (Bagiński [4]);
- p-groups with the center of index p^2 (Drensky [16]); and
- p-groups having a cyclic subgroup of index p^2 (Baginski and Konovalov [5]).

This large number of rather special cases shows the significant interest in the problem, but it also exhibits that the problem is difficult to attack.

There are results on the groups of various small orders and the field with p elements. For example, the MIP holds for

- groups of order dividing p^4 (Passman [29]);
- groups of order 2^5 (Makasikis [26] with remarks by Sandling [34]; alternative proof by Michler, Newman and O'Brien [27]);

- groups of order p^5 (Kovacs and Newman, due to Sandling's remark in [35]; alternative proof by Salim and Sandling [32, 33]);
- groups of order 2^6 (Wursthorn [41, 42] using computers; theoretical proof by Hertweck and Soriano [21]);
- groups of order 2^7 (Wursthorn [9] using computers); and
- groups of orders 2^8 and 3^6 (Eick [17] using computers).

The results on the groups of order dividing 2^8 or 3^6 have been established using computers. As the groups of order dividing 2^8 or 3^6 are classified, this mainly requires an algorithm to check whether two modular group algebras are isomorphic. The first method for this purpose is due to Wursthorn [42]. It has been implemented in the C programming language. The implementation was used on the groups of order dividing 2^7, but this seems to be its limit. Eick [17] has developed a new and independent approach for such an isomorphism test. This is implemented in the ModIsom package [19] of the computational algebra system GAP [20] and proved to be practical for the groups of order 2^8 and 3^6.

We successfully applied Eick's implementation to the 10494213 groups of order 512. This required some improvements as well as a parallelization of the implementation. We report on details of this large-scale computation below.

It is worth mentioning that there is an even stronger conjecture than the MIP: the Modular Isomorphism Problem for Normalized Unit Groups (UMIP) asks whether a finite p-group G is determined by the normalized unit group of its modular group algebra over the field of p elements. Only a few results are known in this direction. For a long time, the positive solution of the UMIP was known only for abelian p-groups. Recently it was solved for 2-groups of maximal class in [6] and for p-groups with the cyclic Frattini subgroup for $p > 2$ in [7]. In [23] the UMIP was verified in GAP [20] for all 2-groups of order at most 32 using the LAGUNA package [10].

2 Invariants

A first step for a computational check of the MIP is the computation of invariants of the considered groups which are known to be determined by the modular group algebra. Hence groups with different such invariants have non-isomorphic modular group algebras. The following lists some such invariants. For a group G we denote by $\mathcal{J}_i(G)$ the i-th term of the Jennings series of G.

(a) The exponent of the group G ([25]; see also [36]).

(b) The isomorphism type of the center of the group G [38, 40].

(c) The isomorphism type of the factor group G/G' ([40]; see also [29, 34]).

(d) The isomorphism type of the factor group $G/\Phi(G)$ [15].

(e) The isomorphism type of the factor group $G/(\gamma_2(G)^p\gamma_3(G))$ [35]. This is also called the *Sandling factor*.

(f) The minimal number of generators $d(G')$ of G' (follows immediately from [39, Prop.III.1.15(ii)]).

(g) The length of the Jennings series and the isomorphism types of the factors

$\mathcal{J}_i(G)/\mathcal{J}_{i+1}(G)$, $\mathcal{J}_i(G)/\mathcal{J}_{i+2}(G)$, and $\mathcal{J}_i(G)/\mathcal{J}_{2i+1}(G)$ [28, 31].

(h) The number of conjugacy classes of elements of the group G and the number of conjugacy classes of all p^n-th powers of elements of the group G for all $n \in \mathbb{N}$ [41].

(i) The number of conjugacy classes of maximal elementary abelian subgroups of given rank [30]. This is also called the *Quillen invariant*.

(j) The *Roggenkamp parameter*

$$R(G) = \sum_{i=1}^{t} \log_p |C_G(g_i)/\Phi(C_G(g_i))|,$$

where $\{g_1, \ldots, g_t\}$ is a set of representatives of the conjugacy classes of the group G (Roggenkamp, see [41]).

Additionally, the nilpotency class of a group G is determined if G has exponent p, or class 2, or G' is cyclic, or G is a group of maximal class and contains an abelian subgroup of index p (see [5]).

3 Isomorphism testing for group algebras

In this section we recall the algorithm by Eick [17] and exhibit some refinements of it which have been necessary to deal with the groups of order 512.

Let $\mathbb{F}$ be the field with p elements and A a finite dimensional $\mathbb{F}$-algebra. The *automorphism group* $Aut(A)$ is the set of all bijective linear maps $\alpha : A \to A$ which are compatible with the multiplication: $\alpha(ab) = \alpha(a)\alpha(b)$ for all $a, b \in A$. The *canonical form* $Can(A)$ is a structure constants table for A which describes A up to isomorphism; that is, $A \cong B$ for two $\mathbb{F}$-algebras A and B if and only if $Can(A) = Can(B)$.

Given a finite p-group G, our aim is to determine $Aut(\mathbb{F}G)$ and $Can(\mathbb{F}G)$. This facilitates an effective check of the modular isomorphism problem for the groups of a given order: we determine the canonical forms $Can(\mathbb{F}G)$ of all considered groups G and then determine isomorphisms by simply comparing the canonical forms.

3.1 A reduction to nilpotent algebras

Let G be a finite p-group and $\mathbb{F}$ the field with p elements. Let $I(G)$ denote the Jacobson radical of the modular group algebra $\mathbb{F}G$. As a first step, we recall the well-known reduction of our given task to the same task for $I(G)$.

Lemma 3.1

1. $I(G)$ *is a nilpotent subalgebra of* $\mathbb{F}G$. *Thus there exists an* $l \in \mathbb{N}$ *with*

$$I(G) > I(G)^2 > \ldots > I(G)^l > I(G)^{l+1} = \{0\}.$$

2. $I(G)$ *coincides with the augmentation ideal of* $\mathbb{F}G$. *Thus* $\{g-1 \mid g \in G, g \neq 1\}$ *is an* $\mathbb{F}$-*basis for* $I(G)$ *and* $\mathbb{F}G = I(G) \oplus \mathbb{F}$.

Lemma 3.1(2) implies that we can readily extend $Aut(I(G))$ and $Can(I(G))$ to $Aut(\mathbb{F}G)$ and $Can(\mathbb{F}G)$. Hence it is sufficient to compute $Aut(I(G))$ and $Can(I(G))$ only. The main advantage of this reduction is that $I(G)$ is a nilpotent algebra by Lemma 3.1(1).

It is well known that the Jennings series of the finite p-group G yields a basis for $I(G)$ which contains bases for all ideals of the power series of $I(G)$. This facilitates an efficient determination of the ideals of the power series of $I(G)$.

3.2   An inductive approach

Let I be a finite dimensional nilpotent associative $\mathbb{F}$-algebra. We describe a method to determine the automorphism group and canonical form of I. The basic idea of this method is to use induction on the quotients of the power series $I > I^2 > \cdots > I^l > I^{l+1} = \{0\}$ of I; that is, we successively determine $C_j = Can(I/I^j)$ and $A_j = Aut(I/I^j)$ from $Can(I/I^{j-1})$ and $Aut(I/I^{j-1})$. Denote $I_j = I/I^j$ and let $d = dim(I_2)$.

The first step: In the initialisation step of the induction we consider the algebra I_2. This algebra satisfies $ab = 0$ for all $a, b \in I_2$. Thus every structure constants table for I_2 is a zero-table. Hence C_2 is the zero-table and $A_2 = GL(d, \mathbb{F})$.

The inductive step: In the inductive step, we have determined C_{j-1} and generators and the order of A_{j-1}. Our aim is to compute C_j and generators and the order of A_j.

Let F be the free nilpotent associative algebra on d generators over $\mathbb{F}$. Then I_{j-1} is a quotient of F, say $I_{j-1} \cong F/R$ for some ideal R. Define $\overline{R}$ be the two-sided ideal of F generated by $FR \cup RF$. Then

$$I_{j-1}^* = F/\overline{R}$$

is the *covering algebra* of I_{j-1}.

Eick [17] provides a detailed investigation of the covering algebra and an effective algorithm to compute a canonical table C_{j-1}^* for this algebra from the canonical table C_{j-1} of I_{j-1}. Here we recall the most important features of the covering algebra only. First, we note that the automorphism group A_{j-1} acts naturally on the covering algebra I_{j-1}^* by directly extending the action on F/R to $F/\overline{R}$.

Theorem 3.2 (Eick [17]) *Let $\rho_j : A_j \to A_{j-1}$ be the natural homomorphism.*
1. *$I_j \cong I_{j-1}^*/U$ for some ideal U in I_{j-1}^*.*
2. *Let V be a canonical element in the orbit $U^{A_{j-1}}$. Then $I_j \cong I_{j-1}^*/V$ and C_j is the table of I_{j-1}^*/V with respect to the basis underlying the table C_{j-1}^*.*
3. *$im(\rho_j) = Stab_{Aut_{j-1}}(V)$ and $ker(\rho_j)$ is the elementary abelian p-group consisting of the automorphisms which fix I/I^{j-1} and I^{j-1}/I^j pointwise.*

Theorem 3.2 is used to reduce the inductive step to an orbit-stabilizer computation. A general algorithm to compute orbits and stabilizers for finite groups is described in [22]. The problem inherent in this algorithm is that if the considered orbit is long, then the computation is time- and space-consuming. In our applications, the arising orbits are often huge. Thus the generic algorithm for finite groups is not going to succeed in most cases.

Eick [17] uses a special orbit-stabilizer algorithm. This exploits the fact that the kernel of the natural homomorphism $A_{j-1} \to A_2$ is a normal p-subgroup of A_{j-1}. Orbit representatives and their stabilizers under the action of a p-group can be determined with a highly effective method due to Schwingel [37] which avoids the explicit computation of the orbits. Using Schwingel's method, our desired orbit-stabilizer computations mainly reduce to an orbit-stabilizer computation under the action of $A_2 \cong GL(d, \mathbb{F})$. This reduction has been sufficient to determine canonical forms for the modular group algebras of the groups of order 2^8.

However, even with this very significant reduction of the problem, the arising orbits and stabilizers in the application to the groups of order 2^9 are frequently too large to be computed. Thus for this new application, we had to reduce the orbit-stabilizer problem further. We exploited an approach which is also used in [18]: we try to reduce the initial group $A_2 \cong GL(d, \mathbb{F})$ *a priori*.

3.3   Fingerprints and precomputing

Let $\varphi : I \to I_2$ denote the natural homomorphism and recall that $I_2 \cong \mathbb{F}^d$. Thus I_2 has $l = (p^d - 1)/(p - 1)$ one-dimensional subspaces. Let $M_1, \ldots, M_l$ denote their preimages under φ and note that $M_1, \ldots, M_l$ are subalgebras of I. In particular, each algebra M_i is nilpotent.

We *fingerprint* each of these subalgebras M; that is, we determine invariants of M. Suitable invariants are, for example, the dimensions of the quotients of the k initial terms of their power series for some given k. That is, given a subalgebra M, we compute $M = M^1 \geq M^2 \geq \ldots \geq M^{k+1}$ and determine the sequence $d_1, \ldots, d_k$ defined by $d_i = dim(M^i/M^{i+1})$. The larger we choose k, the better the resulting fingerprint, but its determination is more time-consuming.

Given a fingerprint for each M in the list $M_1, \ldots, M_l$, we partition the subalgebras according to their fingerprints. For every occurring fingerprint f let L_f be the set of subalgebras with fingerprint f. Define V_f as the sum of all subalgebras in L_f. Then V_f is a subalgebra of I which contains I^2 and is an invariant for $Aut(I)$ and $Can(I)$.

We sort the set of all arising fingerprints and thus obtain a list $f_1, \ldots, f_r$ of fingerprints. Let $V_{f_1}, \ldots, V_{f_r}$ denote the corresponding set of subalgebras of I. Then in the first step of our algorithm we start with a basis for I_2 which exhibits the images of the subalgebras $V_{f_1}, \ldots, V_{f_r}$ under φ and we use the stabilizer in $GL(d, \mathbb{F})$ of these images as initialization for A_2.

As a result we can often reduce *a priori* to a comparatively small subgroup A_2 of $GL(d, \mathbb{F})$. This reduces the subsequent orbit-stabilizer computations significantly, since we only act with a comparatively small subgroup of $GL(d, \mathbb{F})$.

4   The groups of order 512

The complete and non-redundant list of groups of order 512 contains 10494213 groups: these are available in the GAP Small Groups Library [8]. In this section we describe our strategy to check the MIP for these groups and we provide some numerical information on the steps of computation.

Our strategy splits the computation into two steps: first, split the groups of order 512 into possibly small clusters by determining invariants of the groups which are determined by their group algebras and then, secondly, check the MIP for the groups in a cluster for each cluster.

4.1   Computing invariants

We used the invariants listed in Section 2 for the first step. Most of these invariants can be computed readily using available GAP functions; the others are implemented in the LAGUNA package [10]. The computation of these invariants for all groups of order 512 was an initial long-term computation. We outline some more details in the following.

In the initial stage, the following parameters were computed for all groups of order 512 to obtain an initial distribution of groups into clusters: the exponent of G, the number of conjugacy classes of G, orders of $Z(G)$, G', the Frattini subgroup of G and the Sandling factor $G/(\gamma_2(G)^p\gamma_3(G))$, the length of the Jennings series and the Roggenkamp parameter $\sum_{i=1}^{t}\log_p|C_G(g_i)/\Phi(C_G(g_i))|$. These parameters were selected on the ground that they can be computed very effectively and some of them, especially the Roggenkamp parameter, are known to be rather efficient invariants to check that group algebras are non-isomorphic.

As a result of this initial computation, the groups of order 512 were split into 30605 clusters of various sizes. For example, we obtained 5678 clusters of size 1; The groups contained in these need not be considered any further. At the other extreme, there were four clusters containing more than 100000 groups each, with sizes 110248, 112390, 115807 and 118504.

However, the majority of these groups have a Sandling factor of order 512. In this case, the groups are determined by their modular group algebras. After filtering out such groups, and also all clusters of size 1, there remained 1646012 groups in 19877 clusters, including 3373 pairs of groups, and the biggest cluster had size 'only' 9175.

To further refine the set of clusters, the following invariants were computed for the remaining groups: the isomorphism type of $Z(G)$ and G/G', and the number of conjugacy classes of p^n-th powers of elements. This step ruled out only 23222 groups, leaving still 1622790 groups to go, but, however, it increased the number of clusters to 51103 and reduced an average size of the cluster: now we already had 14770 pairs, and the largest cluster contained 5424 groups.

The above-mentioned computations were carried out using the GAP package ParGAP [13] on an 8-core computer.

To split families further, the Quillen invariant (that is, the number of conjugacy classes of maximal elementary abelian subgroups of each rank) was selected. Its computation is rather time-consuming, since it involves computation of the lattice of subgroups. Thus we extended computations to other CIRCA machines, the GILDA cluster available to the second author for the duration of the International Winter School in Grid Computing 2009, and the Beowulf cluster at Heriot-Watt University. Depending on the architecture, we used various technologies: the SC-SCP package [24], the ParGAP package [13] or the Condor job submission system (http://www.cs.wisc.edu/condor/). This computation finished with the following improvement: 1553963 groups in 97116 clusters, including 35486 pairs and the largest cluster of size 1827.

Two more group-theoretical invariants were applied after this stage: the minimal number of generators of G' and the isomorphism types of factors of the Jennings series.

As a final result of the invariant computation, we obtained 345367 clusters containing 1297026 groups mostly in small clusters, including 168486 pairs and the biggest cluster of size 210.

4.2 Isomorphism testing

Now we were able to split each of the clusters using the isomorphism test implemented in the ModIsom package.

This computation consumed about 14000 CPU hours during three weeks of computations on the UK National Grid Service (http://www.ngs.ac.uk/).

At first, Modisom successfully split all clusters, except 293 exceptional clusters containing in total 1660 groups. These clusters needed an improvement of the implementation of Modisom as described in Section 3.3 of this paper.

After implementing this improvement, Modisom was able to split the remaining clusters, and hence returned the result that there are no counterexamples to the modular isomorphism problem among the groups of order 512.

Acknowledgments: Besides the CIRCA computational facilities in St Andrews, we used the research server of the Functional Programming group in the University of St Andrews, the Beowulf cluster at Heriot-Watt University (Edinburgh), the GILDA cluster provided for IWSGC'09, and the UK National Grid Service. We acknowledge all involved organizations for these opportunities.

References

[1] C. Bagiński, The isomorphism question for modular group algebras of metacyclic p-groups, *Proc. Amer. Math. Soc.* **104** (1988), no. 1, 39–42.

[2] C. Bagiński and A. Caranti, The modular group algebras of p-groups of maximal class, *Canad. J. Math.* **40** (1988), no. 6, 1422–1435.

[3] C. Bagiński, Modular group algebras of 2-groups of maximal class, *Comm. Algebra* **20** (1992), no. 5, 1229–1241.

[4] C. Bagiński, On the isomorphism problem for modular group algebras of elementary abelian-by-cyclic p-groups, *Colloq. Math.* **82** (1999), no. 1, 125–136.

[5] C. Bagiński and A. Konovalov, The modular isomorphism problem for finite p-groups with a cyclic subgroup of index p^2, in *Groups St. Andrews 2005. Vol. 1* (C. M. Camp-

bell et al., eds.), London Math. Soc. Lecture Note Ser. **339** (CUP, Cambridge 2007), 186–193.

[6] Zs. Balogh and A. Bovdi, On units of group algebras of 2-groups of maximal class, *Comm. Algebra* **32** (2004), no. 8, 3227–3245.

[7] Zs. Balogh and A. Bovdi, Group algebras with unit group of class p, *Publ. Math. Debrecen* **65** (2004), no. 3-4, 261–268.

[8] H.U. Besche, B. Eick and E. O'Brien, The SmallGroups Library, http://www-public. tu-bs.de:8080/~beick/soft/small/small.html.

[9] F. M. Bleher, W. Kimmerle, K. W. Roggenkamp and M. Wursthorn, Computational aspects of the isomorphism problem, in *Algorithmic algebra and number theory (Heidelberg, 1997)* (Springer, Berlin 1999), 313–329.

[10] V. Bovdi, A. Konovalov, R. Rossmanith and C. Schneider. *LAGUNA – Lie AlGebras and UNits of group Algebras*, Version 3.5.0; 2009, http://www.cs.st-andrews.ac. uk/~alexk/laguna.htm.

[11] J. F. Carlson, Periodic modules over modular group algebras, *J. London Math. Soc. (2)* **15** (1977), no. 3, 431–436.

[12] D. B. Coleman, On the modular group ring of a p-group, *Proc. Amer. Math. Soc.* **15** (1964), 511–514.

[13] G. Cooperman. *ParGAP – Parallel GAP, Version 1.1.2*; 2004, http://www.ccs. neu.edu/home/gene/pargap.html.

[14] W. E. Deskins, Finite Abelian groups with isomorphic group algebras, *Duke Math. J.* **23** (1956), 35–40.

[15] E. M. Dieckmann, Isomorphism of group algebras of p-groups, PhD thesis, Washington University, St. Louis, Missouri, 1967.

[16] V. Drensky, The isomorphism problem for modular group algebras of groups with large centres, in *Representation theory, group rings, and coding theory*, Contemp. Math. **93** (Amer. Math. Soc., Providence, RI 1989), 145–153.

[17] B. Eick, Computing automorphism groups and testing isomorphisms for modular group algebras, *J. Algebra* **320** (2008), 3895–3910.

[18] B. Eick, C. R. Leedham-Green and E. A. O'Brien, Constructing automorphism groups of p-groups, *Comm. Alg.* **30** (2002), 2271–2295.

[19] B. Eick, *ModIsom – Computing automorphisms and checking isomorphisms for modular group algebras of finite p-groups*, Version 1.0, 2009, http://www-public.tu-bs. de:8080/~beick/so.html.

[20] The GAP Group, *GAP – Groups, Algorithms, and Programming, Version 4.4.12*, 2008, http://www.gap-system.org.

[21] M. Hertweck and M. Soriano, On the modular isomorphism problem: groups of order 2^6, in *Groups, rings and algebras*, Contemp. Math. **420** (Amer. Math. Soc., Providence, RI 2006), 177–213.

[22] D. F. Holt, B. Eick and E. A. O'Brien, *Handbook of computational group theory* (CRC Press, 2005).

[23] A. Konovalov and A. Krivokhata, On the isomorphism problem for unit groups of modular group algebras, *Acta Sci. Math. (Szeged)* **73** (2007), no. 1–2, 53–59.

[24] A. Konovalov and S. Linton, *SCSCP – Symbolic Computation Software Composability Protocol, Version 1.1*, 2009, http://www.cs.st-andrews.ac.uk/~alexk/scscp.htm.

[25] B. Külshammer, Bemerkungen über die Gruppenalgebra als symmetrische Algebra. II, *J. Algebra* **75** (1982), no. 1, 59–69.

[26] A. Makasikis, Sur l'isomorphie d'algèbres de groupes sur un champ modulaire, *Bull. Soc. Math. Belg.* **28** (1976), no. 2, 91–109.

[27] G. O. Michler, M. F. Newman and E. A. O'Brien, Modular group algebras. Unpublished report, Australian National Univ., Canberra 1987.

[28] I. B. S. Passi and S. K. Sehgal, Isomorphism of modular group algebras, *Math. Z.* **129** (1972), 65–73.

[29] D. S. Passman, The group algebras of groups of order p^4 over a modular field, *Michigan Math. J.* **12** (1965), 405–415.

[30] D. Quillen, The spectrum of an equivariant cohomology ring: I, *Ann. of Math. (2)* **94** (1971), 549–572.

[31] J. Ritter and S. Sehgal, Isomorphism of group rings, *Arch. Math. (Basel)* **40** (1983), no. 1, 32–39.

[32] M. A. M. Salim and R. Sandling, The modular group algebra problem for groups of order p^5, *J. Austral. Math. Soc. Ser. A* **61** (1996), no. 2, 229–237.

[33] M. A. M. Salim and R. Sandling, The modular group algebra problem for small p-groups of maximal class, *Canad. J. Math.* **48** (1996), no. 5, 1064–1078.

[34] R. Sandling, The isomorphism problem for group rings: a survey, in *Orders and their applications (Oberwolfach, 1984)* Lecture Notes in Math. **1142** (Springer, Berlin 1985), 256–288.

[35] R. Sandling, The modular group algebra of a central-elementary-by-abelian p-group, *Arch. Math. (Basel)* **52** (1989), no. 1, 22–27.

[36] R. Sandling, The modular group algebra problem for metacyclic p-groups, *Proc. Amer. Math. Soc.* **124** (1996), no. 5, 1347–1350.

[37] R. Schwingel, Two matrix group algorithms with applications to computing the automorphism group of a finite p-group, PhD Thesis, Queen Mary, University of London.

[38] S. K. Sehgal, On the isomorphism of group algebras, *Math. Z.* **95** (1967), 71–75.

[39] S. K. Sehgal, *Topics in group rings* (Dekker, New York 1978).

[40] H. N. Ward, Some results on the group algebra of a p-group over a prime field, Seminar on Finite Groups and Related Topics, Mimeographed notes, Harvard Univ., 13–19.

[41] M. Wursthorn, Die modularen Gruppenringe der Gruppen der Ordnung 2^6, Diplomarbeit, Universität Stuttgart, 1990.

[42] M. Wursthorn, Isomorphisms of modular group algebras: an algorithm and its application to groups of order 2^6, *J. Symbolic Comput.* **15** (1993), no. 2, 211–227.

RECENT PROGRESS IN THE SYMMETRIC GENERATION OF GROUPS

BEN FAIRBAIRN

Departamento de Matematicas, Universidad de Los Andes, Carrera 1 N 18A - 12, Bogotá, Colombia
Email: bt.fairbairn20@uniandes.edu.co

Abstract

Many groups possess highly symmetric generating sets that are naturally endowed with an underlying combinatorial structure. Such generating sets can prove to be extremely useful both theoretically in providing new existence proofs for groups and practically by providing succinct means of representing group elements. We give a survey of results obtained in the study of these symmetric generating sets. In keeping with earlier surveys on this matter, we emphasize the sporadic simple groups.

1 Introduction

This article is concerned with groups that are generated by highly symmetric subsets of their elements: that is to say by subsets of elements whose set normalizer within the group they generate acts on them by conjugation in a highly symmetric manner. Rather than investigate the behaviour of known groups we turn this procedure around and ask what groups can be generated by a set of elements that possesses a certain assigned set of symmetries. This enables constructions by hand of a number of interesting groups, including many of the sporadic simple groups. Much of the emphasis of the research project to date has been concerned with using these techniques to construct sporadic simple groups, and this article will emphasize this important special case. Recent work of the author and Müller has been concerned with Coxeter groups, so we shall describe this case too.

This article is intended as an 'update' to the earlier survey article of Curtis [14]. Since [14] appeared several of the larger sporadic groups have succumbed to these techniques and a much wider class of reflections groups have been found to admit symmetric presentations corresponding to symmetric generating sets. We refer the interested reader seeking further details to the recent book of Curtis [15] which discusses the general theory of symmetric generation and many of the constructions mentioned here in much greater detail.

Throughout we shall use the standard ATLAS notation for finite groups as defined in [12].

This article is organized as follows. In Section 2 we shall recall the basic definitions and notation asociated with symmetric generation that are used throughout this article. In Section 3 we shall discuss symmetric generation of some of the sporadic simple groups. In Section 4 our attention will turn to reflection groups and in Section 5 we shall consider symmetric generators more general than cyclic groups of order 2, that earlier parts of the article ignore. After making some concluding remarks in Section 6 we finally give a short appendix describing some of the well-known properties of the Mathieu group M_{24} that are used in several earlier sections.

2   Symmetric Generation

2.1   Definitions and Notation

In this section we shall describe the general theory of symmetric generation giving many of the basic definitions we shall use throughout this survey.

Let $2^{\star n}$ denote the free group generated by n involutions. We write $\{t_1, t_2, \ldots, t_n\}$ for a set of generators of this free product. A permutation $\pi \in S_n$ induces an automorphism of this free product, $\hat{\pi}$, by permuting its generators namely

$$t_i^{\hat{\pi}} := \pi^{-1} t_i \pi = t_{\pi(i)}. \tag{1}$$

Given a group $N \leq S_n$ we can use this action to form a semi-direct product $\mathcal{P} := 2^{\star n} : N$.

When N acts transitively we call $\mathcal{P}$ a *progenitor*. (Note that some of the early papers on symmetric generation insisted that N acts at least 2-transitively.) Elements of $\mathcal{P}$ can all be written as a relator of the form πw where $\pi \in N$ and w is a word in the symmetric generators using (1). Consequently any finitely generated subgroup of $\mathcal{P}$ may be expressed as $H := \langle w_1 \pi_1, \ldots, w_r \pi_r \rangle$ for some r. We shall express the quotient of $\mathcal{P}$ by the normal closure of H, $H^{\mathcal{P}}$, as

$$\frac{2^{\star n} : N}{w_1 \pi_1, \ldots, w_r \pi_r} := G. \tag{2}$$

We say the progenitor $\mathcal{P}$ is factored by the relations $w_1 \pi_1, \ldots, w_r \pi_r$. Whenever we write a relator $w\pi$ we shall tacitly be referring to the relation $w\pi = id$ thus we shall henceforth only refer to relations, when *senso stricto* we mean relators. We call G the *target group*. Often these relations can be written in a more compact form by simply writing $(\pi w)^d$ for some positive integer d. It is the opinion of the author that no confusion should arise from calling both $t \in \mathcal{P}$ and its image in G a *symmetric generator*. Similarly no confusion should arise from calling both $N \leq \mathcal{P}$ and its image in G the *control group*. We define the *length* of the relation πw to be the number of symmetric generators in w.

Henceforth we shall slightly abuse notation in writing t_i both for a generator of $2^{\star n}$ in $\mathcal{P}$ and for its homomorphic image in G. Similarly we shall write N both for the control group in $\mathcal{P}$ and for the homomorphic image of N in G. We shall also assume that N is isomorphic to its image in G as this is often the case in the most interesting examples. Again, it is the opinion of the author that no confusion should arise from this.

We are immediately confronted with the question of how to decide if G is finite or not. To do this we resort to an enumeration of the cosets of N in G. Let $g \in G$.

We have that $gN \subset NgN$. Consequently the number of double cosets of the form NgN in G will be at most the number of single cosets of the form gN in G making them much easier to enumerate. To do this we make the following definition.

Given a word in the symmetric generators, w, we define the *coset stabilizing subgroup* to be the subgroup defined by

$$N^{(w)} := \{\pi \in N \mid Nw\pi = Nw\}.$$

This is clearly a subgroup of N and there are $|N : N^{(w)}|$ right cosets of $N^{(w)}$ in the double coset NwN.

2.2 Double Coset Enumeration

As noted above, to verify a symmetric presentation of a finite group it is usual to perform a double coset enumeration. In early examples of symmetric presentations, the groups involved were sufficiently small for the coset enumeration to be easily performed by hand. However, attention has more recently turned to larger groups and consequently automation of the procedure to enumerate double cosets has been necessary.

Bray and Curtis have produced a double coset enumeration program specially suited to this situation in the MAGMA computer package [3]; it is described in [7]. The program uses an adaptation of the celebrated Todd-Coxeter algorithm first described in [24] and follows an adaptation of an earlier program written by Sayed described in [23, Chapter 4] which worked well with relatively small groups, but could not be made to cope with groups of a larger index or rank [15, p.66].

2.3 Some General Lemmata

Given a particular progenitor we are immediately confronted with the problem of deciding what relations to factor it by. The following lemmata naturally lead us to relations that are often of great interest. In particular these lemmata prove surprisingly effective in naturally leading us to relations to consider.

The following lemma, given a pair of symmetric generators $\{t_i, t_j\}$, tells us which elements of the control group may be expressed as a word in the symmetric generators t_i and t_j inside the target group.

Lemma 2.1
$$\langle t_i, t_j \rangle \cap N \leq C_N(Stab_N(i,j)).$$

This lemma can easily be extended to an arbitrary number of symmetric generators by an obvious induction. Despite its strikingly minimal nature, this lemma proves to be extraordinarily powerful. For a proof see [15, p.58].

Whilst the above lemma tells which elements of the control group may appear in a relation to factor a progenitor by, the precise length of this word remains open. A lemma that proves to be useful in settling this matter is the following.

Lemma 2.2 *Let $\mathcal{P} := 2^{\star n}{:}N$ be a progenitor in which the control group N is perfect. Then any homomorphic image of $\mathcal{P}$ is either perfect or possesses a perfect subgroup to index 2. If w is a word in the symmetric generators of odd length, then the image*

$$\frac{2^{\star n} : N}{\pi w}$$

is perfect.

See [15, Section 3.7] for details. Other general results of this nature may also be found in [15, Chapter 3].

3 The Sporadic Simple Groups

In this section we shall see how the techniques described in the previous section may be used to find symmetric presentations for some of the sporadic simple groups. We shall also see some examples in which symmetric presentations lead to new existence proofs for several of these groups as well as computational applications. We proceed in ascending order of order.

3.1 The Janko Group J_3

In [4] Bradley considered the primitive action of the group $L_2(16){:}4$ on 120 points. Since this action is transitive we can form the progenitor $2^{\star 120} : (L_2(16) : 4)$. Using a single short relation found by Bray, Bradley was able to verify the symmetric presentation

$$\frac{2^{\star 120} : (L_2(16) : 4)}{(\pi t_1)^5} \cong J_3 : 2,$$

where $\pi \in L_2(16){:}4$ is a well chosen permutation of order 12, by performing by hand an enumeration of the double cosets of the form $(L_2(16) : 4)w(L_2(16) : 4)$, where w is a word in the symmetric generators.

The full double coset enumeration in [4] is a fairly long and involved calculation. In [5] Bradley and Curtis state and prove the general lemmata used to perform this calculation and describe the Cayley graph for the symmetric generating set derived from it. They go on to use this symmetric presentation to provide a new existence proof for $J_3{:}2$, by proving that the target group in this symmetric presentation must either have order 2 or the correct order to be $J_3{:}2$. They then exhibit 9×9 matrices over the field of four elements that satisfy the relations of the presentation, thus verifying the above isomorphism.

An immediate consequence of this presentation is that elements of $J_3{:}2$ may be represented as an element of $L_2(16){:}4$ followed by a short word in the symmetric generators. In [4] Bradley gives a program in the algebra package Magma [3] for multiplying elements represented in this form together and expressing their product in this concise form.

3.2   The McLaughlin Group McL

The Mathieu group M_{24} naturally acts on the set of 2576 dodecads (see Appendix). If we fix two of the 24 points that M_{24} acts on, then there are 672 dodecads containing one of these points but not the other. The stabilizer of the two points, the Mathieu group M_{22}, acts transitively on these 672 dodecads. We can therefore form the progenitor $2^{\star 672} : M_{22}$. In [4] Bradley investigated this progenitor and was naturally led to the symmetric presentation

$$\frac{2^{\star 672} : M_{22}}{\pi(t_A t_B)^2} \cong \text{McL:2}$$

where A and B are two dodecads intersecting in eight points and $\pi \in M_{22}$ is a permutation arrived at using Lemma 2.1. As with $J_3{:}2$ the coset enumeration in this case was performed entirely by hand, thus providing a new computer-free existence proof for this group.

3.3   The Conway Group $\cdot 0 = 2{\cdot}\text{Co}_1$

In [8] Bray and Curtis consider the progenitor $2^{\star\binom{24}{4}} : M_{24}$ defined by the action of the Mathieu group M_{24} on subsets of size four of a set of order 24 on which M_{24} naturally acts (see Appendix). After eliminating words of length 2 and other relations of length three they are immediately led to the symmetric presentation

$$\frac{2^{\star\binom{24}{4}} : M_{24}}{\pi t_{ab} t_{ac} t_{ad}} \cong \cdot 0,$$

where a, b, c and d are pairs of points the union of which is a block of the $\mathcal{S}(5,8,24)$ Steiner system on which M_{24} naturally acts (see the ATLAS, [12, p.94]) and $\pi \in M_{24}$ is the unique non-trivial permutation of M_{24} determined by the Lemma 2.1. From this presentation, an irreducible 24 dimensional $\mathbb{Z}$ representation is easily found and considering the action this gives on certain vectors in $\mathbb{Z}^{24}$ the famous Leech lattice effortlessly 'drops out'. Furthermore the symmetric generators are revealed to be essentially the elements of $\cdot 0$ discovered by Conway when he first investigated the group [11].

Using this symmetric presentation of $\cdot 0$ and detailed knowledge of the coset enumeration needed to verify it Curtis and the author have been able to produce a program in MAGMA [3], available from the author's website, that represents elements of $\cdot 0$ as a string of at most 64 symbols and typically far fewer, essentially by representing them as πw where $\pi \in M_{24}$ and w is a word in the symmetric generators [16]. This represents a considerable saving compared to representing an element of $\cdot 0$ as a permutation of 196560 symbols or as a 24×24 matrix (i.e. as a string of $24^2 = 576$ symbols). To date, $\cdot 0$ is the largest group for which a program of this kind has been produced.

3.4   The Janko Group J_4

In [2] Bolt, Bray and Curtis, building on earlier work of Bolt [1], considered the primitive action of the Mathieu group M_{24} on the 3795 triads (see Appendix). Since this action is transitive, we can define the progenitor $2^{\star 3795} : M_{24}$, leading to the presentation

$$\frac{2^{\star 3795} : M_{24}}{t_A t_B t_C, \pi t_A t_D t_A t_E} \cong J_4$$

where the triads A, B and C each have a single octad in common, the remaining octads intersecting in either four or no points, depending on whether they belong to the same triad or not. The triads A, D and E have a more complicated relationship to one another and we merely refer the reader to [2, p.690] for details. Again, the permutation $\pi \in M_{24}$ is naturally arrived at by Lemma 2.1.

Note that in this case the 'excess' of relations is merely an illusion - removing the first of the two relations gives a symmetric presentation of the group $J_4 \times 2$, a relation of odd length being needed to produce a simple group in accordance with Lemma 2.2

3.5   The Fischer Groups

In the earlier survey of Curtis [14] it was noted that the each of the sporadic Fischer groups were homomorphic images of progenitors defined using non-involutory symmetric generators and that relations defining the Fischer groups were at the time of writing being investigated by Bray.

Since that time involutory symmetric presentations for each of the sporadic Fischer groups as well as several of the classical Fischer groups closely related to them have been found. In [9] Bray, Curtis, Parker and Wiedorn proved

$$\frac{2^{\star \binom{10}{4}} : S_{10}}{((45)t_{1234})^3, (12)(34)(56)t_{1234}t_{1256}t_{3456}t_{7890}} \cong Sp_8(2),$$

$$\frac{2^{\star (7+\binom{7}{3})} : S_7}{((12)t_1)^3, ((45)t_{1234})^3, (12)(34)(56)t_{1234}t_{3456}t_{1256}t_7} \cong Sp_6(2),$$

$$\frac{2^{\star 288} : Sp_6(2)}{(r_{4567}t_\infty)^3} \cong 3^{\cdot}O_7(3).$$

In the first case the progenitor is defined by the natural action of the symmetric group S_{10} on subsets of size four. The second 'progenitor', which deviates from the traditional definition of progenitor since the control group does not act transitively, is defined by the natural action of the symmetric group S_7 on seven points and its action on subsets of size three. In the third case the action defining the progenitor is the action of $Sp_6(2)$ on the cosets of a subgroup isomorphic to the symmetric group S_7. The symbol r_{4567} appearing in the third of these presentations corresponds to a symmetric generator defined in the above symmetric presentation of $Sp_8(2)$.

In [10] the same authors, motivated by the above results, were able to exhibit symmetric presentations for the sporadic Fischer groups as follows.

For a symmetric presentation of the sporadic Fischer group Fi_{22}, we consider the action of the group $2^6 : Sp_6(2)$, using its action on the cosets of a copy of a subgroup isomorphic to the symmetric group S_8, which leads us to the presentation

$$\frac{2^{\star 2304} : (2^6 : \mathrm{Sp}_6(2))}{(ts)^3} \cong 3\dot{}\mathrm{Fi}_{22}.$$

Here, t is a symmetric generator and s is a well chosen element of the control group, the choice of which is motivated by the symmetric presentations for the classical Fischer groups given above.

For a symmetric presentation of the sporadic Fischer group Fi_{23}, we consider the natural action of the symmetric group S_{12} on the partitions into three subsets size four of the set $\{1, \ldots, 9, 0, x, y\}$. Using this we can define the progenitor $2^{\star 5775} : S_{12}$ which we factor by the relations

$$\left((15) \left| \begin{array}{c} \underline{1234} \\ \underline{5678} \\ 90xy \end{array} \right| \right)^3$$

and

$$(12)(34)(56)(78) \left| \begin{array}{c} \underline{1234} \\ \underline{5678} \\ 90xy \end{array} \right\| \left. \begin{array}{c} \underline{1256} \\ \underline{3478} \\ 90xy \end{array} \right\| \left. \begin{array}{c} \underline{1278} \\ \underline{3456} \\ 90xy \end{array} \right| .$$

The above presentation was set up to resemble the classical relations satisfied by the bifid maps related to the Weyl groups of type E_6 and E_7 (see [15, Section 4.4] for a discussion relating these to symmetric presentations). In addition to the above presentation, the same authors go on to prove another, substantially simpler, symmetric presentation for Fi_{23} namely

$$\frac{2^{\star 13056} : \mathrm{Sp}_8(2)}{(r_{1234}t_1)^3} \cong \mathrm{Fi}_{23}$$

where the action defining the progenitor in this case is the action of the symplectic group $\mathrm{Sp}_8(2)$ on the cosets of the maximal subgroup isomorphic to the symmetric group S_{10} and r_{1234} is the symmetric generator defined in the symmetric presentation of $\mathrm{Sp}_8(2)$ given earlier.

Finally, for a symmetric presentation of the sporadic Fischer group Fi_{24}, we consider the action of the orthogonal group $O_{10}^-(2) : 2$ on the cosets of a copy of a subgroup isomorphic to the symmetric group $S_{12} \leq O_{10}^- : 2$, which leads us to the symmetric presentation

$$\frac{2^{\star 104448} : (O_{10}^-(2) : 2)}{(ts)^3} \cong 3\dot{}\mathrm{Fi}_{24},$$

where t is a symmetric generator and s is a well chosen element of the control group, the choice of which is motivated by the symmetric presentations for the classical Fischer groups given above.

4 Coxeter Groups

Recall that a *Coxeter diagram* of a presentation is a graph in which the vertices correspond to involutory generators and an edge is labelled with the order of the product of its two endpoints. Commuting vertices are not joined and an edge is left unlabelled if the corresponding product has order three. A Coxeter diagram and its associated group are said to be *simply laced* if all the edges of the graph are unlabelled. In [14] Curtis notes that if such a graph has a "tail" of length at least two, as in Figure I, then we see that the generator corresponding to the terminal vertex, a_r, commutes with the subgroup generated by the subgraph $\mathcal{G}_0$.

Figure I: A Coxeter diagram with a tail.

The author has more recently investigated a slight generalization of this idea to produce extremely succinct symmetric presentations for all of the finite simply laced Coxeter groups. More specifically in [22, 21, Chapter 3] we prove

Theorem 4.1 *Let S_n be the symmetric group acting on n objects and $W(\Phi)$ denote the Weyl group of the root system Φ. Then:*

1.
$$\frac{2^{\star\binom{n}{1}} : S_n}{(t_1(1,2))^3} \cong W(A_n)$$

2.
$$\frac{2^{\star\binom{n}{2}} : S_n}{(t_{12}(2,3))^3} \cong W(D_n) \text{ for } n \geq 4$$

3.
$$\frac{2^{\star\binom{n}{3}} : S_n}{(t_{123}(3,4))^3} \cong W(E_n) \text{ for } n = 6,7,8.$$

In particular, we show that the above symmetric presentations may be naturally arrived at using general results such as Lemma 2.1 without considering the general theory of Coxeter groups.

Subsequently, the author and Müller generalised this result as follows. Let Π be a set of fundamental reflections generating a Coxeter group G. If $s \in \Pi$ is a reflection, the subgroup $N \leq G$ generated by $\Pi \backslash \{s\}$ may be used as a control group for G and the action of N on the set $\{s^N\}$ may be used to define a progenitor, $\mathcal{P}$. Factoring $\mathcal{P}$ by the relations defining G that involve s provides a symmetric presentation of G. The lack of restrictions on G means it is possible that G is infinite. In particular our theorem applies to affine reflection groups and to many of the hyperbolic reflection groups. Our theorem thus provides the first examples of interesting infinite groups for which a symmetric presentation has been found ('interesting' in the sense that any progenitor is an infinite group that is symmetrically generated *ab initio*, but only in unenlightening manner). See [20] for details.

We remark that in [20] various (extremely weak) finiteness assumptions about the control group are made to ensure that the progenitors considered there do actually satisfy the definition of progenitor. The proofs of the results given there, however, nowhere require the progenitor to be finitely generated or the action defining the progenitor to be on a finite number of points. If the definition of progenitor is weakened slightly to allow this wider class of objects to be considered then the results given in [20] and their proofs remain valid.

5 Non-Involutory Symmetric Generation

Whilst many of the applications of symmetric generation have focused on the case where the symmetric generators are involutions, it is also possible to consider groups in which the symmetric generating set contains groups more general than cyclic groups of order 2.

Let H be a group. Instead of considering the free group $2^{\star n}$ we consider the free product of n copies of H, which we shall denote $H^{\star n}$. If N is a subgroup of $S_n \times Aut(H)$ then N can act on $H^{\star n}$ and we can form a progenitor in a similar manner to before, usually writing $H^{\star n} :_m N$ to indicate that the action defining the progenitor is not necessarily just a permutation of the copies of H — the 'm' standing for 'monomial' — see below. (The special case in which H is cyclic of order 2 being the involutory case discussed in earlier sections.)

One common source of actions used for defining progenitors in this way is as follows. Recall that a matrix is said to be *monomial* if it has only one non-zero entry in each row and column. A representation of a group $\rho : G \rightarrow GL(V)$ is said to be monomial if $\rho(g)$ is a monomial matrix for every $g \in G$.

An n-dimensional monomial representation of a control group N may be used to define a progenitor $H^{\star n} :_m N$, each non-zero entry of the monomial matrix acting as a cyclic subgroup of $Aut(H)$.

In [25] Whyte considered monomial representations of decorations of simple groups and their use in forming monomial progenitors. In particular Whyte classified all irreducible monomial representations of the symmetric and alternating groups (reproducing earlier work of Djoković and Malzan [17, 18]) and their covers. Whyte went on to classify the irreducible monomial representations of the sporadic simple groups and their decorations and found a large number of symmetric presentations defined using the actions of these monomial representations on free products of cyclic groups.

In light of Lemma 2.2, it is natural to restrict our attention to control groups that are perfect, and in particular control groups that are simple. In light of the classification of the finite simple groups, the work of Whyte only left open the many families of groups of Lie type. Since the most 'dense' family of groups of Lie type are the groups $L_2(q)$, it is natural to consider monomial representations of these groups. In [19] using only elementary techniques the author classifies the monomial representations for all the natural decorations of $L_2(q)$ and these where put to great use in defining symmetric presentations in [21, Chapter 4].

6 Concluding Remarks

We note that the symmetric presentations discussed so far represent the most 'elegant' symmetric presentations of sporadic simple groups discovered since [14]. Several other symmetric presentations of sporadic groups that have proved more difficult to motivate for various reasons have also been discovered. Examples include Bray's presentation of the Lyons group Ly [6, Chapter 10], Bolt's presentation of the Conway group Co_3 [1, Section 3.3], the author's presentation of the Conway group Co_2 and various authors' work on the Rudvalis group Ru (all unpublished). Furthermore several of the larger sporadic groups, most notably the Thompson group Th, the Baby Monster B and of course the Monster M, still lack any kind of symmetric presentation at all, though various conjectures in these cases do exist (see [14, Section 4.1]).

7 Appendix: The Mathieu Group M_{24}

In this appendix we gather together some well-known facts about the Mathieu group M_{24} that are repeatedly called on in Section 3.

Let X be a set such that $|X| = 24$. Recall that the Steiner system $\mathcal{S}(5,8,24)$ is a certain collection of 759 subsets $O \subset X$ with $|O| = 8$ called *octads* such that any $P \subset X$ with $|P| = 5$ is contained in a unique octad. An automorphism of $\mathcal{S}(5,8,24)$ is a permutation of X such that the set of all octads is preserved. The group of all automorphisms of $\mathcal{S}(5,8,24)$ is the sporadic simple Mathieu group M_{24}. The group M_{24} and the Steiner system $\mathcal{S}(5,8,24)$ have many interesting properties, some of which we summarize below.

- The pointwise stabilizer of two points of X is another sporadic simple group known as the Mathieu group M_{22} — see Section 3.2.
- If A and B are two octads such that $|A \cap B| = 2$ then their symmetric difference $(A \cup B) \backslash (A \cap B)$ is called a *dodecad*. There are 2576 dodecads and M_{24} acts transitively on the set of all dodecads — see Section 3.2.
- The group M_{24} acts 5-transitively on the 24 points of X (so in particular acts transitively on subsets of X of cardinality 4 — see Section 3.3).
- The set X may be partitioned into three disjoint octads. Such a partition is called a *trio*. There are 3795 trios and M_{24} acts primitively on the set of all trios — see Section 3.4.

There are several good discussions of the Mathieu groups and their associated Steiner systems in the literature, many of which contain proofs of the above. See Conway and Sloane [13, Chapters 10 & 11] or the ATLAS [12, p.94] and the references given therein for further details.

References

[1] S. W. Bolt, *Some applications of symmetric generation* (PhD thesis, Birmingham 2002).

[2] S. W. Bolt, J. N. Bray and R. T. Curtis, Symmetric presentation of the Janko group J_4, *J. London Math. Soc. (2)* **76** (2007), 683–701.

[3] W. Bosma, J. Cannon and C. Playoust, The Magma algebra system, I: the user language, *J. Symbolic Comput.* **24** (1997), 235–265.

[4] J. D. Bradley, *Symmetric presentations of sporadic groups* (PhD thesis, Birmingham 2004).

[5] J. D. Bradley and R. T. Curtis, Symmetric generation and existence of $J_3 : 2$, the automorphism group of the third Janko group, *J. Algebra* **304** (2006), 256–270.

[6] J. N. Bray, *Symmetric generation of sporadic groups and related topics* (PhD thesis, Birmingham 1997).

[7] J. N. Bray and R. T. Curtis, Double coset enumeration of symmetrically generated groups, *J. Group Theory* **7** (2004), 167–185.

[8] J.N. Bray and R.T. Curtis, The Leech lattice, Λ and the Conway group $\cdot 0$ revisited, *Trans. Amer. Math. Soc.* to appear.

[9] J. N. Bray, R. T. Curtis, C. W. Parker and C. B. Wiedorn, Symmetric presentations for the Fischer groups I: the classical groups $Sp_6(2)$, $Sp_8(2)$, and $3^{\cdot}O_7(3)$, *J. Algebra* **265** (2003), 171–199.

[10] J. N. Bray, R. T. Curtis, C. W. Parker and C. B. Wiedorn, Symmetric presentations for the Fischer groups II: the sporadic groups, *Geom. Dedicata* **112** (2005), 1–23.

[11] J. H. Conway, A group of order 8,315,553,613,086,720,000, *Bull. Lond. Math. Soc.* **1** (1969), 79–88.

[12] J. H. Conway, R. T. Curtis, S. P. Norton, R. A. Parker and R. A. Wilson, *Atlas of Finite Groups* (OUP, Oxford 1985).

[13] J. H. Conway and N. J. A. Sloane, *Sphere Packing, Lattices and Groups, Third edition*, Grundlehren der mathematischen Wissenschaften **290** (Springer–Verlag, New York 1998).

[14] R. T. Curtis, A survey of symmetric generation of sporadic simple groups, in *The Atlas of Finite Groups: ten years on* (R. T. Curtis and R. A. Wilson, eds.), London Math. Soc. Lecture Note Ser. **249** (CUP, Cambridge 1998), 39–57.

[15] R. T. Curtis *Symmetric generation of groups, with applications to many of the sporadic finite simple groups*, Encyclopedia of Mathematics and Its Applications **111** (CUP, Cambridge 2007).

[16] R. T. Curtis and B. T. Fairbairn, Symmetric representation of the elements of the Conway group $\cdot 0$, *J. Symbolic Comput.* **44** (2009), 1044–1067.

[17] D. Ž. Djoković and J. Malzan, Monomial irreducible characters of the symmetric and alternating groups, *J. Algebra* **35** (1975), 153–158.

[18] D. Ž. Djoković and J. Malzan, Imprimitive, irreducible complex characters of the alternating groups, *Canad. J. Math.* **28** (1976), 1199–1204.

[19] B. T. Fairbairn, A note on monomial representations of linear groups, *Comm. Algebra* to appear.

[20] B. T. Fairbairn and J. Müller, Symmetric generation of Coxeter groups, *Arch. Math (Basel)* to appear, http://arxiv.org/abs/0901.2660.

[21] B. T. Fairbairn, *On the symmetric generation of finite groups* (PhD thesis, Birmingham 2009).

[22] B. T. Fairbairn, Symmetric presentations of Coxeter groups, preprint.

[23] M. S. Mohamed, *Computational methods in symmetric generation of groups* (PhD thesis, Birmingham 1998).

[24] J. A. Todd and H. S. M. Coxeter, A practical method for enumerating cosets of finite abstract groups, *Proc. Edinb. Math. Soc.* **5** (1936), 26–34.

[25] S. Whyte, *Symmetric generation: permutation images and irreducible monomial representations* (PhD thesis, Birmingham 2006).

DISCRIMINATING GROUPS: A COMPREHENSIVE OVERVIEW

BENJAMIN FINE*, ANTHONY M. GAGLIONE†, ALEXEI MYASNIKOV§,
GERHARD ROSENBERGER‡ and DENNIS SPELLMAN*

*Department of Mathematics, Fairfield University, Fairfield, CT 06430, USA
Email: fine@mail.fairfield.edu, dspellman@stagweb.fairfield.edu

†Department of Mathematics, U.S. Naval Academy, Annapolis, MD 21402, USA
Email: amg@usna.edu

§Department of Mathematics, McGill University, Montreal, Quebec, Canada
Email: alexeim@math.mcgill.ca

‡Department of Mathematics, University of Dortmund, Dortmund, Germany
Email: Gerhard.Rosenberger@math.uni-dortmund.de

Abstract

Discriminating groups were introduced by Baumslag, Myasnikov and Remeslennikov as an outgrowth of their theory of algebraic geometry over groups. Algebraic
geometry over groups was the main method of attack used by Kharlampovich and
Myasnikov in their solution of the celebrated Tarski conjectures. The class of discriminating groups, however, has taken on a life of its own and has been an object
of a considerable amount of study. In this paper we survey the large array of results
concerning the class of discriminating groups that have been developed over the
past decade.

1 Introduction

Discriminating groups were introduced by Baumslag, Myasnikov and Remeslennikov as an outgrowth of their theory of algebraic geometry over groups. Algebraic
geometry over groups was the main method of attack used by O. Kharlampovich
and A. Myasnikov in their solution of the celebrated Tarski conjectures. The class
of discriminating groups, however, has taken on a life of its own and has been an
object of a considerable amount of study. In this paper we survey the large array
of results concerning the class of discriminating groups that have been developed
over the past decade.

In Section 1, we define discrimination for groups and describe its ties to other
areas. Also the concept of trivially discriminating (TD) groups is introduced, and
the concept of squarelike groups is defined. It is also indicated how to define discrimination for arbitrary algebraic systems. It is also shown how to generalize the
concept of squarelike to arbitrary algebras. It turns out that in the applications of
discriminating groups it is important not only to know when a group is discriminating, but also to know which groups are not discriminating. These are reviewed
in Sections 2 and 3. In Section 4, examples of nontrivially discriminating groups
are presented. Section 5 gives examples and properties of squarelike groups. In

Section 6, we discuss axiom schema for discriminating and squarelike groups. Also in this section, we outline the proof of the fact that the class of squarelike groups is the axiomatic closure of the class of discriminating groups. Finally in Section 7, we discuss different notions of discrimination and the relations between them.

2 Discrimination and discriminating groups

In the 1940's Tarski conjectured that all nonabelian free groups have the same elementary or first-order theory and that the elementary theory of the nonabelian free groups is decidable. These Tarski conjectures constituted one of the major open problems in both group theory (see [32]) and logic until their positive solution over the past decade by Kharlampovich and Myasnikov [25, 26, 27, 28, 29] and independently by Sela [38, 39, 40, 41, 42, 43]. Although these solutions have not garnered the same public attention as the recent solutions of other major problems such as the Poincaré conjecture, within the infinite group theoretic community they are recognized as major achievements. The solutions were built upon several foundations, one of which was the development of algebraic geometry over groups.

In the 1990's, in an effort to attack the Tarski problems, G. Baumslag, A. Myasnikov and V. Remeslennikov (see [5, 6] and references therein) developed an extensive theory of algebraic geometry over groups, mimicking classical algebraic geometry. Sela used a slightly different but essentially equivalent approach that he called diophantine geometry over groups.

What we look at in this paper is a sidelight to the Tarski solution that has taken on a life of its own and is of independent interest. As a spin-off of algebraic geometry over groups, Baumslag, Myasnikov and Remeslennikov introduced the class of discriminating groups [6]. As a result of the interesting examples and extensive ties this class has with both logic and algebraic geometry, there has been a tremendous amount of study and a wide array of results concerning discriminating groups. In this article, as the title describes, we present a comprehensive overview.

If G and H are groups then G *separates* H provided that to every nontrivial element h of H there is a homomorphism $\varphi_h : H \to G$ such that $\varphi_h(h) \neq 1$. The group G *discriminates* H if to every finite nonempty set S of nontrivial elements of H there is a homomorphism $\varphi_S : H \to G$ such that $\varphi_S(s) \neq 1$ for all $s \in S$. The group G is *discriminating* provided that it discriminates every group it separates.

By analogy with classical algebraic geometry, we may view the discrimination of H by G as an approximation to H, much like the localization of a ring at a prime. (Think of a set of generators for H as a set of variables.)

We note that discrimination can be defined in a much more general algebraic context than just for groups. This was noted by Belegradek in [7]. For complete proofs of the assertions made here about universal algebras see [19]. To do this, we proceed as follows.

Let $\mathcal{V}$ be a variety of algebraic structures of some fixed type, Ω, where Ω is an operator domain (see [10]). For example, $\mathcal{V}$ could be the variety of all groups, in which case Ω could be taken to contain a binary operator $\cdot$, a unary operator $^{-1}$, and a constant symbol. Or $\mathcal{V}$ could be the variety of Lie algebras over an integral domain R (so R could be a field or $\mathbb{Z}$). Even more, $\mathcal{V}$ could be the variety of semigroups, etc.

Now fix Ω and and also fix $\mathcal{V}$. Let $A, B \in \mathcal{V}$. We say that A *separates* B provided for any $(x, y) \in B^2$ with $x \neq y$ there is a homomorphism $\varphi_{(x,y)} : B \to A$ such that $\varphi_{(x,y)}(x) \neq \varphi_{(x,y)}(y)$. We say that A *discriminates* B if to every finite nonempty set $S = \{(x_1, y_1), \ldots, (x_n, y_n)\}$ of elements of B^2 with $x_i \neq y_i$, $i = 1, \ldots, n$, there is a homomorphism $\varphi_S : B \to A$ such that $\varphi_S(x_i) \neq \varphi_S(y_i)$ for all $i = 1, \ldots, n$. The algebra A is *discriminating* if it discriminates every algebra that it separates.

The following is the main criterion for deciding whether a group is discriminating.

Lemma 2.1 ([6]) *A group G is discriminating if and only if G discriminates $G \times G$.*

Since this is fundamental we will give the proof which is very easy:

Proof Since G clearly separates $G \times G$, if it is discriminating it will discriminate $G \times G$. Conversely, suppose G discriminates $G^2 = G \times G$. By induction, then G will discriminate G^n for any positive integer n. To see this, we assume that G discriminates G^n and consider $G^{n+1} = G^n \times G$. Let $g_1, \ldots, g_k$ be finitely many elements of $G^{n+1} \backslash \{1\}$. Without loss of generality, we may assume that $g_i = (g_{i1}, g_{i2})$ where $g_{i1} \in G^n$, $g_{i2} \in G$ and $g_{i2} = 1$ for $1 \leq i \leq m$, but $g_{i2} \neq 1$ for $m + 1 \leq i \leq k$. Here $0 \leq m \leq k$ and when $m = 0$ all $g_{i1} = 1$ and necessarily all $g_{i2} \neq 1$. But when $m = k$ all $g_{i1} \neq 1$. Then since G discriminates G^n, there is a homomorphism $\theta : G^n \to G$ such that $\theta(g_{i1}) \neq 1$ for $1 \leq i \leq m$, assuming $m \neq 0$. If $m = 0$, then take $\theta(g) = 1$ for all $g \in G^n$. Consider the homomorphism $\phi : G^{n+1} \to G^2$ defined by $\phi(g, h) = (\theta(g), h)$ where $g \in G^n$ and $h \in G$. Clearly, $\phi(g_1), \ldots, \phi(g_k)$ are then nontrivial elements of G^2. So since G discriminates G^2 there exists another homomorphism $\varphi : G^2 \to G$ which does not annihilate any of the $\phi(g_i)$ for $1 \leq i \leq k$. Taking $\delta = \varphi \circ \phi$ completes the induction. Suppose that G separates a group H; we want to show that it discriminates H. Let $h_1, \ldots, h_n$ be nontrivial elements in H. Then for each h_i there is a homomorphism $\varphi_i : H \to G$ such that $\varphi_i(h_i) \neq 1$. Taking $\varphi = \varphi_1 \times \cdots \times \varphi_n$ and using the fact that G discriminates G^n gives us that G discriminates H. $\square$

We note that the above proof translates over exactly to show that if A is an algebra of type Ω, then A is discriminating if and only if A discriminates A^2.

It is clear from Lemma 2.1 that if $G \times G$ embeds in G, then G is discriminating. If this is the case we say that G is *trivially discriminating* which we abbreviate as *TD*. In [12], it was asked whether there exist finitely generated nonabelian, nontrivially discriminating groups. We will give some examples of these later in the survey.

2.1 Some ties to logic

We now present some ties between discriminating groups and logic.

Let L be the first order language with equality and a binary operation symbol $\cdot$ (usually omitted in favor of juxtaposition), a unary operation symbol $^{-1}$ and a constant symbol 1. We call L the language of group theory. A *universal sentence* of L is one of the form $\forall \overline{x} \varphi(\overline{x})$ where $\overline{x}$ is a tuple of distinct variables and $\varphi(\overline{x})$ is a quantifier free formula containing at most the variables in $\overline{x}$. Similarly, an existential sentence of L is one of the form $\exists \overline{x} \varphi(\overline{x})$ where $\overline{x}$ and $\varphi(\overline{x})$ are as before. Vacuous quantifications are permitted and so a quantifier free sentence of L is considered a special case both of universal sentences and existential sentences. Observe that the negation of a universal sentence is logically equivalent to an existential sentence and vice-versa. If G is a group, then the *theory* (respectively, *universal theory*) of G, written $\text{Th}(G)$ (respectively, $\text{Th}_\forall(G)$) is the set of all sentences of L (respectively, all universal sentences of L) true in G. Two groups G and H are *elementarily equivalent* (respectively, *universally equivalent*) provided they have the same theory, i.e., $\text{Th}(G) = \text{Th}(H)$ (respectively, $\text{Th}_\forall(G) = \text{Th}_\forall(H)$). We denote this by $G \equiv H$ (respectively, $G \equiv_\forall H$). Observe that two groups satisfy precisely the same universal sentences if and only if they satisfy precisely the same existential sentences. Notice that it is easy to prove that all finitely generated free groups of rank $r \geq 2$ have the same universal theory.

Theorem 2.2 ([12]) *If G is discriminating, then $G \times G \equiv_\forall G$.*

In order to better capture the axiomatic properties of discriminating groups, we give:

Definition 2.3 ([13]) *The group G is* squarelike *provided $G \times G \equiv_\forall G$.*

Hence, every discriminating group is squarelike, but (as we shall see) the converse does not hold.

Corresponding to each operator domain Ω is the first order language L_Ω. If A is an algebra of type Ω, and if $\forall \overline{x} \varphi(\overline{x})$ is a universal sentence of L_Ω holding in a direct square A^2, then it must hold in A, since the diagonal map $\delta : A \to A^2$, $a \mapsto (a, a)$ embeds A as a subalgebra of A^2 and universal sentences are preserved in subalgebras. If A is discriminating, then it is not hard to show that $A^2 \equiv_\forall A$. We can then introduce the concept of *squarelike* just as in Definition 2.3 and say that the algebra A is squarelike provided $A^2 \equiv_\forall A$.

2.2   Some ties to the algebraic geometry of groups

As mentioned, discriminating groups arose out of the study of algebraic geometry over groups. Here we present some of the ties.

Algebraic geometry over groups is most useful in its relativized version, where a group H plays the role of the ring of scalars. In this theory an H-group G is a just a group containing a distinguished copy of H. For our purposes, we consider the case $H = 1$, so that an H-group G is just a group. Let $n \in \mathbb{N}$ and let $X_n = \{x_1, \ldots, x_n\}$ be a set of distinct variables. Let S be a set of words on $X_n \cup X_n^{-1}$. If G is a group, then by *a solution in G to the system $S = 1$* is meant an ordered n-tuple $\mathbf{g} = (g_1, \ldots, g_n) \in G^n$ such that $s(g_1, \ldots, g_n) = 1$ for all $s(x_1, \ldots, x_n) \in S$. We write $S(\mathbf{g}) = 1$ in that event. The solution set of the system is then denoted $V_G(S) = \{\mathbf{g} \in G^n : S(\mathbf{g}) = 1\}$. These are the *affine algebraic subsets* of G^n and they form a closed subbase for a topology (the *Zariski topology*) on G^n. The Zariski topology will be Noetherian if every descending chain

$$A_0 \supseteq A_1 \supseteq \cdots \supseteq A_n \supseteq \cdots$$

of closed sets stabilizes after finitely many steps, i.e., $A_N = A_{N+1} = \cdots$. A necessary and sufficient condition that the Zariski topology be Noetherian for all n is given in the following:

Definition 2.4 The group G is *equationally Noetherian* if for every natural number $n \in \mathbb{N}$, and every system $S = 1$ of equations in n unknowns, there is a finite subset $S_o \subseteq S$ such that $V_G(S) = V_G(S_o)$.

Examples of equationally Noetherian groups are groups linear over a commutative Noetherian ring with 1. In particular, every group linear over a field is equationally Noetherian. Also, according to a theorem of Sela, torsion free word-hyperbolic groups are equationally Noetherian.

A first order expression $\forall \overline{x} \, (\wedge_i(w_i(\overline{x}) = 1) \to (u(\overline{x}) = 1))$, where $u(\overline{x})$ and the $w_i(\overline{x})$ are words in the variables in $\overline{x}$ and their formal inverses, is a *quasi-identity*.

Observe that every identity $\forall \overline{x}(A(\overline{x}) = a(\overline{x}))$ is equivalent to a quasi-identity $\forall \overline{x}((1 = 1) \to (A(\overline{x}) = a(\overline{x})))$. So laws and the group axioms themselves may be viewed as instances of quasi-identities. If Φ is a consistent set of sentences of L containing the group axioms, then the class $\mathrm{M}(\Phi)$ of all groups G satisfying every sentence φ in Φ is the *model class* of Φ. Note that every such class is assumed to be nonempty and also is closed under isomorphism.

Definition 2.5 The model class of a set of quasi-identities including the group axioms is a *quasivariety* of groups. If G is a group then the quasivariety generated by G is denoted qvar(G).

Quasivarieties are examples of *axiomatic classes*. That is, they are model classes of sets of sentences of L, the language of group theory. Quasivarieties are closed under (unrestricted) direct products. They are also closed under subgroups, since they have a set of universal axioms. Put another way, quasivarieties are axiomatic *prevarieties*. The class of all groups is a quasivariety of groups, and the intersection of any family of quasivarieties is again a quasivariety. For that matter, the intersection of any family of universally axiomatizable model classes (i.e. axiomatic classes having a set of universal axioms) is again a universally axiomatizable model class. It follows that if $\mathcal{X}$ is any class of groups then there is a least quasivariety, denoted qvar($\mathcal{X}$), containing $\mathcal{X}$, and a least universally axiomatizable model class, denoted ucl($\mathcal{X}$), containing $\mathcal{X}$. We call qvar($\mathcal{X}$) the *quasivariety generated by $\mathcal{X}$* and ucl($\mathcal{X}$) the *universal closure of $\mathcal{X}$*. If $\mathcal{X} = \{G\}$ is a singleton we write qvar(G) and ucl(G) for qvar($\mathcal{X}$) and ucl($\mathcal{X}$) respectively. Then qvar(G) is the model class of the quasi-identities satisfied by G and ucl(G) is the model class of the universal sentences satisfied by G. In general ucl(G) is a proper subclass of qvar(G). (The more properties a structure is required to satisfy, the fewer structures can satisfy them in general.)

Tying all of these concepts together is:

Theorem 2.6 ([12]) *Let G be a finitely generated equationally Noetherian group. Then G is discriminating if and only if $\mathrm{qvar}(G) = \mathrm{ucl}(G)$.*

Now, for the operator domain $\Omega = (F, C, d)$, we always have the trivial Ω-algebra (unique up to isomorphism) $\mathbb{A} = (\{\theta\}, (f_{\mathbb{A}})_{f \in F}, (c_{\mathbb{A}})_{c \in C})$, where $c_{\mathbb{A}} = \theta$ for all $c \in C$ and $f_{\mathbb{A}}(\theta, \ldots, \theta) = \theta$ for all $f \in F$ (i.e., $\mathbb{A}$ is the one element Ω-algebra). Now let $\mathcal{Q}$ be a nonempty class of Ω-algebras closed under isomorphism. Then $\mathcal{Q}$ is a quasivariety of Ω-algebras provided that $\mathcal{Q}$ is the model class of a set of quasi-identities, i.e., a set of universal sentences of the form

$$\forall \overline{x} \left(\bigwedge_i (u_i(\overline{x}) = v_i(\overline{x})) \rightarrow (u(\overline{x}) = v(\overline{x})) \right)$$

where the u_i, v_i, u. and v are terms of L_Ω. This is equivalent to being an axiomatic class of Ω-algebras closed under taking subalgebras, direct products, and containing the trivial algebra. For groups, one does not have to postulate the condition about containing the trivial algebra because the trivial group is a subgroup of any group. However, for semigroups for example, one needs that extra condition. This fact that quasivarieties must contain the trivial algebra forces us to restrict ourselves to the following slightly less general structures. If we do so, however, all our results will go through with the same proofs as for groups.

Here is how we restrict our Ω-algebras. Let $\Omega = (F, C, d)$ be an operator domain with the set C of constant symbols a singleton $C = \{\theta\}$. Let $\mathcal{V}$ be any variety of Ω-algebras satisfying at least the laws $f(\theta, \ldots, \theta) = \theta$ as f varies over F. We note then that some of our results will fail for semigroups, but still hold for monoids.

2.3   Abelian discriminating groups

In the original paper by Baumslag, Myasnikov and Remeslennikov [6] the question was raised as to whether the abelian discriminating groups can be characterized. In [13] this was extended to ask whether the abelian squarelike groups could be characterized. Also [13] settled this by giving such a characterization. If there is any torsion this can be answered immediately.

Lemma 2.7 *Suppose the set of nontrivial elements of finite order in the group G is finite and nonempty. Then G is not squarelike. Hence, G is not discriminating.*

We immediately deduce that no nontrivial finitely generated nilpotent group with torsion can be squarelike. In particular, no nontrivial finitely generated abelian group with torsion, and no nontrivial finite group, can be squarelike. In the torsion free case we always get discrimination.

Lemma 2.8 *Every torsion free abelian group is discriminating. Hence, every torsion free abelian group is squarelike.*

So, among finitely generated abelian groups, the discriminating groups and the squarelike groups are precisely the torsion free ones. A moment's reflection produces nontrivial discriminating abelian groups (necessarily not finitely generated) with torsion. For example, every group free of infinite rank in the variety of abelian groups of exponent dividing n for any fixed integer $n > 1$ is trivially discriminating. Baumslag, Myasnikov and Remeslennikov in [6] give a partial answer to the total characterization of abelian discriminating groups. In order to state their result, we must introduce some terminology and notation.

Let A be an abelian group written additively. As a reference for all facts about abelian groups see [30]. We say that an element $a \in A$ has *infinite p-height* if the equation $p^k x = a$ has a solution in A for every positive integer k. We also must introduce the *Szmielew invariants* of A. Given an integer $m > 0$ and a family of elements (a_i) in A, (a_i) is *linearly independent modulo m* provided $\Sigma_i n_i a_i = 0$ implies $n_i \equiv 0 \pmod{m}$ for all i; (a_i) is *linearly independent modulo m in the stronger sense* provided $\Sigma_i n_i a_i \in mA$ implies the coefficients $n_i \equiv 0 \pmod{m}$ for all i. For each prime p and positive integer k we define three ranks, each of which is a nonnegative integer or the symbol ∞.

1. $\rho_{[p,k]}^{(1)}(A)$ is the maximum number of elements of A of order p^k and linearly independent modulo p^k.

2. $\rho_{[p,k]}^{(2)}(A)$ is the maximum number of elements of A linearly independent modulo p^k in the stronger sense.

3. $\rho_{[p,k]}^{(3)}(A)$ is the maximum number of elements of A of order p^k and linearly independent modulo p^k in the stronger sense.

Proposition 2.9 ([44]) *Let A and B be abelian groups. Then A and B are elementarily equivalent if and only if the following two properties are satisfied.*

1. *A and B either both have the same finite exponent or both have infinite exponent.*

2. *For all primes p and positive integers k, $\rho_{[p,k]}^{(i)}(A) = \rho_{[p,k]}^{(i)}(B)$ for $i = 1, 2, 3$.*

Proposition 2.10 ([6]) *Let A be a torsion abelian group such that, for each prime p, the p-primary component of A modulo its maximal divisible subgroup contains no nontrivial elements of infinite p-height. Then A is discriminating if and only if, for each prime p, the following two properties are satisfied.*

1. *For all positive integers k, $\rho_{[p,k]}^{(1)}(A)$ is either 0 or ∞.*

2. *The rank of the maximal divisible subgroup of the p-primary component of A is either 0 or infinite.*

The question of a complete characterization of all abelian discriminating groups remains open. Extending the ideas of [6], Fine, Gaglione, Mayasnikov and Spellman [13] developed a complete characterization of abelian squarelike groups.

Proposition 2.11 ([13]) *Let A be an abelian group. Then A is squarelike if and only if, for each prime p and positive integer k, $\rho_{[p,k]}^{(1)}(A)$ is either 0 or ∞.*

3   Positive examples: discriminating groups

In the theory as applied to the algebraic geometry of groups it is important to know both positive examples of discriminating groups and negative examples. The original positive examples of discriminating groups were either abelian or trivially discriminating. Recall that a group is trivially discriminating if it embeds its direct square. In this section we present these positive examples. We will return later to examples of nontrivially discriminating groups.

Most of the first-discovered positive examples were also "universal type groups".

Example 3.1 (Abelian groups) Both torsion free abelian groups, and the torsion abelian groups described above [6, 12].

Example 3.2 (Higman's universal group H) Higman constructed this group [20], which is a finitely presented group which embeds every other finitely presented group. Since this group is finitely presented its direct square is also finitely presented. Therefore it embeds its direct square and hence is trivially discriminating.

Example 3.3 (Thompson's group F) This group consists of all orientation preserving piecewise linear homeomorphisms from the unit interval $[0, 1]$ onto itself that are differentiable except at finitely many dyadic rational numbers, and such that on intervals of differentiability the derivatives are powers of 2. As in Higman's group, this group embeds its direct square and hence is trivially discriminating [12]. This group is also finitely presented.

Example 3.4 (The commutator subgroup of the Gupta–Sidki–Grigorchuk Group) For each odd prime p, Gupta and Sidki and independently Grigorchuk constructed a group H_p which is a subgroup of the group of automorphisms of a rooted tree. H_p is a 2-generator infinite p-group. Consider the commutator subgroup H_p' of H_p. It can be shown that H_p', while finitely generated, is not finitely presentable. It can also be shown that H_p' is trivially discriminating.

Example 3.5 (Groups that are isomorphic to their direct squares)
Examples of nontrivial finitely generated groups isomorphic to their direct squares were first constructed by Tyrer Jones [23] and subsequently by Hirshon and Meier [22]. The question of whether or not there exists a finitely presented group isomorphic to its direct square remains open.

Example 3.6 (The Grigorchuk group of intermediate growth)
Grigorchuk developed a group of intermediate growth [12]. It was shown in that paper, due to conversations with Grigorchuk, that this group is discriminating. Whether it is trivially discriminating or not is open.

Example 3.7 In [2], the following was given without proof, so we prove it now.

Lemma 3.8 *A simple discriminating group is trivially discriminating.*

Proof Suppose that G is a simple discriminating group. We wish to show that $G \times G$ embeds into G. Let g be any nontrivial element of G and consider the elements $x = (g, 1), y = (1, g) \in G \times G$. Since G is discriminating there exists a homomorphism

$$\phi \colon G \times G \to G$$

such that both $\phi(x)$ and $\phi(y)$ are nontrivial elements of G. Let $G_1 = G \times \{1\}$, $G_2 = \{1\} \times G$, $H_1 = \phi(G_1)$ and $H_2 = \phi(G_2)$. Since G is simple, it follows that $G \cong G_1 \cong \phi(G_1) = H_1$ and $G \cong G_2 \cong \phi(G_2) = H_2$. The elements in H_1 and H_2 commute element-wise and hence their intersection is central in H_1. Since H_1 is simple, this intersection must be trivial. It follows that the subgroup of G generated by H_1 and H_2 is their direct product $H_1 \times H_2$. But from the above, this is isomorphic to $G \times G$. Therefore $G \times G$ embeds into G and hence G is TD. □

For each integer $n \geq 2$ and each integer $r \geq 1$, Higman defined a group $G_{n,r}$ (see [21]) as follows. Let $\mathcal{V}_n$ be the variety of all algebras with one n-ary operation λ and n unary operations $\alpha_1, \ldots, \alpha_n$ subject to the laws

$$\lambda(\alpha_1(x), \ldots, \alpha_n(x)) = x \text{ and } \alpha_i(\lambda(x_1, \ldots, x_n)) = x_i, \ 1 \leq i \leq n.$$

If $\mathcal{V}_{n,r}$ is an algebra free on r generators in $\mathcal{V}_n$, then we let $G_{n,r} = Aut(\mathcal{V}_{n,r})$ be its group of automorphisms. Higman proved that the $G_{n,r}$ are finitely presented, when n is even $G_{n,r}$ is simple, and when n is odd $G_{n,r}$ contains a simple subgroup $G^+_{n,r}$ of index 2. Setting $G^+_{n,r} = G_{n,r}$ when n is even, he showed that, for fixed r, $G_{m,r} \cong G_{n,r}$ implies $m = n$. Thus he found an infinite family of finitely presented infinite simple groups. The $G_{n,r}$ are all trivially discriminating.

Example 3.9 The group S_ω of all permutations of $\mathbb{N}$ which move only finitely many integers, and its subgroup A_ω of even permutations, are trivially discriminating.

Example 3.10 The existence of a finitely generated but not finitely presentable trivially discriminating group G_1 (e.g. the commutator subgroup of a Gupta–Sidki–Grigorchuk group) together with the existence of a universal finitely presented group G_2 (Higman's group) allowed us to construct in [17] a group $G = G_1 \times G_2$ which is proven in that paper to be a finitely generated discriminating group which is neither finitely presented nor equationally Noetherian.

This led to the obvious question: *Does there exist a finitely presented nonabelian nontrivially discriminating group?* We will return to this shortly.

4   Negative examples: nondiscriminating groups

In the application of algebraic geometry it is as important to know whether a group is discriminating, as whether it is not discriminating. This raised further general questions on the nondiscrimination of various classes of groups in various varieties.

The general idea to show nondiscrimination of a group G is to produce some universal property which is true in G but cannot be true in $G \times G$, and then apply

Theorem 2.2 or to find a number — dimension etc. — which is additive so cannot hold in $G \times G$ and in G.

Definition 4.1 The group G is *commutative transitive* or *CT* provided the centralizer of every nontrivial element is abelian.

We observe that the group G is CT if and only if it satisfies the universal sentence

$$\forall x, y, z(((y \neq 1) \wedge (xy = yx) \wedge (yz = zy)) \rightarrow (xz = zx))$$

If G is nonabelian then $G \times G$ cannot satisfy this sentence. Therefore the following lemma follows directly.

Lemma 4.2 *A nonabelian CT group is nondiscriminating.*

In particular we obtain the negative examples:
1. Any torsion free hyperbolic group and in particular any nonabelian free group is nondiscriminating.
2. Nonabelian free solvable groups and their nonabelian subgroups are nondiscriminating.
3. The free product of two nondiscriminating groups is nondiscriminating.

In another direction, if we think of discrimination as having the direct square $G \times G$ almost embed in the group G, then this would be impossible for a finite group. In particular it is impossible if the torsion elements form a finite subgroup.

Lemma 4.3 *If the torsion elements of a group G form a nontrivial finite subgroup then G is nondiscriminating.*

This leads to the further negative examples:
4. Any finitely generated nilpotent group with nontrivial torsion is nondiscriminating
5. Any nontrivial finite group is nondiscriminating.

One of the major questions raised in the early work on discriminating groups [6, 12] was whether or not a nonabelian nilpotent group can be discriminating. The full result was:

Theorem 4.4 *A finitely generated nilpotent group is discriminating if and only if it is torsion free abelian.*

There were several stages to obtaining the final result Theorem 4.4, although the final proof was rather direct. In [12] it was shown that nonabelian finitely generated free nilpotent groups are nondiscriminating. To do this an extension of commutative transitivity was introduced. Myasnikov and Shumyatsky [35] introduced the concept of centralizer dimensions (see [35]) and showed that a group with finite centralizer dimension is nondiscriminating. As a consequence of this it follows that a finitely generated nilpotent group is discriminating if and only if it is torsion free abelian. Their method yielded the following further result.

Theorem 4.5 *A finitely generated linear group is discriminating only if it is torsion free abelian.*

Independently, Baumslag, Fine, Gaglione and Spellman, using vector space dimension, proved that a nilpotent group is discriminating if and only if its Malcev completion is discriminating. Theorem 4.4 can then be recovered from this result.

Theorem 4.6 ([3]) *The Malcev completion of a finitely generated torsion free nilpotent group is discriminating if and only if it is abelian.*

The nondiscrimination of nonabelian nilpotent groups leads to the question as to whether any nonabelian group in a variety must be nondiscriminating. There is some further evidence for this. Kassabov [24] proved it for solvable groups.

Theorem 4.7 ([24]) *A finitely generated solvable group is discriminating only if it is torsion free abelian.*

Further in this direction, Baumslag, Fine, Gaglione and Spellman [2] proved some general results on nondiscrimination of relatively free groups.

Theorem 4.8 ([2]) *Let F be a nonabelian free group and let R be a nontrivial, normal subgroup of F such that F/R is torsion free and contains a free abelian group of rank two. Then $F/V(R)$ is not discriminating.*

Theorem 4.9 ([2]) *Let V and U be two varieties. If the variety V contains the infinite cyclic group, then the nonabelian free groups in the product variety VU are not discriminating.*

Theorem 4.10 ([2]) *For all sufficiently large primes p, the nonabelian free groups G in the variety of all groups of exponent p are not discriminating.*

We pose the following question: *Does there exist a nonabelian, discriminating, relatively free group?*

5 Nontrivially discriminating groups

Recall the question of whether there exists a finitely generated nonabelian nontrivially discriminating group. That is, a finitely generated discriminating group which does not embed its direct square. One possibility (still open) was the Grigorchuk groups of intermediate growth. Here we give other examples — done with Peter Neumann and Gilbert Baumslag. We present two classes of examples. The groups in the first class are finitely presented. The second is a class of groups studied by B. H. Neumann which are nontrivially discriminating and finitely generated but not finitely presented (see [4]).

Example 5.1 To describe the first class of groups let X be a nonabelian, finitely generated, torsion free, nilpotent group and let Y be one of the infinite simple groups $G_{n,r}$, as described by Higman. The following properties of X and Y are stated without proof. As a reference for properties of X see [1] and for properties of Y see [21].

(a) X and Y are finitely presented;

(b) X is residually finite;

(c) every finite group is embeddable in Y;

(d) X is not embeddable in Y; in fact, a torsion free nilpotent subgroup of Y is abelian;

(e) $Y \times Y \times Y$ is embeddable in Y. (This follows from the fact that $V_{n,r} \cong V_{n,s}$ in the notation of [21] if and only if $r \equiv s \pmod{n-1}$.)

Our class $\mathcal{G}_1$ consists of the groups $X \times Y$.

Theorem 5.2 ([17]) *Each group $G \in \mathcal{G}_1$ is a nonabelian, finitely presented nontrivially discriminating group.*

We present the proof to exhibit the techniques involved.

Proof Clearly each group $G \in \mathcal{G}_1$ is nonabelian and finitely presented. Then suppose $G \in \mathcal{G}_1$. We must show that G is discriminating but that G does not embed its direct square.

Since $G = X \times Y$, with X and Y as described above, to prove that G is discriminating it suffices to show that Y discriminates any group of the form $W \times Y \times Y$, where W is residually finite. This is sufficient, for then $G = X \times Y$ will discriminate its direct square and hence be discriminating. For this purpose, let $h_1, \ldots, h_n$ be finitely many non-identity elements of $W \times Y \times Y$. Write $h_i = (a_i, b_i)$ where $a_i \in W$ and $b_i \in Y \times Y$. Without loss of generality, we may suppose that $a_i \neq 1$ for $1 \leq i \leq m$ and $a_i = 1$ for $m + 1 \leq i \leq n$. Since W is residually finite, there exists a finite group V and a homomorphism $\alpha : W \to V$ such that $\alpha(a_i) \neq 1$ for $1 \leq i \leq m$. By (c) above we may embed V into Y and so we get a homomorphism $\beta : W \to Y$ such that $\beta(a_i) \neq 1$ for $1 \leq i \leq m$. Let γ be the homomorphism $\gamma : W \times Y \times Y \to Y \times Y \times Y$ defined by $\gamma(w, y, z) = (\beta(w).y, z)$. By (e) above, let δ be an embedding of $Y \times Y \times Y$ into Y. If we put $\varphi = \delta \circ \gamma$, then clearly φ is a homomorphism fron $W \times Y \times Y$ into Y such that $\varphi(h_i) \neq 1$ for all $i = 1, \ldots, n$. Thus Y discriminates $W \times Y \times Y$ and therefore it follows that G discriminates $G \times G$. Thus G is discriminating.

To show that G is nontrivially discriminating we must show that $G \times G$ is not embeddable into G. Assume, to deduce a contradiction, that $G \times G \hookrightarrow G$. Restricting the embedding we get an embedding $X \times X \hookrightarrow X \times Y$. Let A be the projection of $X \times X$ into Y. Then $X \times X \leq X \times A$, so $X' \times X' \leq X' \times A'$. But this tells us that if h denotes the Hirsch length of the commutator subgroup X', then $h \leq 2h \leq h$ since A' is finite. This contradicts the fact that X is nonabelian and torsion free nilpotent. Therefore $G \times G$ is not embeddable into G and hence is nontrivially discriminating. $\square$

Example 5.3 The second class of nontrivally discriminating groups, which we will denote $\mathcal{G}_2$, was introduced by B. H. Neumann in a different context. Let

$\mathbf{n} = (n_1, n_2, \dots)$ be a strictly increasing sequence of odd positive integers with $n_1 \geq 5$. Let $\Lambda_r = \{1, 2, \dots, n_r\}$ and let G_r be the alternating group on Λ_r. In G_r consider the two elements

$$x_r = (123), \quad y_r = (12 \dots n_r)$$

In the unrestricted direct product $\Pi_r G_r$ define

$$x = (x_1, x_2, \dots), \quad y = (y_1, y_2, \dots)$$

and let

$$G = G_\mathbf{n} = \langle x, y \rangle \subseteq \prod_r G_r.$$

B. H. Neumann (see [36]) proved that the restricted direct product $D = G_1 \times G_2 \times \cdots$ is a subgroup of G (see also [31]). The class $\mathcal{G}_2$ consists of all the groups $G = G_\mathbf{n}$ constructed as above.

Theorem 5.4 *Each group $G \in \mathcal{G}_2$ is a finitely generated nontrivially discriminating group.*

The methods used in constructing these non-TD examples can be generalized via the use of ascending chains of subgroups.

Theorem 5.5 ([2]) *Let U be the union of a properly ascending chain of subgroups*

$$U_1 < U_2 < \cdots < U_n < \cdots$$

and let P be the unrestricted direct product of the U_i:

$$P = \prod_{i=1}^{\overline{\infty}} U_i \ .$$

Furthermore let Q be the restricted direct product

$$Q = \prod_{i=1}^{\infty} U_i.$$

Then every subgroup G of P containing Q is discriminating (where here we view Q as a subgroup of P).

Proof The proof of this result is straightforward. To this end, let $D = G \times G$ and let $d_1 = (a_1, b_1), \dots, d_k = (a_k, b_k)$ be finitely many nontrivial elements of D. Each of the elements of G can be viewed as a sequence of elements whose n-th term is contained in U_n. Since G contains Q and the series of subgroups U_j is increasing, there exists an integer α such that all of $a_1, b_1, \dots, a_k, b_k \in U_\alpha$. We now define a homomorphism of D into G, by projecting the first coordinates of the elements of D to U_α and the second coordinates to $U_{\alpha+1}$. The upshot of this is that we have defined a homomorphism θ of $G \times G$ into $U_\alpha \times U_{\alpha+1}$. If we view $U_\alpha \times U_{\alpha+1}$ as a subgroup of G, it follows that θ is a homomorphism of $G \times G$ into G mapping the given elements $d_1, \dots, d_k$ nontrivially. This completes the proof. $\square$

This final theorem can be used to recover the proofs of the preceding examples and to obtain the fact that there exist continuously many 2-generator non-TD discriminating groups. In particular, let A_i denote the alternating group of degree i and let $G_{\mathbf{n}}$ be the subgroup defined before the statement of Theorem 5.4. Then B. H. Neumann not only proved that $G_{\mathbf{n}}$ contains the restricted direct product of the A_{n_i} but also that

$$G_{\mathbf{n}} \cong G_{\mathbf{s}} \text{ if and only if } \mathbf{n} = \mathbf{s}.$$

This implies that there are continuously many non-isomorphic 2-generator groups. Theorem 5.5 then applies to these groups. Further, B. H. Neumann proved that the only finite normal subgroups of $G_{\mathbf{n}}$ are direct products of finitely many of the A_{n_i}. This implies, remembering that the alternating groups involved here are all simple, that none of the $G_{\mathbf{n}}$ are TD. We remark here that it has recently been proved that none of the groups $G_{\mathbf{n}}$ is finitely presented (see [4]).

Corollary 5.6 *The groups $G_{\mathbf{n}}$ are discriminating and further are not TD. Thus there exist continuously many 2-generator nontrivially discriminating groups.*

6   Squarelike groups

In order to better capture the axiomatic properties of discriminating groups, the class of squarelike groups was introduced. These are groups which share the same universal theory as their direct squares. The concept of a squarelike group was defined in Definition 2.3. Hence, every discriminating group is squarelike but (as we shall see) not conversely.

Theorem 6.1 *The class of discriminating groups is a proper subclass of the class of squarelike groups.*

To prove this we explicitly construct a squarelike which is provably nondiscriminating. This can also be proved with an axiomatic argument done below.

Example 6.2 Let H be the subgroup of the group of all permutations of the set $\mathbb{Z}$ of integers generated by the 3-cycle $\xi = (012)$ and the translation $\eta(n) = n + 1$ for all $n \in \mathbb{Z}$. The group $H = \langle \xi, \eta \rangle$ can also be described as the semidirect product of the group M, of all even parity permutations within the group N of all permutations of the set $\mathbb{Z}$ of integers which move only finitely many integers, by an infinite cyclic group $C = \langle c; \rangle$ where the automorphism $\alpha(c) : M \to M$ acts by $\alpha(c)(\pi)(n) = \pi(n-1) + 1$. (We say that $\alpha(c)$ acts by translation by 1.) Note that any bijection between $\mathbb{N}$ and $\mathbb{Z}$ induces an isomorphism between the infinite alternating group A_ω and M. The group H first appeared in print in the same paper of B. H. Neumann (see [36]) in which the uncountably many nontrivially discriminating groups $G_{\mathbf{n}}$ exhibited before were introduced. B. H. Neumann observed that (independent of $\mathbf{n}$) if K_0 is the restricted direct product of the family $\{A_{n_r}\}_{r \in \mathbb{N}}$ of alternating groups, then the quotient of $G = G_{\mathbf{n}}$ modulo K_0 is isomorphic to H. It can be proved that H is nondiscriminating, however, H is squarelike. (See [17].)

In an earlier paper this was done differently. Recall that a class of groups is *axiomatic* if it is the model class of a set of first order sentences (axioms). It can be shown that a class of groups is axiomatic if and only if it is closed under ultraproducts and elementary equivalence. (See [13].)

Theorem 6.3 ([13]) *The class of squarelike groups is axiomatic but the class of discriminating groups is not axiomatic.*

To prove that the discriminating groups are nonaxiomatic, a specific example of a nondiscriminating group (however squarelike) was constructed which was elementarily equivalent to a discriminating group. Hence the discriminating groups are not closed under elementary equivalence.

Originally the fact that the squarelike groups are axiomatic was proved by using the closure properties noted above. However subsequently V. H. Dyson [11] and Oleg Belegradek [7] discovered explicit axiom schema for the class of squarelike groups. Both of these are basically the same. We now state these.

To each ordered pair $(\mathbf{w}, \mathbf{u})$ of finite tuples of words on a fixed but arbitrary finite set $\{x_1, \ldots, x_n\}$ of distinct variables and their formal inverses, which we indicate by $\overline{x}$, we assign the following sentence $\sigma(\mathbf{w}, \mathbf{u})$ of L:

$$\forall \overline{x} \left(\bigwedge_i (w_i(\overline{x}) = 1) \to \bigvee_j (u_j(\overline{x}) = 1) \right) \to$$
$$\bigvee_j \forall \overline{x} \left(\bigwedge_i (w_i(\overline{x}) = 1) \to (u_j(\overline{x}) = 1) \right).$$

The contrapositive of $\sigma(\mathbf{w}, \mathbf{u})$ is (up to logical equivalence) the sentence $\tau(\mathbf{w}, \mathbf{u})$ asserting

$$\bigwedge_j \exists \overline{x} \left(\bigwedge_i (w_i(\overline{x}) = 1) \wedge (u_j(\overline{x}) \neq 1) \right) \to \exists \overline{x} \left(\bigwedge_i (w_i(\overline{x}) = 1) \wedge \bigwedge_j (u_j(\overline{x}) \neq 1) \right).$$

Theorem 6.4 ([11, 16]) *The class of squarelike groups is the model class of the group axioms and the sentences $\sigma(\mathbf{w}, \mathbf{u})$. Hence, the class of squarelike groups is axiomatic.*

Although the squarelike groups properly contain the discriminating groups they are very close and in fact correspond in the presence of finite presentation. In particular:

Theorem 6.5 ([13]) *Let G be a finitely presented group. Then G is discriminating if and only if G is squarelike.*

We give the proof because it's very pretty.

Proof Suppose G is a finitely presented group. If it is discriminating, it is squarelike. Now we suppose that G is squarelike and we must show that it is discriminating.

Let
$$G = \langle x_1, \ldots, x_n; \ R_1, \ldots, R_m \rangle$$
be a finite presentation for G where $R_i = R_i(x_1, \ldots, x_n)$ are words in $x_1, \ldots, x_n$. To show that G is discriminating we show that G discriminates $G \times G$.

A finite presentation for $G \times G$ is then given by

$$\begin{aligned}
G \times G \ = \ & \langle x_1, \ldots, x_n, y_1, \ldots, y_n; R_1(x_1, \ldots, x_n) = 1, \ldots, \\
& R_m(x_1, \ldots, x_n) = 1, \ R_1(y_1, \ldots, y_n) = 1, \ldots, \\
& R_m(y_1, \ldots, y_n) = 1, \ [x_i, y_j] = 1, \quad i, j = 1, \ldots, n \rangle.
\end{aligned}$$

Now suppose $W_1, \ldots, W_k$ are nontrivial elements of $G \times G$. Then each W_i, $i = 1, \ldots, k$ is given by $W_i = W_i(x_1, \ldots, x_n, y_1, \ldots, y_n)$, a word in the given generators of $G \times G$. Consider now the existential sentence

$$\exists \overline{x} \exists \overline{y} \left(\left(\bigwedge_{i=1}^m R_i(\overline{x}) = 1 \right) \wedge \left(\bigwedge_{i=1}^m R_i(\overline{y}) = 1 \right) \wedge \left(\bigwedge_{i,j} [x_i, y_j] = 1 \right) \wedge \left(\bigwedge_{i=1}^k W_i(\overline{x}, \overline{y}) \neq 1 \right) \right)$$

This existential sentence is clearly true in $G \times G$. Since G is squarelike, G and $G \times G$ have the same universal theory. Hence they have the same existential theory and therefore the above existential sentence is true in G. Therefore there exist elements $a_1, \ldots, a_n, b_1, \ldots, b_n$ in G such that $R_i(a_1, \ldots, a_n) = 1$ for $i = 1, \ldots, m$; $R_i(b_1, \ldots, b_n) = 1$ for $i = 1, \ldots, m$; $[a_i, b_j] = 1$ for $i, j = 1, \ldots, n$ and $W_i(a_1, \ldots, a_n, b_1, \ldots, b_n) \neq 1$ for $i = 1, \ldots, k$. The map from $G \times G$ to G given by mapping x_i to a_i and y_i to b_i for $i = 1, \ldots, n$ defines a homomorphism for which the images of $W_1, \ldots, W_k$ are nontrivial. Hence G discriminates $G \times G$ and therefore G is discriminating. $\qquad \square$

Finally recall Theorem 2.6: Let G be a finitely generated equationally Noetherian group. Then G is discriminating if and only if qvar(G) =ucl(G). For squarelike groups we do not need the equationally Noetherian condition.

Theorem 6.6 ([13]) *Let G be a group. The following are pairwise equivalent:*

(1) G is squarelike.

(2) ucl(G) = qvar(G).

(3) There is a discriminating group H such that $G \equiv_\forall H$.

Therefore the squarelike groups are the universal closure of the discriminating groups. In fact they are the axiomatic closure.

7 Axiomatics and the Axiomatic Closure Property

The last axiomatic class containing the discriminating groups is the class of those groups G for which there exists a discriminating group H with $G \equiv H$. Here $\equiv$ denotes elementary equivalence. We call this class the *axiomatic closure* of G.

In [15] it was established that the squarelike groups are precisely the axiomatic closure of the discriminating groups.

Theorem 7.1 ([15]) *Let G be a group. Then G is squarelike if and only if there is a discriminating group H with $G \equiv H$. Hence the class of squarelike groups is the axiomatic closure of the class of discriminating groups.*

Proof We outline the proof, which uses ultrapowers and ultralimits.

If $G \equiv H$, then $G \equiv_\forall H$; so, G elementarily equivalent to a discriminating group implies G is squarelike.

Assume now that G_0 is a squarelike group. Then, in the notation of[9], $G_0 \forall (G_0 \times G_0)$ so there is an ultrapower $G_1 = G_0^I/D$ admitting an embedding $\varphi_1 : G_0 \times G_0 \to G_1$. Let $d_{0,1} : G_0 \to G_1$ be the canonical embedding and let $G_2 = G_1^I/D$. Then φ_1 induces $\varphi_2 : G_1 \times G_1 \to G_2$ and a diagram chase convinces one that the square

$$
\begin{array}{ccc}
 & \varphi_1 & \\
G_0 \times G_0 & \to & G_1 \\
d_{0,1} \times d_{0,1} \downarrow & & \downarrow d_{1,2} \\
G_1 \times G_1 & \to & G_2 \\
 & \varphi_2 &
\end{array}
$$

is commutative, where $d_{1,2} : G_1 \to G_2$ is the canonical embedding.

We may iterate a countable infinity of times. Taking G_ω to be the ultralimit of G_0 with respect to the constant sequence of ultrafilters $(D, D, \ldots)$, we get an embedding $G_\omega \times G_\omega \to G_\omega$ so that G_ω is discriminating. But the limit map $d_{0,\omega} : G_0 \to G_\omega$ is elementary. Hence, $G_0 \equiv G_\omega$. $\qquad\square$

There is a great deal of further axiomatic information about both squarelike and discriminating groups (see [16]). As we have seen in the previous sections the class of squarelike groups is axiomatic, that is, is given by a set of first order axioms. while the class of discriminating groups is not. Originally this was proved in [13] using Malcev's conditon for axiomability. Subsequently an explicit set of axioms was given by V. H. Dyson (see [16]) and by O. Belegradek (see [7]). These were previously stated here. P. Schupp raised the question of whether the theory of squarelike groups was finitely axiomatizable, that is, whether the axiom set can be taken to be finite. This was answered in the negative:

Theorem 7.2 ([16]) *The theory of squarelike groups is not finitely axiomatizable.*

Recall that if $\mathcal{X}$ is a class of groups then the first order theory of $\mathcal{X}$ is *decidable* if there exists a recursive algorithm which, given a sentence ϕ of L, decides whether or not ϕ is true in every group in $\mathcal{X}$. As a consequence of the undecidability of the universal theory of groups (see [16]) it was shown that the theory of the squarelike groups is undecidable.

Theorem 7.3 ([16]) *The theory of squarelike groups is undecidable.*

If we restrict, however, ourselves to the abelian squarelike groups, then we have the following positive result of Oleg Belegradek.

Theorem 7.4 ([8]) *The first order theory of the abelian squarelike groups is decidable.*

8   Varietal discrimination

There is an older notion of discrimination, which we call *varietal discrimination*, that was also introduced by G. Baumslag jointly with Bernard, Hanna and Peter Neumann. The properties of this type of discrimination are described in the book of Hanna Neumann [37] and play a role in the structure of product varieties.

In [18], the relationship between these notions of discrimination was examined. Let $\mathcal{D}$ denote the class of discriminating groups, $\mathcal{S}$ the class of squarelike groups and $\mathcal{VD}$ the class of varietally discriminating groups. The main result is that

$$\mathcal{D} \subset \mathcal{S} \subset \mathcal{VD}$$

and all inclusions are proper.

In order to prove that $\mathcal{S} \subset \mathcal{VD}$, a further notion of discrimination called q-*discriminating* was introduced, and then it was proved that a group is squarelike if and only if it is q-discriminating.

Definition 8.1 ([37]) Let $\mathcal{V}$ be a variety of groups, and $G \in \mathcal{V}$. Then G discriminates $\mathcal{V}$ if, to every finite set $w_i(\overline{x})$ of words in the variables $X \cup X^{-1}$, with none of the equations $w_i(\overline{x}) = 1$ being a law in $\mathcal{V}$, there is a tuple $\overline{g}$ of elements of G, such that $w_i(\overline{x}) \neq 1$ in G for all i. A group G is *varietally discriminating* if it discriminates the variety it generates.

Theorem 8.2 $\mathcal{D} \subset \mathcal{S} \subset \mathcal{VD}$ *and all inclusions are proper.*

The proper inclusion $\mathcal{D} \subset \mathcal{S}$ has already been described. The proper inclusion $\mathcal{D} \subset \mathcal{VD}$ can be proved independently of the middle. In order to prove the proper inclusion $\mathcal{S} \subset \mathcal{VD}$ we introduce a quasivarietal version of varietal discrimination.

Definition 8.3 Let $\mathcal{Q}$ be a quasivariety and let $G \in \mathcal{Q}$. Then G q-*discriminates* $\mathcal{Q}$ provided that, given finitely many quasilaws

$$\forall \overline{x} \left(\bigwedge_i (u_i(\overline{x}) = 1) \rightarrow (w_j(\overline{x}) = 1) \right)$$

with the same antecedents, none of which holds in $\mathcal{Q}$, there exists a tuple $\overline{g}$ from G, such that $u_i(\overline{g}) = 1$ and $w_j(\overline{g}) \neq 1$ for all i, j. A group G is q-*discriminating* if G q-discriminates $\mathrm{qvar}(G)$.

This almost, except for having the same antecedents, appears to be the translation of varietal discrimination in terms of quasivarieties. The condition of having the same antecedents is necessary for the next theorem. Without this restriction the only q-discriminating group would be the trivial group. The above notion of q-discrimination characterizes squarelike groups.

Definition 8.4 A group G is q-*algebraically closed* if and only if whenever a finite system

$$u_i(x_1, \ldots, x_n) = 1, \quad w_j(x_1, \ldots, x_n) \neq 1$$

of equations and inequations has a solution in some group $H \in \text{qvar}(G)$, it also has a solution in G.

We can now characterize squarelike groups in terms of q-discrimination.

Theorem 8.5 ([18]) *Let G be a group. The following are pairwise equivalent.*
(1) G is q-discriminating.
(2) G is q-algebraically closed.
(3) G is squarelike.

Acknowledgements: This paper was partially prepared while two of the authors (B. Fine and A. Myasnikov) were visitors at the CRM in Barcelona. We'd like to thank the CRM for its hospitality.

References

[1] G. Baumslag, *Lecture notes on nilpotent groups* (Amer. Math. Soc., Providence, RI 1969).

[2] G. Baumslag, B. Fine, A. M. Gaglione and D. Spellman, Reflections on discriminating groups, *J. Group Theory* **10** (2007), 87–99.

[3] G. Baumslag, B. Fine, A. M. Gaglione and D. Spellman, A note on nondiscrimination of nilpotent groups and Malcev completions, *Contemp. Math.* **421** (2006), 29–34.

[4] G. Baumslag and C. F. Miller III, *Reflections on some groups of B. H. Neumann*, preprint.

[5] G. Baumslag, A. G. Myasnikov and V. N. Remeslennikov, Algebraic geometry over groups 1, *J. Algebra* **219** (1999), 16–79.

[6] G. Baumslag, A. G. Myasnikov and V. N. Remeslennikov, Discriminating and co-discriminating groups, *J. Group Theory* **3** (2000), 467–479.

[7] O. Belegradek, Discriminating and square-like groups, *J. Group Theory* **7** (2004), 521–532.

[8] O. Belegradek, The theory of square-like abelian groups is decidable, in *Algebra, logic, set theory*, Stud. Log. (Lond.) **4** (Coll. Publ., London 2007), 33–46.

[9] J. L. Bell and A. B. Slomson, *Models and ultraproducts: an introduction (Second revised printing)* (North-Holland, Amsterdam 1972).

[10] P. M. Cohn, *Universal algebra* (Harper and Row, New York, NY 1965).

[11] V. H. Dyson, *private communication.*

[12] B. Fine, A. M. Gaglione, A. G. Myasnikov and D. Spellman, Discriminating groups, *J. Group Theory* **4** (2001), 463–474.

[13] B. Fine, A. M. Gaglione, A. G. Myasnikov and D. Spellman, Groups whose universal theory is axiomatizable by quasi-identities, *J. Group Theory* **5** (2002), 365–381.

[14] B. Fine, A. M. Gaglione and D. Spellman, Every abelian group universally equivalent to a discriminating group is elementarily equivalent to a discriminating group, *Contemp. Math.* **296** (2002), 129–137.

[15] B. Fine, A. M. Gaglione and D. Spellman, The axiomatic closure of the class of discriminating groups, *Arch. Math. (Basel)* **83** (2004), 106–112.

[16] B. Fine, A. M. Gaglione and D. Spellman, Discriminating and squarelike groups I: axiomatics, *Contemp. Math* **360** (2004), 35–46.

[17] B. Fine, A. M. Gaglione and D. Spellman, Discriminating and squarelike groups II: examples, *Houston J. Math.* **31** (2005), 649–673.

[18] B. Fine, A. M. Gaglione and D. Spellman, Notions of discrimination, *Comm. Algebra* (2006), 2175–2182.

[19] B. Fine, A. M. Gaglione and D. Spellman, Discrimination properties in universal algebras, *in preparation*.

[20] G. Higman, Subgroups of finitely presented groups, *Proc. Royal Soc. London Ser. A* **264** (1961), 455–475.

[21] G. Higman, *Finitely presented infinite simple groups*, Notes on Pure Math. **8**, (I.A.S., Austral. Nat. Univ., Canberra 1974).

[22] R. Hirshon and D. Meier, Groups with a quotient that contains the original group as a direct factor, *Bull. Austral. Math. Soc.* **45** (1992), 513–520.

[23] J. M. Tyrer Jones, Direct products and the Hopf property, *J. Austral. Math. Soc.* **17** (1974), 174–196.

[24] M. Kassabov, On discriminating solvable groups, *preprint*.

[25] O. Kharlampovich and A. Myasnikov, Irreducible affine varieties over a free group: I. Irreducibility of quadratic equations and Nullstellensatz, *J. Algebra* **200** (1998), 472–516.

[26] O. Kharlampovich and A. Myasnikov, Irreducible affine varieties over a free group: II. Systems in triangular quasi-quadratic form and a description of residually free groups, *J. Algebra* **200** (1998), 517–569.

[27] O. Kharlampovich and A. Myasnikov, Description of fully residually free groups and irreducible affine varieties over free groups, in *Summer school in Group Theory in Banff, 1996*, CRM Proc. Lecture Notes **17** (Amer. Math. Soc., Providence, RI 1999), 71–81.

[28] O. Kharlampovich and A. Myasnikov, Hyperbolic groups and free constructions, *Trans. Amer. Math. Soc.* **350(2)** (1998), 571–613.

[29] O. Kharlampovich and A. Myasnikov, The elementary theory of free nonabelian groups, *J. Algebra* **302** (2006), 451–552.

[30] A. G. Kurosh, *The Theory of Groups, Vol. I* (Chelsea, New York 1956).

[31] A. G. Kurosh, *The Theory of Groups, Vol. II* (Chelsea, New York 1956).

[32] R. C. Lyndon, Problems in combinatorial group theory, in *Annals of Mathematical Studies* **111** (Princeton Univ. Press 1987), 3–33.

[33] W. Magnus, A. Karass and D. Solitar, *Combinatorial Group Theory* (Interscience, New York 1966).

[34] A. I. Mal'cev, On free soluble groups, *Dokl. Akad. Nauk SSR* **130** (1960), 495–498. English transl. *Soviet Math. Dokl.* (1960), 65–68.

[35] A. M. Myasnikov and P. Shumyatsky, Discriminating groups and c-dimension, *J. Group Theory* **7(1)** (2004), 135–142.

[36] B. H. Neumann, Some remarks on infinite groups, *J. London Math. Soc.* **12** (1937), 120–127.

[37] H. Neumann, *Varieties of Groups* (Springer-Verlag, New York 1968).

[38] Z. Sela, Diophantine geometry over groups. I. Makanin-Razborov diagrams, *Publ. Math. Inst. Hautes Études Sci.* **93** (2001), 31–105.

[39] Z. Sela, Diophantine geometry over groups. II. Completions, closures, and formal solutions, *Israel J. Math.* **104** (2003), 173–254.

[40] Z. Sela, Diophantine geometry over groups. III. Rigid and solid solutions, *Israel J. Math.* **147** (2005), 1–73.

[41] Z. Sela, Diophantine geometry over groups. IV. An iterative procedure for validation of a sentence, *Israel J. Math.* **143** (2004), 1–130.

[42] Z. Sela, Diophantine geometry over groups. V_1. Quantifier elimination, *Israel J. Math.* **150** (2005), 1–97.

[43] Z. Sela, Diophantine geometry over groups. VI. The elementary theory of a free group, *Geom. Funct. Anal.* **16** (2006), 707–730.

[44] W. Szmielew, Elementary properties of abelian groups, *Fund. Math.* **41** (1955), 203–271.

EXTENDING THE KEGEL WIELANDT THEOREM THROUGH π-DECOMPOSABLE GROUPS

L. S. KAZARIN*, A. MARTÍNEZ-PASTOR[†] and M. D. PÉREZ-RAMOS[§]

*Department of Mathematics, Yaroslavl P. Demidov State University, Sovetskaya Str 14, 150000 Yaroslavl, Russia
Email: Kazarin@uniyar.ac.ru

[†]Instituto Universitario de Matemática Pura y Aplicada IUMPA-UPV, Universidad Politécnica de Valencia, Camino de Vera, s/n, 46022 Valencia, Spain
Email: anamarti@mat.upv.es

[§]Departament d'Àlgebra, Universitat de València, C/ Doctor Moliner 50, 46100 Burjassot (València), Spain
Email: dolores.perez@uv.es

Abstract

A celebrated theorem of Kegel and Wielandt asserts the solubility of a finite group which is the product of two nilpotent subgroups. In this survey we report on some extensions of this result by considering π-decomposable subgroups, for a set of primes π, instead of nilpotent groups.

1 Introduction

The study of groups which can be factorised as the product of two subgroups has developed extensively in recent decades. The general aim is to obtain information about the structure of the whole group from the structure of the subgroups in the factorization, and vice versa. An example is the well known theorem of Kegel and Wielandt which establishes the solubility of a finite group factorised as the product of two nilpotent subgroups. This result has been the motivation for a wide variety of results in the literature. In particular some of them consider the situation when either one or both of the factors are π-decomposable, for a set of primes π. This paper is a survey article containing a detailed account of recent achievements which extend the Kegel–Wielandt theorem in this direction.

Only finite groups are considered in this paper.

Let us start with an explicit statement of the starting point of our development:

Theorem 1.1 (Kegel [14] and Wielandt [19]) *If the group $G = AB$ is the product of two nilpotent subgroups A and B, then G is soluble.*

This theorem was proved first by Wielandt in 1958 for the case when the subgroups A and B have coprime orders, i.e., $(|A|, |B|) = 1$, and later it was extended

This research has been supported by Proyecto MTM2007-68010-C03-03, Ministerio de Educación y Ciencia and FEDER, Spain.

to the general case by Kegel in 1961. Many authors have studied generalisations of this result from many different points of view. We focus our attention in this paper by regarding nilpotent groups as a special case of π-decomposable groups, for a set of primes π. First we settle the notation and basic concepts involved in our study.

Notation For any group X and any set of primes π, we denote by $O_\pi(X)$ the π-radical of X, that is, the largest normal π-subgroup in X. Moreover, X_π represents a Hall π-subgroup of X. We write $\mathrm{Hall}_\pi(X)$ for the set of all Hall π-subgroups of X and, as usual, $\mathrm{Syl}_p(X)$ is the set of all Sylow p-subgroups of X, for a prime p. Finally $\pi(X)$ denotes the set of all prime divisors of $|X|$, the order of X.

Definition 1.2 A group X is said to be π-decomposable if

$$X = O_\pi(X) \times O_{\pi'}(X) = X_\pi \times X_{\pi'}$$

is the direct product of a π-subgroup and a π'-subgroup, where π' stands for the complement of π in the set of all prime numbers.

We recall also that a group X is π-separable if its composition factors are either π-groups or π'-groups. Furthermore, a π-separable group X satisfies the so-called property D_π, that is, X has a unique conjugacy class of Hall π-subgroups and every π-subgroup is contained in a Hall π-subgroup.

Inspired by the Kegel-Wielandt theorem, since a group is soluble if and only if it is p-separable, for all primes p, it is natural to settle the following question:

Question 1 Let the group $G = AB$ be the product of two π-decomposable subgroups $A = A_\pi \times A_{\pi'}$ and $B = B_\pi \times B_{\pi'}$.
  Is it true that G is π-separable?

The answer to this question is, in general, negative, as can be seen in the following example:

Example 1.3 Let $G = A_5$, the alternating group of degree 5, which can be factorised in the following way: $G = AB$, $A = G_{\{2,3\}} \in \mathrm{Hall}_{\{2,3\}}(G)$, $B = G_5 \in \mathrm{Syl}_5(G)$. Let $\pi = \{2,3\}$. Then $G = AB$ is the product of a π-group and a π'-group, but G is not π-separable.

So, for products of π-decomposable groups we have considered the existence of Hall π-subgroups as a preliminary step to finding conditions for π-separability. More concretely, we have asked ourselves the following question:

Question 2 Let the group $G = AB$ be the product of two π-decomposable subgroups $A = A_\pi \times A_{\pi'}$ and $B = B_\pi \times B_{\pi'}$.
  Is it true that

$$A_\pi B_\pi = B_\pi A_\pi$$

and so G posseses Hall π-subgroups?

2   Preliminaries

In this section we collect some preliminary results which have turned out to be significant in our work.

First it is interesting to notice that in any factorised group and for each prime p, there always exists a Sylow p-subgroup of the group which is the product of Sylow p-subgroups of the factors (see [1, Corollary 1.3.3]).

Lemma 2.1 *Let the group $G = AB$ be the product of the subgroups A and B. Then for each prime p there exist Sylow p-subgroups A_p of A and B_p of B such that $A_p B_p$ is a Sylow p-subgroup of G.*

In particular, for the problem we are dealing with, this result allows us to restrict our study to the situation when the set of primes π contains at least two primes. On the other hand, an extension of this result for Hall π-subgroups (see [1, Lemma 1.3.2]) provides an easy positive answer to Question 2 when we impose the additional hypotheses that all groups involved (i.e., the factorised group and the factors) satisfy the property D_π.

The next basic result is a reformulation of a very useful criterion for the non-simplicity of a finite group in terms of some suitable permutability conditions on subgroups, found by Kegel ([14, Satz 3]), for subgroups of coprime orders, and later improved by Wielandt in [20, Satz 1]. (See also [1, Lemmas 2.4.1, 2.5.1].)

Lemma 2.2 *Let the group $G = AB$ be the product of the subgroups A and B and let A_0 and B_0 be normal subgroups of A and B, respectively. If $A_0 B_0 = B_0 A_0$, then $A_0^g B_0 = B_0 A_0^g$ for all $g \in G$.*

Assume in addition that A_0 and B_0 are π-groups for a set of primes π. If $O_\pi(G) = 1$, then $[A_0^G, B_0^G] = 1$.

If $G = AB$ is the product of nilpotent subgroups A and B, then the hypotheses of this result hold for $A_0 = A_p$ and $B_0 = B_p$, the Sylow p-subgroups of A and B, respectively, and for any prime p. This fact is in the core of proof of the solubility of the group G. We note also that this result is applicable in particular if $A = A_\pi \times A_{\pi'}$ and $B = B_\pi \times B_{\pi'}$ are π-decomposable, by considering $A_0 = A_\pi$ and $B_0 = B_\pi$.

Finally, it is to be noticed that our results depend heavily on the Classification of Finite Simple Groups. In particular, we have often used the information about the maximal factorisations of the finite simple groups and their automorphism groups, due to Liebeck, Praeger and Saxl (see [16]). Moreover, the information about the centralizers of Sylow 2-subgroups in such groups, which appears in the paper of Kondratiev and Mazurov [15], has also been useful.

3   The case when one of the subgroups is just a π-group

To tackle the questions posed in the introduction we have proceeded step by step. First, we handle the situation when one of subgroups is a π-group. In this particular case, Question 2 about the existence of Hall subgroups can be expressed in an equivalent way:

Lemma 3.1 *Let the group $G = AB$ be the product of a π-decomposable subgroup A and a π-subgroup B. Then the following statements are equivalent:*
(i) $A_\pi = O_\pi(A) \le O_\pi(G)$;
(ii) *G possesses Hall π-subgroups.*
In this case $A_\pi B$ is a Hall π-subgroup of G.

And we have proved in [11] that the above-mentioned question has a positive answer when π is a set of odd primes.

Theorem 3.2 *Let π be a set of odd primes. Let the group $G = AB$ be the product of a π-decomposable subgroup A and a π-subgroup B. Then $O_\pi(A) \le O_\pi(G)$.*

Notice that this result provides, in particular, criteria for the existence of non-trivial soluble normal subgroups in a factorised group G and so non-simplicity criteria for such a group.

Moreover, it extends the following result proved by Kazarin in 1983:

Theorem 3.3 (Kazarin [10]) *Let $G = AB$ be a group satisfying:*
(1) $A = A_2 \times A_{2'}$ *is 2-decomposable.*
(2) $|B|$ *is odd.*
Then $A_{2'} = O_{2'}(A) \le O_{2'}(G)$.

Next we present some examples showing that the result analogous to Theorem 3.2 is not true when the prime 2 belongs to the set of primes π:

Examples 3.4 (a) Let G be a group isomorphic to $L_2(2^n)$ where n is a positive integer such that $2^n + 1$ is divisible by two distinct primes (this happens if $n \ne 3$ and $2^n + 1$ is not a Fermat prime). Set $q = 2^n$. Then $G = AB$ where $A \cong C_{q+1}$ is a cyclic group of order $q + 1$ and $B = N_G(G_2)$, $G_2 \in \mathrm{Syl}_2(G)$. Let r be a prime dividing $q + 1$ and take $\pi = \pi(N_G(G_2)) \cup \{r\}$. Then $A = O_\pi(A) \times O_{\pi'}(A)$ is a π-decomposable group and B is a π-group, but $O_\pi(A) \not\le O_\pi(G)$; note that $2 \in \pi$.

(b) Consider now a group G isomorphic to $PGL_2(q)$, $q > 3$ odd, where q is not a Mersenne prime. (We note that $L_2(4) \cong L_2(5)$.) Thus $|G| = q(q^2-1)$ and it is known that this group has cyclic subgroups of orders $(q-1)$ and $(q+1)$. Then $G = AB$ where $A \cong C_{q+1}$ is a cyclic group of order $q + 1$ and $B = N_G(G_p)$, $G_p \in \mathrm{Syl}_p(G)$, is a subgroup of order $q(q-1)$. Clearly $\pi(A) \cap \pi(B) = \{2\}$. Set $\pi = \pi(N_G(G_p))$ and note that $2 \in \pi$. Then $A = O_\pi(A) \times O_{\pi'}(A)$ and B is a π-group, but $O_\pi(A) \not\le O_\pi(G)$ (except when $q+1$ is a power of 2, that is, q is a Mersenne prime, in which case G is a π-group).

From Theorem 3.2 together with the results in [3] we obtain our first extension of the Kegel–Wielandt theorem:

Corollary 3.5 *Let π be a set of odd primes. Let the group $G = AB$ be the product of a π-decomposable subgroup A and a π-subgroup B. Then the composition factors of G belong to one of the following types:*

(1) *π-groups,*

(2) *π'-groups,*

(3) *groups in the list of Arad-Fisman [3, Theorem 1.1], that is:*

   (i) *A_r with $r \geq 5$ a prime,*

   (ii) *M_{11},*

   (iii) *M_{23},*

   (iv) *$L_2(q)$ where either $q = 29$ or $3 < q \not\equiv 1 \, (mod \, 4)$,*

   (v) *$L_r(q)$ with r an odd prime such that $(r, q - 1) = 1$.*

In particular, if none of these groups is involved in G, then the group is π-separable.

4   Products of π-decomposable groups of coprime orders

We state in this section another case for which Question 2 has a positive answer, which follows from Theorem 3.2 and appears in [12].

Theorem 4.1 *Let π be a set of odd primes. Let the group $G = AB$ be the product of two π-decomposable subgroups $A = A_\pi \times A_{\pi'}$ and $B = B_\pi \times B_{\pi'}$. Assume in addition that $(|A_{\pi'}|, |B_{\pi'}|) = 1$. Then $A_\pi B_\pi = B_\pi A_\pi$.*

It is worthwhile emphasizing that the desired conclusion holds in the case when $(|A|, |B|) = 1$. This case is quite a significant one, since we can find in the literature a variety of extensions of the Kegel-Wielandt theorem which consider this hypothesis. In particular, we report some of them here involving products of a 2-decomposable group and a group of odd order, with coprime orders.

First, Berkovich obtained in 1966 the following:

Theorem 4.2 (Berkovich [4]) *Let $G = AB$ be a group satisfying:*

(1) *$A = A_2 \times A_{2'}$ is 2-decomposable.*

(2) *B is nilpotent and $|B|$ is odd.*

(3) *$(|A|, |B|) = 1$.*

Then G is soluble.

Later, Rowley in 1977 generalised this result by proving:

Theorem 4.3 (Rowley [18]) *Let $G = AB$ be a group satisfying:*

(1) *$A = A_2 \times A_{2'}$ is 2-decomposable.*

(2) *B is metanilpotent and $|B|$ is odd.*

(3) *$(|A|, |B|) = 1$.*

Then G is $\pi(A_{2'})$-separable.

Moreover, Arad and Chillag showed in 1981 that the same conclusion is true without any restriction on the nilpotent length of B:

Theorem 4.4 (Arad and Chillag [3]) *Let $G = AB$ be a group satisfying:*

(1) $A = A_2 \times A_{2'}$ *is 2-decomposable.*

(2) $|B|$ *is odd.*

(3) $(|A|, |B|) = 1$.

Then G is $\pi(A_{2'})$-separable.

Previously Kazarin had shown that $A_{2'} = O_{2'}(A) \leq O_{2'}(G)$ under the above hypothesis (see [9]) as a preliminary step in the proof of Theorem 3.3.

We should point out that the proofs by Berkovich, Rowley and Kazarin do not depend on the classification of finite simple groups. Notice that all the above-mentioned results can be derived from our Theorem 4.1 by considering π the set of all odd prime numbers in this theorem and taking into account the particular hypotheses in each of these results.

5   Products of soluble π-decomposable groups

As a next step towards our objective, in this section we consider the case of a group factorised as the product of two *soluble* π-decomposable subgroups. Under these additional hypotheses we have analyzed in [12] what happens in both cases: when π is a set of odd primes and when $2 \in \pi$.

5.1   The soluble case with π a set of odd primes

In the case of soluble factors we have again proved that the desired result about the permutability of the Hall subgroups holds when π is a set of odd primes:

Theorem 5.1 *Let π be a set of odd primes. Let the group $G = AB$ be the product of two π-decomposable soluble subgroups $A = A_\pi \times A_{\pi'}$ and $B = B_\pi \times B_{\pi'}$. Then $A_\pi B_\pi = B_\pi A_\pi$ and this is a Hall π-subgroup of G.*

5.2   The soluble case with $2 \in \pi$

For the case when $2 \in \pi$ we have shown in Examples 3.4 that the result corresponding to Theorem 5.1 is false. But we have been able to give a detailed description of what happens in this situation. Indeed we have shown that groups involving those examples are the only exceptions to the desired result.

Theorem 5.2 *Let π be a set of primes with $2 \in \pi$. Let the group $G = AB$ be the product of two soluble π-decomposable subgroups $A = A_\pi \times A_{\pi'}$ and $B = B_\pi \times B_{\pi'}$. Assume that the following simple groups are not involved in G:*

(i) $L_2(2^n)$, *where $n \geq 2$, $n \neq 3$ and $q = 2^n + 1 > 5$ is not a Fermat prime,*

(ii) $L_2(q)$, *where $q > 3$ is odd and q is not a Mersenne prime.*

Then $A_\pi B_\pi = B_\pi A_\pi$ and this is a Hall π-subgroup of G.

5.3 The soluble case for an arbitrary set of primes

As a consequence of Theorems 5.1 and 5.2 we deduce the following extensions of the Kegel–Wielandt theorem for products of soluble π-decomposable groups, for an arbitrary set of primes π:

Corollary 5.3 *Let π be a set of primes. Let the group $G = AB$ be the product of two soluble π-decomposable subgroups $A = A_\pi \times A_{\pi'}$ and $B = B_\pi \times B_{\pi'}$. Assume that the following simple groups are not involved in G:*
  (i) $L_2(2^n)$, *where $n \geq 2$, $n \neq 3$ and $q = 2^n + 1 > 5$ is not a Fermat prime,*
  (ii) $L_2(q)$, *where $q > 3$ is odd and q is not a Mersenne prime.*
Then the composition factors of G belong to one of the following types:
  (1) π-*groups,*
  (2) π'-*groups,*
  (3) *the following groups in the list of Fisman [7, Theorem 1.1]:*
      (i) $L_2(2^n)$, *$n \geq 2$, where either $n = 3$ or $q = 2^n + 1 > 5$ is a Fermat prime,*
      (ii) $L_2(q)$ *where $q > 3$ and q is a Mersenne prime,*
      (iii) $L_3(3)$,
      (iv) M_{11}.

Corollary 5.4 *Let π be a set of primes. Let the group $G = AB$ be the product of two soluble π-decomposable subgroups $A = A_\pi \times A_{\pi'}$ and $B = B_\pi \times B_{\pi'}$ and assume that the simple groups*
$$L_2(q), \; q > 3, \; L_3(3) \; and \; M_{11}$$
are not involved in G.
  Then the group G is π-separable.

As an easy consequence of our results and the list of Fisman [7, Theorem 1.1] or the classification of Guralnick [8, Theorem 1] of non-abelian simple groups having subgroups of prime power index, we can deduce another extension of the Kegel-Wielandt theorem.

Corollary 5.5 *Let the group $G = AB$ be the product of two 2-decomposable subgroups $A = A_2 \times A_{2'}$ and $B = B_2 \times B_{2'}$ and assume that the simple group $L_2(r)$, where $r > 3$ is a Mersenne prime, is not involved in G. Then G is soluble.*

In particular, this corollary generalizes a result of Monakhov (see [17]) which proves the solubility of a group $G = AB$ under the assumption that the subgroups A and B are both 2-decomposable and 3-decomposable, and, in addition, they are p-closed (i.e., their Sylow p-subgroups are normal) for all primes $p \in \pi(A) \cap \pi(B)$.

6 A conjecture for the general case

Regarding our initial Question 2, and motivated by the above developments, we have conjectured the following result:

Conjecture *Let π be a set of odd primes. Let the group $G = AB$ be the product of two π-decomposable subgroups $A = A_\pi \times A_{\pi'}$ and $B = B_\pi \times B_{\pi'}$. Then*

$$A_\pi B_\pi = B_\pi A_\pi$$

and this is a Hall π-subgroup of G.

The partial results we have already obtained provide strong evidence in favour of a positive answer to this conjecture. Indeed, in [12] and [13] we have studied carefully a minimal counterexample to the desired result. In particular, we have succeed in proving that this minimal counterexample should be an almost simple group (that is, a group having a unique minimal normal subgroup which is a non-abelian simple group). Currently a case by case analysis is being carried out to discard any possible non-abelian simple group involved.

As in our previous cases, a positive answer to this conjecture will provide conditions for the π-separability of a group factorised as the product of two π-decomposable groups, in the general situation.

Note added in proof: The above conjecture has been proved by the authors since the submission of this paper.

References

[1] B. Amberg, S. Franciosi and F. de Giovanni, *Products of Groups* (Clarendon Press, Oxford 1992).

[2] Z. Arad and D. Chillag, Finite groups containing a nilpotent Hall subgroup of even order, *Houston J. Math.* **7** (1981), 23–32.

[3] Z. Arad and E. Fisman, On finite factorizable groups, *J. Algebra* **86** (1984), 522–548.

[4] Ya. G. Berkovich, Generalization of the theorems of Carter and Wielandt, *Sov. Math. Dokl.* **7** (1966), 1525–1529.

[5] R. Carter, *Simple groups of Lie type* (Wiley, London 1972).

[6] J. H. Conway, R. T. Curtis, S. P. Norton, R. A. Parker and R. A. Wilson, *Atlas of Finite Groups* (OUP, Oxford 1985).
http://brauer.maths.qmul.ac.uk/Atlas/v3/

[7] E. Fisman, On the product of two finite solvable groups, *J. Algebra* **80** (1983), 517–536.

[8] R. Guralnick, Subgroups of prime power index in a simple group, *J. Algebra* **81** (1983), 304–311.

[9] L. S. Kazarin, Criteria for the nonsimplicity of factorable groups, *Izv. Akad. Nauk SSSR, Ser. Mat.* **44** (1980), 288–308.

[10] L. S. Kazarin, The product of a 2-decomposable group and a group of odd order, *Problems in group theory and homological algebra* (1983), 89–98. (Russian)

[11] L. S. Kazarin, A. Martínez-Pastor and M. D. Pérez-Ramos, On the product of a π-group and a π-decomposable group, *J. Algebra* **315** (2007), 640–653.

[12] L. S. Kazarin, A. Martínez-Pastor and M. D. Pérez-Ramos, On the product of two π-decomposable soluble groups, *Publ. Mat.* **53** (2) (2009), 439–456.

[13] L. S. Kazarin, A. Martínez-Pastor and M. D. Pérez-Ramos, On the product of two π-decomposable groups, preprint.

[14] O. H. Kegel, Produkte nilpotenter Gruppen, *Arch. Math.* **12** (1961), 90–93.

[15] A. S. Kondratiev and V. D. Mazurov, 2-signalizers of finite simple groups, *Algebra Logic* **42** (2003), 333–348.

[16] M. Liebeck, C. E. Praeger, and J. Saxl, *The maximal factorizations of the finite simple groups and their automorphism groups*, Mem. Amer. Math. Soc. **86**, No. 432, (Amer. Math. Soc., Providence, RI, 1990).

[17] V. S. Monakhov, Solvability of a factorable group with decomposable factors, *Mat. Zametki* **34** (1983), 337–340. (Russian)

[18] P. J. Rowley, The π-separability of certain factorizable groups, *Math. Z.* **153** (1977), 219–228.

[19] H. Wielandt, Über Produkte von nilpotenten Gruppen, *Ill. J. Math.* **2** (1958), 611–618.

[20] H. Wielandt, Vertauschbarkeit von Untergruppen und Subnormalität, *Math. Z.* **133** (1973), 275–276.

ON THE PRIME GRAPH OF A FINITE GROUP

BEHROOZ KHOSRAVI

Dept. of Pure Math., Faculty of Math. and Computer Sci., Amirkabir University of Technology (Tehran Polytechnic), 424, Hafez Ave., Tehran 15914, Iran

School of Mathematics, Institute for Research in Fundamental Sciences (IPM), P. O. Box: 19395-5746, Tehran, Iran

Email: khosravibbb@yahoo.com

Abstract

Let G be a finite group. The prime graph $\Gamma(G)$ of G is defined as follows. The vertices of $\Gamma(G)$ are the primes dividing the order of G and two distinct vertices p, p' are joined by an edge if there is an element in G of order pp'. In this paper we give a survey about the question of which groups have the same prime graph. It is proved that some finite groups are uniquely determined by their prime graph. Applications of this result to the problem of recognition of finite groups by the set of element orders are also considered.

1 Introduction

If n is an integer, then we denote by $\pi(n)$ the set of all prime divisors of n. If G is a finite group, then the set $\pi(|G|)$ is denoted by $\pi(G)$. Also the set of order elements of G is denoted by $\pi_e(G)$. Obviously $\pi_e(G)$ is partially ordered by divisibility. Therefore it is uniquely determined by $\mu(G)$, the subset of its maximal elements. We construct the prime graph of G as follows:

Definition 1.1 *The prime graph* $\Gamma(G)$ *of a group* G *is the graph whose vertex set is* $\pi(G)$, *and two distinct primes* p *and* q *are joined by an edge (we write* $p \sim q$) *if and only if* G *contains an element of order* pq. *Let* $t(G)$ *be the number of connected components of* $\Gamma(G)$ *and let* $\pi_1(G), \pi_2(G), \ldots, \pi_{t(G)}(G)$ *be the connected components of* $\Gamma(G)$. *Sometimes we use the notation* π_i *instead of* $\pi_i(G)$. *If* $2 \in \pi(G)$, *then we always suppose* $2 \in \pi_1$.

The concept of prime graph arose during the investigation of certain cohomological questions associated with integral representations of finite groups. It turns out that $\Gamma(G)$ is not connected if and only if the augmentation ideal of G is decomposable as a module [13]. Also non-connectedness of $\Gamma(G)$ has relations with the existence of isolated subgroups of G. A proper subgroup H of G is isolated if $H \cap H^g = 1$ or H for every $g \in G$ and $C_G(h) \leq H$ for all non-trivial $h \in H$. It was proved in [19] that G has a nilpotent isolated Hall π-subgroup whenever G is non-solvable and $\pi = \pi_i$ ($i > 1$). In fact we have the following equivalences:

Theorem 1.2 (see [13]) *If* G *is a finite group, then the following are equivalent:*
(i) *the augmentation ideal of* G *is decomposable as a module,*

(ii) *the group G contains an isolated subgroup,*

(iii) *the prime graph of G has more than one component.*

It is therefore interesting to discuss the prime graph of finite groups. It has been proved that for every finite group G we have $t(G) \le 6$ [5, 13, 19] and the diameter of $\Gamma(G)$ is at most 5 [15].

Let G be a finite group such that $G \cong H$ if and only if $\pi_e(G) = \pi_e(H)$. Then G is called *recognizable by the set of element orders*, see [14, 16, 17, 20]. Also a nonabelian finite simple group P is called *quasirecognizable by the set of element orders*, if every finite group G with $\pi_e(G) = \pi_e(P)$ has a composition factor isomorphic to P, see [2]. Alekseeva and Kondrat'ev in [2] proved that every finite simple group with at least three connected components (except A_6) is quasirecognizable by the set of element orders.

Following these concepts we introduce similar concepts for the prime graph.

Definition 1.3 (see [11]) A finite group G is called *recognizable by the prime graph* if $H \cong G$ for every finite group H with $\Gamma(H) = \Gamma(G)$. Also a finite simple nonabelian group P is called *quasirecognizable by the prime graph* if every finite group G with $\Gamma(G) = \Gamma(P)$ has a composition factor isomorphic to P.

We note that recognizability by prime graph implies recognition by element orders, but the converse is not true in general. Also recognizability by prime graph is in general harder to establish than recognizability by element orders since some methods fail in the former case.

2   *n*-Recognizable Groups by Prime Graph

Hagie in [4] determined finite groups G satisfying $\Gamma(G) = \Gamma(S)$, where S is a sporadic simple group. The author considered almost sporadic simple groups and determined the structure of finite groups with the same prime graph as $\Gamma(A)$, where A is an almost sporadic group. For example it is proved that $\mathrm{Aut}(O'N)$ is uniquely determined by its prime graph (see [9]).

Theorem 2.1 (see [7, 21]) *If $q = 3^{2n+1}$ $(n > 0)$, then the simple group $^2G_2(q)$ is uniquely determined by its prime graph.*

Theorem 2.2 (see [11]) *If $p > 11$ is a prime number and $p \not\equiv 1$ (mod 12), then $PSL(2,p)$ is uniquely determined by its prime graph.*

We denote by $k(\Gamma(G))$ the number of isomorphism classes of finite groups H satisfying $\Gamma(G) = \Gamma(H)$. Hence if G is a finite group, then $k(\Gamma(G)) \ge 1$. By using this function, we introduce the following definition:

Definition 2.3 Given a natural number n, a finite group G is called *n-recognizable by prime graph* if $k(\Gamma(G)) = n$. In fact a 1-recognizable group is called a *characterizable group*. If there exist infinitely many non-isomorphic finite groups H such that $\Gamma(G) = \Gamma(H)$, we call G a *non-recognizable group by prime graph*.

If p is a prime number, then $k(\Gamma(\mathbb{Z}_p)) = \infty$ and hence every p-group is a non-recognizable group. Also in [9] it is proved that $k(\Gamma(\text{Aut}(Suz))) = 3$, since Fi_{22}, $\text{Aut}(Fi_{22})$ and $\text{Aut}(Suz)$ have the same prime graph and there is not any other finite group with the same prime graph.

Theorem 2.4 (see [8]) *Let p be a prime number. If G is a finite group and $\Gamma(G) = \Gamma(PSL(2, p^2))$, then by using the notations of [3] we have*

(i) *if $p = 2$ or $p = 3$, then $G \cong PSL(2, 9)$ or $G \cong PSL(2, 9).2_3$, the non-split extension of $PSL(2, 9)$ by $\mathbb{Z}_2$, or $G/O_2(G) \cong PSL(2, 4)$;*

(ii) *if $p = 7$, then $G \cong PSL(2, 49)$, $PSL(2, 49).2_3$, the non-split extension of $PSL(2, 49)$ by $\mathbb{Z}_2$, or $G/O_{\{2,3\}}(G)$ is isomorphic to $PSL(3, 4)$, $PSL(3, 4).2_1$, $PSU(4, 3)$ or A_7;*

(iii) *if $p \neq 2$, 3, 7, then $G \cong PSL(2, p^2)$ or $G \cong PSL(2, p^2).2_3$, the non-split extension of $PSL(2, p^2)$ by $\mathbb{Z}_2$. Hence $PSL(2, p^2)$ is 2-recognizable by prime graph.*

Theorem 2.5 (see [12]) *Let $q = p^k$ and p be an odd prime number, $k \geq 3$. Then $PSL(2, q)$ is n-recognizable by prime graph where $n = (2, k)$. In fact if G is a finite group such that $\Gamma(G) = \Gamma(PSL(2, q))$, and k is odd, then $G \cong PSL(2, q)$ and if k is even, then $G \cong PSL(2, q)$ or $G \cong PSL(2, q).2$, the non-split extension of $PSL(2, q)$ by $\mathbb{Z}_2$.*

3 Quasirecognizable Groups by Prime Graph

In [9] it is proved that almost simple groups $\text{Aut}(Fi'_{24})$ or $\text{Aut}(M_{22})$ are quasirecognizable by prime graph.

Theorem 3.1 (see [10]) *Let $M = Sz(q)$, where $q = 2^{2m+1}$ and $m \geq 1$. If $\Gamma(G) = \Gamma(M)$, then $G/O_2(G) \cong Sz(q)$. Also if $O_2(G) \neq 1$, then $O_2(G)$ is elementary abelian.*

Theorem 3.2 (see [10]) *Let G be a finite group and $\Gamma(G) = \Gamma(PSL(2, 2^m))$, where $m \geq 2$.*

(i) *If $m = 2$, then $G/O_2(G) \cong PSL(2, 4) \cong A_5$; $G \cong PSL(2, 9) \cong A_6$ or $G \cong M_{10} \cong PSL(2, 9).2$, the non-split extension of $PSL(2, 9)$ by $\mathbb{Z}_2$.*

(ii) *If $m = 3$, then $G \cong PSL(2, 7)$ or $G/O_2(G) \cong PSL(2, 8)$.*

(iii) *If $m \geq 4$, then $G/O_2(G) \cong PSL(2, 2^m)$.*

If $G/O_2(G) \cong PSL(2, 2^m)$, where $m \geq 2$ and $O_2(G) \neq 1$, then $O_2(G)$ is elementary abelian and is the direct product of minimal normal subgroups of G, each of order 2^{2m}.

Theorem 3.3 (see [1]) *The simple group $^2F_4(q)$, where $q = 2^{2m+1}$ ($m \geq 1$), is quasirecognizable by prime graph.*

Theorem 3.4 (see [9]) *Let G be a finite group such that $\Gamma(G) = \Gamma(\text{Aut}(McL))$. Then $G/O_2(G)$ is isomorphic to McL, Aut(McL), $U_6(2)$, $U_6(2):2$, HS or Aut(HS). Therefore Aut(McL) is not quasirecognizable by prime graph.*

In [6] similar discussion for $PGL(2, p)$ is considered.

4 Some Related Results

We note that $\Gamma(\mathbb{Z}_6)$ is a graph with two vertices, i.e., $V = \{2, 3\}$ and there exists an edge between 2 and 3. But $\Gamma(\mathbb{Z}_3 \times \mathbb{Z}_{2^k}) = \Gamma(\mathbb{Z}_6)$ for every $k > 0$. Also $S_3 \times \mathbb{Z}_{2^k}$, where $k > 0$, is a non-abelian group and $\Gamma(S_3 \times \mathbb{Z}_{2^k}) = \Gamma(\mathbb{Z}_6)$. Therefore there exist infinitely many non isomorphic groups G such that $\Gamma(G) = \Gamma(\mathbb{Z}_6)$. Hence $\mathbb{Z}_6$ is a non-recognizable group. Also note that even if $|G| = |M|$ and $\Gamma(G) = \Gamma(M)$, we cannot conclude that $G \cong M$.

Remark 4.1 W. Shi and J. Bi in [18] put forward the following conjecture:
Conjecture. Let G be a group and M be a finite simple group. Then $G \cong M$ if and only if (i) $|G| = |M|$, (ii) $\pi_e(G) = \pi_e(M)$.

This conjecture is valid for sporadic simple groups, alternating groups and some simple groups of Lie type. Following Shi and Bi we put forward the following question:

Question 4.2 For which finite groups A is it true that if $|G| = |A|$ and $\Gamma(G) = \Gamma(A)$, then $G \cong A$?

We answer the above question for many finite simple groups as a consequence of our recognition and quasirecognition by prime graph. For example we have

Theorem 4.3 (see [9]) *Let G be a finite group and A be the automorphism group of a sporadic simple group, except $\text{Aut}(J_2)$ and $\text{Aut}(McL)$. If $|G| = |A|$ and $\Gamma(G) = \Gamma(A)$, then $G \cong A$.*

Question 4.4 Let G be a finite group satisfying $\Gamma(G) = \Gamma(\text{Aut}(J_2))$. What can we say about the structure of G?

Acknowledgement

The author express his gratitude to the referee for valuable remarks. The author was supported in part by a grant from IPM (No. 88200038). This paper is dedicated to the memory of Professor Maria Silvia Lucido.

References

[1] Z. Akhlaghi, M. Khatami and B. Khosravi, Quasirecognition by prime graph of the simple group $^2F_4(q)$, *Acta Math. Hungar.* **122** (2009), no. 4, 387–397.

[2] O. A. Alekseeva and A. S. Kondrat'ev, Quasirecognizability of a class of finite simple groups by the set of element orders, *Siberian Math. J.* **44** (2003), no. 2, 195–207.

[3] J. H. Conway, R. T. Curtis, S. P. Norton, R. A. Parker and R. A. Wilson, *Atlas of Finite Groups*, Oxford University Press, Oxford (1985).

[4] M. Hagie, The prime graph of a sporadic simple group, *Comm. Algebra* **31** (2003), no. 9, 4405–4424.

[5] N. Iiyori and H. Yamaki, Prime graph components of the simple groups of Lie type over the field of even characteristic, *J. Algebra* **155** (1993), no. 2, 335–343.

[6] M. Khatami, B. Khosravi and Z. Akhlaghi, NCF-distinguishablity by prime graph of $PGL(2,p)$, where p is a prime, *Rocky Mountain J. Math.*, to appear.

[7] A. Khosravi and B. Khosravi, Quasirecognition by prime graph of the simple group $^2G_2(q)$, *Siberian Math. J.* **48** (2007), no. 3, 570–577.

[8] A. Khosravi and B. Khosravi, 2-Recognizability of $PSL(2,p^2)$ by the prime graph, *Siberian Math. J.* **49** (2008), no. 4, 749–757.

[9] B. Khosravi, A Characterization of the automorphism groups of sporadic groups by the set of orders of maximal abelian subgroups, *Kumamoto J. Math.* **22** (2009), 17–34.

[10] B. Khosravi, B. Khosravi and B. Khosravi, Groups with the same prime graph as a CIT simple group, *Houston J. Math.* **33** (2007), no. 4, 967–977.

[11] B. Khosravi, B. Khosravi and B. Khosravi, On the prime graph of $PSL(2,p)$ where $p > 3$ is a prime number, *Acta. Math. Hungar.* **116** (2007), no. 4, 295–307.

[12] B. Khosravi, n-Recognition by prime graph of the simple group $PSL(2,q)$, *J. Algebra Appl.* **7** (2008), no. 6, 735–748.

[13] A. S. Kondrat'ev, On prime graph components of finite simple groups, *Math. USSR-Sb.* **67** (1990), no. 1, 235–247.

[14] A. S. Kondrat'ev and V. D. Mazurov, Recognition of alternating groups of prime degree from their element orders, *Siberian Math. J.* **41** (2000), no. 2, 294–302.

[15] M. S. Lucido, The diameter of the prime graph of a finite group, *J. Group Theory* **2** (1999), no. 2, 157–172.

[16] V. D. Mazurov, Characterization of finite groups by sets of element orders, *Algebra and Logic* **36** (1997), no. 1, 23–32.

[17] V. D. Mazurov and W. Shi, Groups whose elements have given orders, in *Groups St Andrews 1997 in Bath, II*, London Math. Soc. Lecture Note Ser. **261** (CUP 1999), 532–537.

[18] W. Shi and J. Bi, A characteristic property for each finite projective special linear group, in *Groups—Canberra 1989*, Lecture Notes in Math. **1456** (Springer, Berlin, 1990), 171–180.

[19] J. S. Williams, Prime graph components of finite groups, *J. Algebra* **69** (1981), no. 2, 487–513.

[20] A. Zavarnitsin and V. D. Mazurov, Element orders in coverings of symmetric and alternating groups, *Algebra and Logic* **38** (1999), 159–170.

[21] A. V. Zavarnitsin, Recognition of finite groups by the prime graph, *Algebra and Logic* **43** (2006), no. 4, 220–231.

APPLICATIONS OF LIE RINGS WITH FINITE CYCLIC GRADING

E. I. KHUKHRO

Sobolev Institute of Mathematics, Novosibirsk, 630090, Russia
Email: khukhro@yahoo.co.uk

Abstract

Various results on graded Lie rings are discussed along with their applications in group theory.

1 Introduction

Graded Lie rings appear naturally in the study of automorphisms or derivations, with grading components defined by eigenspace decomposition. For example, the well-known theorems on solubility and nilpotency of Lie rings with a fixed-point-free automorphism of finite order n are essentially about $(\mathbb{Z}/n\mathbb{Z})$-graded Lie rings $L = \bigoplus_{i=0}^{n-1} L_i$ with trivial zero-component $L_0 = 0$. These results have been generalized in several directions: for the case of bounded, rather than trivial, L_0, for the case of few non-trivial components, etc. In this survey we show how these results on graded Lie rings, including some recent generalizations, find applications both in the theory of Lie rings (algebras) and in group theory.

Lie ring methods. The very invention of Lie algebras was motivated by applications to Lie groups. Later other applications in abstract group theory emerged. For example, the Mal'cev correspondence makes use of the exponential and logarithmic maps and the Baker–Campbell–Hausdorff formula (BCHf) to establish the category equivalence between locally nilpotent radicable torsion-free groups and locally nilpotent Lie algebras over $\mathbb{Q}$. Under rather restrictive conditions a similar correspondence of Lazard holds for certain p-groups and Lie rings (or p-algebras). BCHf is also used in the theory of pro-p-groups, including the coclass conjectures.

In general the logarithmic/exponential map and BCHf cannot be applied, say, to finite groups. Other constructions are used instead. The *associated Lie ring* $L(G)$ is defined for any group G on the additively written direct sum of the factors of the lower central series

$$L(G) = \bigoplus_i \gamma_i/\gamma_{i+1}.$$

The Lie products (commutators) are defined for homogeneous elements $\bar{a} = a + \gamma_{i+1} \in \gamma_i/\gamma_{i+1}$, $\bar{b} = b + \gamma_{j+1} \in \gamma_j/\gamma_{j+1}$ via group commutators as

$$[\bar{a}, \bar{b}]_{\text{Lie ring}} = [a, b]_{\text{group}} + \gamma_{i+j+1} \in \gamma_{i+j}/\gamma_{i+j+1}$$

and then extended to the direct sum by linearity.

On the plus side: $L(G)$ always exists, and if the group G is nilpotent, then the nilpotency class of G is equal to the nilpotency class of $L(G)$. However, $L(G)$ obviously can only reflect the properties of $G/\bigcap \gamma_i(G)$, so it only makes sense for (residually) nilpotent groups. But even for these, some information may be lost: for example, the derived length of $L(G)$ may become smaller than that of G.

Any Lie ring method can be described by the following informal diagram:

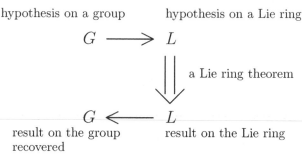

hypothesis on a group   hypothesis on a Lie ring

a Lie ring theorem

result on the group recovered   result on the Lie ring

$\rightarrow$ A hypothesis on a group is translated into a hypothesis on a Lie ring constructed from the group in some way.

$\Downarrow$ Then a theorem on Lie rings is proved (or used).

$\leftarrow$ Finally, a result about the group must be recovered from the Lie ring information obtained.

As we shall see, there may be difficulties both in translating a group problem to Lie rings and in recovering the information obtained.

Notation and terminology. The Lie subring generated by a subset U is denoted by $\langle U \rangle$, and the ideal generated by U by $_{\mathrm{id}}\langle U \rangle$. Products in a Lie ring are called commutators. For subsets X, Y we denote by $[X, Y]$ the additive subgroup generated by all commutators $[x, y]$; this is an ideal if X, Y are ideals. Terms of the derived series of a Lie ring (or a group) L are defined as $L^{(0)} = L$; $L^{(k+1)} = [L^{(k)}, L^{(k)}]$. Then L is soluble of derived length at most n (sometimes called "solubility index") if $L^{(n)} = 0$. Terms of the lower central series of L are defined as $\gamma_1(L) = L$; $\gamma_{k+1}(L) = [\gamma_k(L), L]$ (sometimes denoted as powers of L). Then L is nilpotent of class at most c (sometimes called "nilpotency index") if $\gamma_{c+1}(L) = 0$. A simple commutator $[X_1, X_2, \ldots, X_s]$ (of elements or subsets) is by definition the commutator $[\ldots[[X_1, X_2], X_3], \ldots, X_s]$.

Let A be an additively written abelian group. A Lie ring L is A-*graded* if

$$L = \bigoplus_{a \in A} L_a, \qquad [L_a, L_b] \subseteq L_{a+b},$$

where the *grading components* L_a are additive subgroups of L.

The results on Lie rings stated here are also valid for Lie algebras, with minimal modifications, like replacing index with codimension, additive subgroups with subspaces, etc.

Throughout the paper we use the abbreviation "(m, n)-bounded" for "bounded above in terms of m, n only".

2 Finite cyclic grading with $L_0 = 0$

Let $L = \bigoplus_{i=0}^{n-1} L_i$ be a $(\mathbb{Z}/n\mathbb{Z})$-graded Lie ring, where the L_i are additive subgroups satisfying $[L_i, L_j] \subseteq L_{i+j \,(\mathrm{mod}\, n)}$.

Theorem 2.1 (Higman [14], Kostrikin–Kreknin [37, 39]) *If $n = p$ is a prime and $L_0 = 0$, then L is nilpotent of p-bounded class at most $h(p)$.*

Theorem 2.2 (Kreknin [37]) *If $L_0 = 0$, then L is soluble of derived length at most $2^n - 2$.*

Another way of stating these results is $L^{(2^n - 2)} \subseteq \mathrm{id}\langle L_0 \rangle$ and $\gamma_{h(p)+1}(L) \subseteq \mathrm{id}\langle L_0 \rangle$, respectively. Higman's proof [14] was a pure existence theorem for $h(p)$; Kostrikin and Kreknin [37, 39] produced a new short proof, with an explicit upper bound for $h(p)$. However, this bound for $h(p)$ is in the region of p^{2^p}, a far cry from Higman's [14] lower bound $(p^2 - 1)/4$, which was confirmed to be the true value of $h(p)$ for several first primes.

Problem 2.3 Is the best-possible Higman's function given by $h(p) = (p^2 - 1)/4$? (for $p > 2$; and we know that $h(2) = 1$).

Problem 2.4 What is the best-possible bound for the derived length in Kreknin's theorem?

Theorems 2.2 and 2.1 are virtually equivalent to statements about fixed-point-free automorphisms. Indeed, let L be a Lie ring with an automorphism φ of finite order n: after adjoining a primitive nth root of unity ω we obtain in a 'typical' case $L = L_0 \oplus L_1 \oplus \cdots \oplus L_{n-1}$ for $L_i = \{x \in L \mid \varphi(x) = \omega^i x\}$. It is easy to see that $[L_i, L_j] \subseteq L_{i+j \,(\mathrm{mod}\, n)}$, so this is a $(\mathbb{Z}/n\mathbb{Z})$-grading. In the general case the sum is not direct and it only contains nL, but this is not a major issue in deriving the following corollary.

Corollary 2.5 (Higman [14], Kostrikin–Kreknin [37, 39]) *Suppose that a Lie ring L admits an automorphism $\varphi \in \mathrm{Aut}\, L$ of prime order p with $C_L(\varphi) = 0$. Then L is nilpotent of class $\leqslant h(p)$.*

Corollary 2.6 (Kreknin [37]) *Suppose that a Lie ring L admits an automorphism $\varphi \in \mathrm{Aut}\, L$ of finite order n with $C_L(\varphi) = 0$. Then L is soluble of derived length $\leqslant 2^n - 2$.*

Earlier Jacobson [15] (for n being a prime) and Borel–Mostow [4] (for any n and simisimple φ) proved these results for finite-dimensional Lie algebras, without upper bounds for the nilpotency class or derived length. We also mention Kreknin's paper [38], where he proved the solubility of a finite-dimensional Lie algebra with a fixed-point-free automorphism (possibly of infinite order, not necessarily semisimple).

Groups with fixed-point-free automorphisms. Corollaries 2.5 and 2.6 for Lie algebras must have corresponding consequences for connected and simply connected

Lie groups with fixed-point-free automorphisms of finite order. For arbitrary finite or nilpotent groups there is an immediate application for an automorphism of prime order.

Corollary 2.7 (Higman [14]) *If a locally nilpotent group G admits an automorphism $\varphi \in \mathrm{Aut}\, G$ of prime order p such that $C_G(\varphi) = 1$, then G is nilpotent of class $\leqslant h(p)$.*

In Higman's paper [14], the passage "$G \to L$" in the Lie ring method is to the associated Lie ring $L = L(G)$ with the induced automorphism φ: it can be shown that $C_L(\varphi) = 0$ (with extra care for infinite nilpotent groups, where the isolators of the γ_i must be taken in the definition of $L(G)$ instead of the γ_i). Then "$\Downarrow$" by Corollary 2.5 L is nilpotent of class $\leqslant h(p)$. Then "$G \leftarrow L$" so is G, since the nilpotency class of G coincides with that of $L(G)$.

Remark 2.8 Corollary 2.7 is actually true for any finite group G, with nilpotency provided by the celebrated theorem of Thompson [71]. As for infinite groups, some additional conditions are unavoidable: for example, a free group $F = \langle x, y \rangle$ admits the automorphism τ of order 2 interchanging the free generators, for which $C_F(\tau) = 1$.

It is worth recalling the above results of more than 40 years ago, because there are still unsolved problems in relation to group-theoretic applications.

Problem 2.9 Does an analogue of Kreknin's theorem hold for a finite (nilpotent) group G with a fixed-point-free automorphism φ of coprime order n? that is, is the derived length of G bounded by a function of n alone?

Even assuming G nilpotent, if we try to apply the same Lie ring method "$G \to L(G)$", then the induced fixed-point-free automorphism of $L(G)$ is fixed-point-free, and "$\Downarrow$" therefore $L(G)$ is soluble of derived length $\leqslant 2^n - 2$ by Kreknin. But this does not yield anything for the derived length of G, that is, the recovery "$G \leftarrow L$" does not work here. The only case where Problem 2.9 is solved, apart from φ of prime order, is that of $|\varphi| = 4$, due to Kovács [36]: then G is centre-by-metabelian.

Remark 2.10 For arbitrary finite groups Problem 2.9 is already reduced to nilpotent groups: solubility is by the classification (even for non-coprime order of the automorphism as shown by Rowley [61]), and the Fitting height is bounded by Hall–Higman-type theorems, starting from Thompson [72], (even without the coprimeness condition as shown by Dade [6]). Moreover, in these papers not just cyclic groups of automorphisms were considered.

Of course, as soon as a better Lie ring method can be applied, we have a partial solution to Problem 2.9. Apart from Lie groups, in abstract group theory we can mention the following case of locally nilpotent torsion-free groups.

Corollary 2.11 (Folklore) *If a locally nilpotent torsion-free group G admits an automorphism $\varphi \in \mathrm{Aut}\, G$ of finite order n such that $C_G(\varphi) = 1$, then G is soluble of derived length $\leqslant 2^n - 2$.*

Proof Embed G into its Mal'cev completion $\hat{G}$ by adjoining all roots of non-trivial elements; then φ extends to $\hat{G}$ with $C_{\hat{G}}(\varphi) = 1$. Let L be the Lie algebra over $\mathbb{Q}$ in the Mal'cev correspondence with $\hat{G}$, given by the inverse BCHf on the same set. Then φ can be regarded as an automorphism of L with $C_L(\varphi) = 0$. By Kreknin's theorem, L is soluble of derived length $\leqslant 2^n - 2$; hence so is $\hat{G}$, and so is G.   □

'Modular situation'. Nevertheless, the Higman–Kreknin–Kostrikin theorems were very successfully applied to finite p-groups P with an automorphism of order p^n. Of course, such an automorphism cannot be fixed-point-free; the problem is to restrict the structure of P in terms of p^n and the number of fixed points p^m. A crucial advantage here is a bound for the rank of P: by counting the Jordan blocks of φ as a linear transformation of an elementary abelian φ-invariant section we obtain that its rank is at most mp^n, which implies a bound for the rank of any section of P. The theory of powerful p-groups comes into play providing certain 'linear' tools, like Shalev's lemma $[M^p, N] = [M, N]^p$ for powerfully embedded subgroups, or various constructions of Lie rings that have a better grasp of the power-commutator structure than the ordinary associated Lie ring.

Theorem 2.12 ([1, 18, 46]) *If a finite p-group P admits an automorphism φ of prime order p with exactly p^m fixed points, then P has a subgroup of (p, m)-bounded index that is nilpotent of class $\leqslant h(p)$.*

Alperin [1] proved that the derived length of P is (p, m)-bounded; the bound $h(p) + 1$ for the class of a subgroup of bounded index was obtained in [18]; this bound was improved to $h(p)$ by Makarenko [46]. In the proofs, Higman's theorem is applied to the associated Lie rings L of certain groups in the form $\gamma_{h(p)+1}(pL) \subseteq {}_{\mathrm{id}}\langle C_L(\varphi)\rangle$.

Theorem 2.13 ([20, 63]) *If a finite p-group P admits an automorphism of order p^n with exactly p^m fixed points, then P has a subgroup of (p, m, n)-bounded index that is soluble of p^n-bounded derived length.*

Shalev [63] proved that the derived length of P is (p, m, n)-bounded; in the present form this is the result of [20]. In Shalev's proof, Kreknin's theorem was applied in the form $(p^n L)^{(k(p^n))} \subseteq {}_{\mathrm{id}}\langle C_L(\varphi)\rangle$ to a certain Lie ring L (not the associated Lie ring) constructed for uniformly powerful sections in the manner developed in his papers on coclass conjectures, when a new Lie ring product is defined by 'lifting' the old one by the inverse of taking a certain power.

In [20] Kreknin's theorem was applied first to the associated Lie rings of uniformly powerful sections S to yield that $T = S^{(k(p^n))}$ is nilpotent of 'weakly' (p, m, n)-bounded class $c = c(p, m, n)$. Then Kreknin's theorem was again applied to the Lie algebra L over $\mathbb{Q}$ that is in the Mal'cev correspondence with the Mal'cev completion $\hat{F}$ of a free nilpotent $\langle\varphi\rangle$-operator group F covering those sections T. The Lie algebra inclusion $L^{(k(p^n))} \subseteq {}_{\mathrm{id}}\langle C_L(\varphi)\rangle$ was then translated into the group language as $\hat{F}^{(k(p^n))} \subseteq \langle C_{\hat{F}}(\varphi)^{\hat{F}}\rangle$. Due to a bound for the nilpotency class, this yields the inclusion $(F^{g(p,m,n)})^{(k(p^n))} \subseteq \langle C_F(\varphi)^F\rangle$, where a certain power $F^{g(p,m,n)}$

replaces $\hat{F}$. This in turn implies a similar inclusion for those sections T. Quotients like $T/T^{p^{f(p,m,n)}}$ have bounded order because the rank of P is bounded; all such quotients can be 'pushed-up', so that a subgroup of bounded index appears that is soluble of p^n-bounded derived length.

The extreme case, where $|C_P(\varphi)| = p$, corresponds to the semidirect product $P\langle\varphi\rangle$ being a p-group of maximal class (if $|\varphi| = p$) or of given coclass (in the so-called 'uncovered' case). Shepherd [66] and Leedham-Green–McKay [42] proved that if $|\varphi| = p$ and $|C_P(\varphi)| = p$, then P has a subgroup of p-bounded index that is nilpotent of class 2. If $|\varphi| = p^n$ and $|C_P(\varphi)| = p$, Kiming [35] and McKay [57] proved that P has a subgroup of (p, n)-bounded index that is nilpotent of class 2. These results were generalized to the case of $|C_P(\varphi)| = p^m$ aiming at subgroups with derived length or nilpotency class bounded in terms of m only. For $|\varphi| = p$ Medvedev [59] proved the following.

Theorem 2.14 ([59]) *If a finite p-group P admits an automorphism of prime order p with p^m fixed points, then P has a subgroup of (p, m)-bounded index that is nilpotent of m-bounded class.*

There is an easy reduction to Lie rings based on Theorem 2.12 and Lazard's correspondence. For a Lie ring L one of the tricks is to define a new Lie ring structure: one can assume that the additive group of L is homocyclic, say, of exponent p^e; if $[L, L] \subseteq p^s L$, then the new product $[[a, b]] = (1/p^s)[a, b]$ induces a Lie ring structure on $L/p^{e-2s}L$. Roughly speaking, Higman's theorem is applied to this new Lie ring.

For $|\varphi| = p^n$ Jaikin-Zapirain [16] proved the following.

Theorem 2.15 ([16]) *If a finite p-group P admits an automorphism of order p^n with p^m fixed points, then P has a subgroup of (p, m, n)-bounded index that is soluble of m-bounded derived length.*

The method of proof can be viewed as a development of the earlier methods of Shalev [64] and Shalev–Zelmanov [65], where Lie rings were constructed from a p-group equipped with the action of a transcendental variable ϑ reflecting taking the pth powers in the group, and the above-mentioned method of Medvedev, with a new Lie ring multiplication defined by 'lifting' the old one. The assumption that there is no desired bound for the derived lengths leads to a certain limit Lie ring in zero characteristic, where Kreknin's theorem is applied, which leads to a contradiction. At the final stage, a subring N of (p, m, n)-bounded index is produced such that $\gamma_3(\gamma_3(\ldots\gamma_3(N)\ldots)) = 0$, where γ_3 appears $2^m - 1$ times; this is relevant to the m-bounded estimate for the derived length of a subgroup in Theorem 2.15. In this context, Jaikin-Zapirain stated the following problem.

Problem 2.16 Can $2^m - 1$ be reduced to m? Or at least a linear function of m?

Similar methods (ultimately based on Kreknin's theorem) were applied by Jaikin-Zapirain [17] to prove the following.

Theorem 2.17 ([17]) *If a finite group G of rank r admits an automorphism φ with exactly m fixed points, then G has a soluble subgroup of (r,m)-bounded index that has r-bounded derived length.*

Note that the bounds do not depend on $|\varphi|$. Earlier this result was proved in [62] and [21] in the case of $(|G|, |\varphi|) = 1$ by using different methods, see §4.

Coclass conjectures. A finite p-group P of nilpotency class c is said to have coclass n if $|P| = p^{c+n}$ (thus, p-groups of maximal class are those of coclass 1). Leedham-Green and Newman [44] stated a series of conjectures, the strongest of which said that a p-group of coclass n must have a class 2 nilpotent subgroup of (p,n)-bounded index. There are also natural extensions to the realm of pro-p-groups. These conjectures have been proved in a series of papers by Leedham-Green, Donkin, McKay, Plesken, Shalev, and Zel'manov [7, 40, 41, 43, 64, 65]. Without going into too much detail, here we only mention applications of graded Lie rings (the Higman–Kreknin–Kostrikin theorems).

First, one of the strongest results is that every pro-p-group of finite coclass is abelian-by-finite. The first proof of this fact was given by Donkin [7] for $p > 3$ and used the Iwahori–Matsumoto theory of p-adic Chevalley groups. A short proof of this result for all p was given by Shalev and Zel'manov in [65], where they used various Lie algebras and arrived at a final contradiction by using Kreknin's theorem.

Second, an effective proof of the coclass conjectures was given by Shalev in [64], where, among other techniques, he used the fact that a Lie algebra of characteristic $p > 0$ admitting a derivation D satisfying $D^{p-1} = 1$ is nilpotent. The eigenspace decomposition for D makes L a $(\mathbb{Z}/p\mathbb{Z})$-graded Lie algebra with $L_0 = 0$; hence Higman's theorem can be applied. (Actually, the argument by contradiction in [64] does not need a bound for the nilpotency class of L, only the nilpotency as such, which follows by an earlier result of Jacobson [15], since here L is also finite-dimensional.)

3 Finite cyclic grading with bounded L_0

If in a $(\mathbb{Z}/n\mathbb{Z})$-graded Lie ring L the zero-component L_0 is almost trivial, then L is almost soluble or nilpotent in the following precise sense.

Theorem 3.1 ([19, 48, 53]) *Let L be a $(\mathbb{Z}/n\mathbb{Z})$-graded Lie ring with $|L_0| = m$ (or $\dim L_0 = m$).*

(a) *Then L contains a soluble ideal of n-bounded derived length and of (m,n)-bounded index (codimension).*

(b) *If is addition n is a prime, then L contains a nilpotent ideal of n-bounded nilpotency class and of (m,n)-bounded index (codimension).*

Proof is a difficult calculation of complicated commutators in elements of the grading components, using induction on several parameters, where the Higman–Kreknin–Kostrikin theorems are applied repeatedly. Note that in general there does not have to be a subring of bounded index (or codimension) trivially intersecting L_0.

The so-called *method of graded centralizers* was developed in these calculations. To give a flavour of this method, consider the simplest case of $n = 2$.

Example 3.2 Let $L = L_0 \oplus L_1$ be $(\mathbb{Z}/2\mathbb{Z})$-graded with $|L_0| = m$. We can assume that $L = \langle L_1 \rangle$. 'Freeze' some expressions for all generators of the additive group L_0, we need at most m of them: $[x_1, x_2], \ldots, [x_{2m-1}, x_{2m}]$, where $x_i \in L_1$. Each 'graded centralizer' $C(x_j) = \{y \in L_1 \mid [x_j, y] = 0\}$ has index at most m in L_1, since its cosets correspond to distinct elements of L_0. Then $Z = \bigcap_{i=1}^{m} C(x_i)$ has bounded index in L_1 and therefore in L. The subring $\langle Z \rangle$ is nilpotent of class 2. Indeed, for any $z_i \in Z$, any $[z_1, z_2]$ is a linear combination of the $[x_j, x_{j+1}]$; by the definition of Z (and by Jacobi identity), $[[x_j, x_{j+1}], z_3] = 0$; hence, $[[z_1, z_2], z_3] = 0$. Thus a subring is constructed; further effort is required to produce an ideal.

This case of $n = 2$ really does not reflect the difficulty of the general case; even for $p = 3$ (which is still easy) many more steps are required. One can draw a comparison with the Burnside problem, where groups of exponent 2 are easily shown to be abelian, while for large exponents all those celebrated results of Adian–Novikov, Ol'shansky, Kostrikin, Zel'manov, et al. are required.

As before, a virtually equivalent formulation is in terms of automorphisms of finite order: almost fixed-point-free implies almost solubility or nilpotency.

Corollary 3.3 ([19, 48, 53]) *Suppose that a Lie ring L admits an automorphism φ of finite order n such that $|C_L(\varphi)| = m$ (or $\dim C_L(\varphi) = m$).*
(a) *Then L contains a soluble ideal of n-bounded derived length and of (n, m)-bounded index (codimension).*
(b) *If in addition n is a prime, then L contains a nilpotent ideal of n-bounded nilpotency class and of (n, m)-bounded index (codimension).*

This result is non-trivial even for finite-dimensional Lie algebras, because of the bound for the codimension.

Remark 3.4 In a broader context, Bahturin, Zaitsev, and Linchenko [2, 45] proved that if a Lie algebra L admits a finite group of automorphisms G of order coprime to the characteristic of the ground field, then a polynomial identity for the fixed-point subalgebra $C_L(G)$ implies some polynomial identity for L. But one cannot hope for (almost) solubility if G is non-cyclic: the simple three-dimensional Lie algebra with basis e_1, e_2, e_3 and structure constants $[e_i, e_{i+1}] = e_{i+2}$ (indices mod 3) admits the non-cyclic group F of order 4 generated by the linear transformations given in this basis by $\mathrm{diag}(-1, -1, 1)$, $\mathrm{diag}(-1, 1, -1)$, for which $C_L(F) = 0$.

However, nilpotency results hold for associative rings with arbitrary finite automorphism groups. Somewhat deviating from the theme, we mention a problem modelled on the above almost nilpotency results for Lie rings.

Problem 3.5 Suppose that an associative algebra A admits a finite group of automorphisms G of order coprime to the characteristic of the ground field with finite-dimensional fixed-point subalgebra $C_A(G)$. Must A have a nilpotent ideal of $|G|$-bounded class and of codimension bounded in terms of $|G|$ and $\dim C_A(G)$?

The Bergman–Isaacs theorem [3] deals with the case $C_A(G) = 0$: then A is nilpotent of $|G|$-bounded class. Kharchenko informed the author that in Problem 3.5 it can be shown that the algebra A contains an ideal of bounded codimension that is nilpotent of class 'weakly' bounded in terms of $|G|$ *and* $\dim C_A(G)$. Interestingly, for an associative algebra graded by any finite group the corresponding question has an affirmative answer, as shown by Makarenko (unpublished) by using analogues of graded centralizers; but it is not clear how it may help in Problem 3.5, except, may be, for a soluble G. One situation where Problem 3.5 is rather easily solved is that of a cyclic group G. In fact, Makarenko generalized Theorem 3.1 to so-called algebras of Lie type as follows (products are not necessarily associative but denoted by concatenation).

Theorem 3.6 ([49]) *Let* $L = \bigoplus_{i=0}^{n} L_i$ *be a* $(\mathbb{Z}/n\mathbb{Z})$-*graded algebra of Lie type over a field* F, *that is,* $L_s L_t \subseteq L_{s+t \,(\mathrm{mod}\, n)}$ *and for any integers* g, h, k *there are* $\alpha \neq 0, \beta \in F$ *such that* $(ab)c = \alpha a(bc) + \beta (ac)b$ *for all* $a \in L_g$, $b \in L_h$, $c \in L_k$. *If* $\dim L_0 = m < \infty$, *then* L *has a soluble ideal of* n-*bounded derived length and of* (m, n)-*bounded codimension.*

The proof is a generalization of the proof of Theorem 3.1 by the method of graded centralizers (the proof is much simpler if L is associative). Corollaries include facts on automorphisms and (colour) Lie superalgebras.

Groups admitting an automorphism with bounded $C_G(\varphi)$. Bearing in mind that even the fixed-point-free case is open for arbitrary order of automorphism (apart from those nice cases where a Lie ring method works for derived length), for the moment we can only expect an application for groups with an automorphism of prime order. Again, almost fixed-point-free implies almost nilpotency.

Theorem 3.7 ([19], [58]) *If a nilpotent group* G *admits an automorphism of prime order* p *with* $|C_L(\varphi)| = m$, *then* G *contains a subgroup of* (p, m)-*bounded index that is nilpotent of* p-*bounded class.*

The case of finite groups was dealt with in [19]; Medvedev [58] extended the result to infinite nilpotent groups by a reduction to finite. The proof in [19] is based on the Lie ring method with the associated Lie ring for "$G \to L$", and Corollary 3.3(b) for "$\Downarrow$". But the recovery "$G \leftarrow L$" is a difficult task here, as there is no good correspondence between subgroups of G and subrings or ideals of $L(G)$.

Remark 3.8 The result of Theorem 3.7 is also true if G is any finite group: Fong [8] proved, based on the classification, that then G has a soluble subgroup of (p, m)-bounded index, and for soluble groups Hartley–Meixner [12] and Pettet [60] proved that the index of the Fitting subgroup is (p, m)-bounded. The result for $p = 2$ was also proved earlier by Hartley–Meixner [11] (moreover, by Shunkov's theorem [70], for $p = 2$ the result holds even for any periodic group G).

Problem 3.9 It is of course natural to conjecture that if a finite group G admits an automorphism φ of coprime order n with $|C_G(\varphi)| = m$, then it must have a subgroup of (m, n)-bounded index that is soluble of n-bounded derived length.

...But surely the case of $C_G(\varphi) = 1$ must be settled first (Problem 2.9).

Remark 3.10 It is nice to know, though, that Problem 3.9, too, is already reduced to nilpotent groups: Hartley [9] proved that there is a soluble subgroup of (m, n)-bounded index, and Hartley–Isaacs [10] (using Turull's [73]) produced a subgroup of (m, n)-bounded index that has n-bounded Fitting height.

The only case of composite order where satisfactory results were obtained is that of $|\varphi| = 4$, which can be summed as following.

Theorem 3.11 ([54]) *If a finite group G admits an automoprphism φ of order 4 with $|C_G(\varphi)| = m$, then G contains a subgroup H of m-bounded index such that $\gamma_3(H)$ is nilpotent of '4-bounded' (by a constant) class.*

Reduction to soluble groups is due to Fong [8]. Then Hall–Higman type theorems are applied in [54], using Hartley–Turau [13], to reduce to nilpotent groups. Then the Lie ring method with the associated Lie ring is used in [51, 54], with "⇓" being the results for $(\mathbb{Z}/4\mathbb{Z})$-graded Lie rings in [52]. Makarenko's result [47] in the 'modular' case where G is a 2-group is especially strong, giving a centre-by-metabelian subgroup of bounded index.

Another way to regard an automorphism φ as 'almost fixed-point-free' is to seek restrictions on G depending on $|\varphi|$ and the rank of the fixed-point subgroup $C_G(\varphi)$. (Recall that a finite group has rank $\leqslant r$ if each of its subgroups can be generated by r elements.) Almost fixed-point-free in this sense also implies almost nilpotency in respective terms, for $|\varphi|$ being a prime.

Theorem 3.12 ([24]) *If a finitely generated, or periodic, nilpotent group G admits an automorphism φ of prime order p with centralizer $C_G(\varphi)$ of finite rank r, then G has a characteristic nilpotent subgroup C of p-bounded class such that G/C has finite (p, r)-bounded rank.*

Examples show that one cannot drop both additional conditions of being finitely generated or periodic. Earlier the case of $p = 2$ was done by Shumyatsky [67]. The proof in [24] uses powerful p-groups, the Lie ring method with associated Lie ring, and Corollary 3.3(b). Note that here even the transition "$G \to L$" is difficult, since the rank of fixed points can increase. As before, the recovery "$G \leftarrow L$" is also difficult.

Remark 3.13 If in Theorem 3.12 G is an arbitrary finite group, then it is proved in [31] that there are characteristic subgroups $G > N > R > 1$ with G/N and R of (p, r)-bounded rank and N/R nilpotent — of p-bounded class due to Theorem 3.12 (and examples show that R cannot be removed). This result was also extended in [32] to non-cyclic groups of automorphisms A of coprime order with $C_G(A)$ of given rank r by induction based on Thompson's [72].

4 Gradings with few non-trivial components

Shalev [62] noticed that if there are only few non-trivial components in a cyclic grading, then the bound in Kreknin's theorem for the derived length can be improved. A similar improvement holds for the nilpotency class in Higman's theorem.

Theorem 4.1 ([23, 62]) *Suppose that $L = L_0 \oplus \cdots \oplus L_{n-1}$ is a $(\mathbb{Z}/n\mathbb{Z})$-graded Lie ring in which only r of the components L_i are non-trivial, and $L_0 = 0$.*
(a) Then L is soluble of r-bounded derived length.
(b) If in addition n is a a prime, then L is nilpotent of r-bounded class.

The proof in [62] followed Kreknin's proof [37], 'skipping' the non-existent components; the case of n being a prime was similarly derived in [23] from the Kreknin–Kostrikin proof [37, 39] of Higman's theorem.

Theorem 4.1 found natural applications for groups of given rank with an almost fixed-point-free automorphism; cf. Problem 2.9.

Theorem 4.2 ([21, 23, 62]) *Suppose that a finite group G of rank r admits an automorphism φ of coprime order with $|C_G(\varphi)| = m$.*
(a) Then G has a soluble subgroup of r-bounded derived length and of (m, r)-bounded index. If $m = 1$, then G itself is soluble of r-bounded derived length.
(b) If in addition $|\varphi|$ is a prime, then G has a nilpotent subgroup C of r-bounded class and of (m, r)-bounded index, and if $m = 1$, then G itself is nilpotent of r-bounded class.

Note that the bounds are independent of $|\varphi|$. Shalev [62] obtained a soluble subgroup of bounded index with (m, r)-bounded derived length; the stronger bounds for the derived length and nilpotency class in terms of r only were obtained in [21] and [23].

Theorem 4.1 was generalized in a familiar manner: from trivial to bounded L_0.

Theorem 4.3 ([30, 50]) *Suppose that $L = L_0 \oplus \cdots \oplus L_{n-1}$ is a $(\mathbb{Z}/n\mathbb{Z})$-graded Lie ring in which only r of the components L_i are non-trivial, and $|L_0| = m$ (or $\dim L_0 = m$).*
(a) Then L contains a soluble ideal of r-bounded derived length and of (m, r)-bounded index (codimension).
(b) If in addition n is a prime, then L contains a nilpotent ideal of r-bounded class and of (m, r)-bounded index (codimension).

It is important that the bounds in this theorem are independent of n. As before, there is an application for an almost fixed-point-free semisimple automorphism φ giving bounds in terms of the number of eigenvalues (rather than $|\varphi|$).

We also mention another result on few grading components, where Engel-type conditions are involved. The essence of the theorem is in the bound for the nilpotency class. For finite-dimensional L, nilpotency, without a bound, follows from the well-known results of Engel–Jacobson.

Theorem 4.4 ([34]) *Let A be an additively written abelian group, and let $L = \bigoplus_{a \in A} L_a$ be an A-graded Lie ring. Suppose that there are at most d non-trivial grading components among the L_a. If $[L, \underbrace{L_i, \ldots, L_i}_{t}] = 0$ for all $i \in A$, then L is nilpotent of (d, t)-bounded class.*

The proof uses the following result of Makarenko [50], which generalized, to the case of few non-trivial components, the criterion for solubility in [22, 33], which in turn was an effectivization of Winter's [74] generalization of Kreknin's theorem. This result is used in the situation where $[L, \underbrace{L_0, \ldots, L_0}_{t}] = 0$, so that L is then soluble of (d, t)-bounded derived length.

Theorem 4.5 ([50]) *Let L be a $(\mathbb{Z}/n\mathbb{Z})$-graded Lie ring with exactly d non-trivial components among the L_i. There exists a function $f : \mathbb{N} \times \mathbb{N} \to \mathbb{N}$ such that the $f(d, t)$-th term of the derived series $L^{(f(d,t))}$ is contained in the subalgebra generated by the set $[L, \underbrace{L_0, \ldots, L_0}_{t}]$.*

An application to almost constant-free derivations. Let us see how the fact that the bound for the nilpotency class in Theorem 4.3(b) is independent of n (as long as it is a prime) is used in the proof of the following theorem. A classical result of Jacobson [15] states that if a finite-dimensional Lie algebra L of characteristic 0 admits a nilpotent Lie algebra of derivations D that is constant-free (that is, $x\delta = 0$ for all $\delta \in D$ only if $x = 0$), then L is nilpotent. Now if D is almost constant-free, then L is almost nilpotent in the following precise sense.

Theorem 4.6 ([30]) *If a finite-dimensional Lie algebra L over an algebraically closed field of characteristic 0 admits a nilpotent Lie algebra of derivations D with exactly d weights in L and the dimension of the Fitting null-component for D is m, then L has an ideal of (m, d)-bounded codimension that is nilpotent of d-bounded class. If $m = 0$, then L itself is nilpotent of d-bounded class.*

The last part also gives a bound for the nilpotency class in Jacobson's theorem (where $m = 0$).

Proof The d weights of D are elements of the dual space D^*. By Zassenhaus's theorem,

$$L = \bigoplus_{a \in W} L_a,$$

for the weight spaces

$$L_a = \{x \in L \mid x.(y - a(y)\mathbf{Id})^N = 0 \text{ for all } y \in D \text{ for some } N\}.$$

The latter satisfy

$$[L_a, L_b] \subseteq L_{a+b} \quad \text{if } a + b \in W, \qquad [L_a, L_b] = 0 \quad \text{if } a + b \notin W.$$

Recall that $m = \dim L_0$ by hypothesis.

Since W is a finite subset of the dual D^* and char $= 0$, the subgroup $A = {}_+\langle W \rangle$ is a finitely generated free abelian group. Hence there is a (possibly large!) prime p and a subgroup B of index p such that $W \cap B = \{0\}$.

We identify $A/B = \mathbb{Z}/p\mathbb{Z}$ and define a $(\mathbb{Z}/p\mathbb{Z})$-grading of L by setting

$$L_i = \bigoplus_{a+B=i} L_a.$$

Then still $\dim L_0 = m$ and there are at most d non-trivial components. It remains to apply Theorem 4.3(b).   □

It may be worth mentioning that the above argument can be applied for a Lie ring graded by any abelian torsion-free group.

Corollary 4.7 *Let A be an abelian torsion-free group. Suppose that an A-graded Lie ring L has only r non-trivial grading components. If $L_0 = 0$, then L is nilpotent of r-bounded class. If $|L_0| = m$ (or $\dim L_0 = m$), then L has an ideal of (m, r)-bounded index (codimension) that is nilpotent of r-bounded class.*

5 Gradings with many commuting components

Yet another variation on the Higman–Kreknin–Kostrikin theorem may seem artificial at first glance, but it actually emerged in answering a question on 2-Frobenius groups (see below).

Theorem 5.1 ([25]) *Let L be a $(\mathbb{Z}/n\mathbb{Z})$-graded Lie ring. Suppose that $L_0 = 0$ and for some m each component L_k commutes with all but at most m components, that is, $|\{i \mid [L_k, L_i] \neq 0\}| \leqslant m$.*
  (a) *Then L is soluble (for n prime, nilpotent) of m-bounded derived length.*
  (b) *If in addition n is a prime, then L is nilpotent of m-bounded class.*

It is important that the bounds are independent of n. The proof uses Theorem 4.1.

In a recent paper Makarenko and Shumyatsky [55] proved a more general result giving a bound for the nilpotency class of a $(\mathbb{Z}/p\mathbb{Z})$-graded Lie ring L with $L_0 = 0$, p being a prime, under a certain condition, which has the nature of generalized 'many commuting components', in a certain precise sense. This condition, however, is rather technical and too cumbersome to be discussed here. They also applied this criterion to 2-Frobenius groups.

The following two theorems were also used recently in the study of Frobenius groups of automorphisms (see below). First, in familiar vein, there is an extension to bounded, rather than trivial, L_0, with Theorem 4.3 used in the proof.

Theorem 5.2 ([26]) *Let L be a $(\mathbb{Z}/n\mathbb{Z})$-graded Lie ring. Suppose that $|L_0| = r$ (or $\dim L_0 = r$) and for some m each component L_k for $k \neq 0$ commutes with all but at most m components, that is, $|\{i \mid [L_k, L_i] \neq 0\}| \leqslant m$ for $k \neq 0$.*

(a) *Then L contains a soluble ideal of m-bounded derived length and of (n, r)-bounded index (codimension).*

(b) *If in addition n is a prime, then L contains a nilpotent ideal of m-bounded class and of (n, r)-bounded index (codimension).*

Problem 5.3 Is it possible to replace "(n, r)-bounded index (codimension)" in Theorem 5.2 by "(m, r)-bounded index (codimension)"?

In the second theorem, the Lie ring is graded by an arbitrary abelian group. The proof uses Theorem 4.4; as in that theorem, the essence is in the bound for the nilpotency class (without a bound, the nilpotency follows for finite-dimensional L from Engel–Jacobson).

Theorem 5.4 ([34]) *Let A be an additively written abelian group, and let $L = \bigoplus_{a \in A} L_a$ be an A-graded Lie ring. Suppose that for some m every grading component L_k for $k \neq 0$ commutes with all but at most m components, that is, $|\{i \mid [L_k, L_i] \neq 0\}| \leqslant m$ for each $k \neq 0$. If $[L, \underbrace{L_i, \ldots, L_i}_{t}] = 0$ for all $i \in A$, then L is nilpotent of (m, t)-bounded class.*

Frobenius groups of automorphisms. First we have an application to so-called 2-Frobenius groups. In a recent paper Makarenko and Shumyatsky [55] obtained the following remarkable result.

Theorem 5.5 ([55]) *Let BC be a finite Frobenius group with kernel B and complement C of order $|C| = t$ acting on a finite group G so that GB is also a Frobenius group with kernel G and complement B. If $C_G(C)$ is nilpotent of class c, then G is nilpotent of (c, t)-bounded class.*

Of course, the group G is nilpotent by Thompson's theorem [71], of nilpotency class bounded in terms of the least prime divisor of $|B|$ by Higman's theorem [14]. Theorem 5.5 gives a better bound if $|C|$ is small. 2-Frobenius groups arise, in particular, in the study of the Gruenberg–Kegel prime graphs of finite groups. Earlier the result of Theorem 5.5 was proved by Mazurov [56] for $c = 1$ and $t = 2, 3$. In [25] the case of $c = 1$ was considered for any t. The proof of Theorem 5.5 uses the associated Lie ring $L(G)$, with grading given by the eigenspaces of B. This grading is shown to have many commuting components in a certain generalized sense due to $C_G(C)$ being nilpotent of class c. (In a simpler situation of $c = 1$ this is the condition of Theorem 5.1.)

Problem 5.6 Let BC be a Frobenius group with kernel B and complement C. Suppose that BC acts on a finite group G so that GB is also a Frobenius group with kernel G and complement B. Is the exponent of G bounded in terms of $|C|$ and the exponent of $C_G(C)$?

In a more general situation we have a Frobenius group of automorphisms.

Theorem 5.7 ([26]) *Let BC be a Frobenius group with kernel B of prime order p and with complement C of order t acting on a finite group G so that $|C_G(B)| = s$ and $(|G|, |BC|) = 1$. If $C_G(C)$ is abelian, then G contains a subgroup of (p, s)-bounded index that is nilpotent of t-bounded class.*

By Remark 3.8 and Theorem 3.7, G has a subgroup of (p, s)-bounded index that is nilpotent of p-bounded class. Theorem 5.7 gives a better bound for the class if t is small.

Problem 5.8 (a) In Theorem 5.7, can one replace "(p, s)-bounded index" by "(t, s)-bounded index"?

(b) If in Theorem 5.7, "$C_G(C)$ is abelian" is replaced by "$C_A(C)$ is nilpotent of class c", will G contain a subgroup of (p, s)-bounded (or even (t, s)-bounded) index that is nilpotent of (c, t)-bounded class?

Theorems 5.5 and 5.7 are essentially about the case where the kernel B of a Frobenius group of automorphisms is cyclic. Now consider a Frobenius group of automorphisms BC with non-cyclic kernel B.

Theorem 5.9 ([34]) *Let BC be a Frobenius group with non-cyclic abelian kernel B and with complement C of order t. Suppose that BC acts coprimely on a finite group G so that $C_G(C)$ is abelian and $[\underbrace{C_G(u), C_G(v), \ldots, C_G(v)}_{k}] = 1$ for any $u, v \in B \setminus \{1\}$. Then G is nilpotent of (k, t)-bounded class.*

The proof uses Theorem 5.4. Earlier Shumyatsky [68] proved that if a finite p'-group G admits an elementary abelian group of automorphisms B of order p^2 such that $[\underbrace{C_G(u), C_G(v), \ldots, C_G(v)}_{k}] = 1$ for any $u, v \in B \setminus \{1\}$, then G is nilpotent of (p, k)-bounded class. In Theorem 5.9 the additional condition on the fixed points of the Frobenius complement C yields a bound for the nilpotency class that is independent of the order of B.

If B is abelian of rank at least three, then the result becomes stronger.

Theorem 5.10 ([34]) *Let BC be a Frobenius group with abelian kernel B of rank at least three and with complement C of order t. Suppose that BC acts coprimely on a finite group G so that $C_G(C)$ is abelian and $C_G(b)$ is nilpotent of class at most c for every $b \in B \setminus \{1\}$. Then G is nilpotent of (c, t)-bounded class.*

The proof uses Theorem 5.4. Shumyatsky [69] proved that if a finite p'-group G admits an elementary abelian group of automorphisms B of order p^3 such that $C_G(b)$ is nilpotent of class c for every $b \in B \setminus \{1\}$, then G is nilpotent of (p, c)-bounded class. In Theorem 5.10 the additional condition on the fixed points of the Frobenius complement C yields a bound for the nilpotency class that is independent of the order of B.

Problem 5.11 If the condition of $C_G(C)$ being abelian in Theorems 5.9 and 5.10 is replaced by $C_G(C)$ being nilpotent of class d, will similar results hold, with the bounds for the nilpotency class in terms of k, t, d and c, t, d, respectively?

6 Remark on characteristic subgroups

In all the above results, where a subgroup of (bounded) index s is constructed that is nilpotent of class t, or soluble of derived length t, one can claim that there is also a characteristic subgroup of (s, t)-bounded index which is nilpotent of class at most t, or soluble of derived length at most t, respectively. This is due to the results in [5, 28]. (The result for nilpotent subgroups of finite index is due to B. Bruno and F. Napolitani [5, Lemma 3]; unfortunately, this was not known to the authors of [28] earlier. I am grateful to E. Jabara for bringing the paper [5] to my attention.)

The same applies to the results in which a normal subgroup H of a p-group G or of a torsion-free group G is constructed with G/H of (bounded) rank r such that H is nilpotent of class t, or soluble of derived length t. Then one can claim that there is also a characteristic subgroup C of G with (r, t)-bounded rank of the quotient G/C such that C is nilpotent of class at most t, or soluble of derived length at most t, respectively. This is due to the results in [27] and [29].

Likewise, whenever an ideal in a Lie algebra is produced that has codimension s and is nilpotent of class t, or soluble of derived length t, one can claim that there is also an automorphically-invariant ideal of (s, t)-bounded codimension which is nilpotent of class at most t, or soluble of derived length at most t, respectively. This is due to the result in [29].

Actually in [28, 29, 27] there were considered not just subgroups or ideals satisfying nilpotency or solubility laws, but arbitrary outer-commutator laws (multilinear laws). Moreover, in [27] not only finiteness of the order, or of the rank, or of dimension, but certain general 'smallness' properties of the quotient by a normal subgroup or ideal were considered, for general groups with multioperators.

References

[1] J. L. Alperin, Automorphisms of solvable groups, *Proc. Amer. Math. Soc.* **13** (1962), 175–180.

[2] Yu. A. Bahturin and M. V. Zaicev, Identities of graded algebras, *J. Algebra* **205** (1998), 1–12.

[3] G. M. Bergman and I. M. Isaacs, Rings with fixed-point-free group actions, *Proc. London Math. Soc.* **27** (1973), 69–87.

[4] A. Borel and G. D. Mostow, On semi-simple automorphisms of Lie algebras, *Ann. Math. (2)* **61** (1955), 389–405.

[5] B. Bruno and F. Napolitani, A note on nilpotent-by-Černikov groups, *Glasgow Math. J.* **46** (2004) 211–215

[6] E. C. Dade, Carter subgroups and Fitting heights of finite solvable groups, *Illinois J. Math.* **13** (1969), 449–514.

[7] S. Donkin, Space groups and groups of prime power order. VIII. Pro-p-groups of finite coclass and p-adic Lie algebras, *J. Algebra* **111** (1987), 316–342.

[8] P. Fong, On orders of finite groups and centralizers of p-elements, *Osaka J. Math.* **13** (1976), 483–489.

[9] B. Hartley, A general Brauer–Fowler theorem and centralizers in locally finite groups, *Pacific J. Math.* **152** (1992), 101–117.

[10] B. Hartley and I. M. Isaacs, On characters and fixed points of coprime operator groups, *J. Algebra* **131** (1990), 342–358.

[11] B. Hartley and T. Meixner, Periodic groups in which the centralizer of an involution has bounded order, *J. Algebra* **64** (1980), 285–291.

[12] B. Hartley and T. Meixner, Finite soluble groups containing an element of prime order whose centralizer is small, *Arch. Math. (Basel)* **36** (1981), 211–213.

[13] B. Hartley and V. Turau, Finite soluble groups admitting an automorphism of prime power order with few fixed points, *Math. Proc. Cambridge Philos. Soc.* **102** (1987), 431–441.

[14] G. Higman, Groups and rings which have automorphisms without non-trivial fixed elements, *J. London Math. Soc. (2)* **32** (1957), 321–334.

[15] N. Jacobson, A note on automorphisms and derivations of Lie algebras, *Proc. Amer. Math. Soc.* **6** (1955), 281–283.

[16] A. Jaikin-Zapirain, On almost regular automorphisms of finite p-groups, *Adv. Math.* **153** (2000), 391–402.

[17] A. Jaikin-Zapirain, Finite groups of bounded rank with an almost regular automorphism, *Israel J. Math.* **129** (2002), 209–220.

[18] E. I. Khukhro, Finite p-groups admitting an automorphism of order p with a small number of fixed points, *Mat. Zametki* **38** (1985), 652–657 (Russian); English transl., *Math. Notes.* **38** (1986), 867–870.

[19] E. I. Khukhro, Groups and Lie rings admitting an almost regular automorphism of prime order, *Mat. Sbornik* **181**, no. 9 (1990), 1207–1219; English transl., *Math. USSR Sbornik* **71**, no. 9 (1992), 51–63.

[20] E. I. Khukhro, Finite p-groups admitting p-automorphisms with few fixed points, *Mat. Sb.* **184**, no. 12 (1993), 53–64; English transl., *Russ. Acad. Sci., Sb., Math.* **80** (1995), 435–444.

[21] E. I. Khukhro, Almost regular automorphisms of finite groups of bounded rank, *Sibirsk. Mat. Zh.* **37** (1996), 1407–1412; English transl., *Siberian Math. J.* **37** (1996), 1237–1241.

[22] E. I. Khukhro, On the solvability of Lie rings with an automorphism of finite order, *Sibirsk. Mat. Zh.* **42** (2001), 1187–1192; *Siberian Math. J.* **42** (2001) 996–1000.

[23] E. I. Khukhro, Finite groups of bounded rank with an almost regular automorphism of prime order, *Sibirsk. Mat. Zh.* **43** (2002), 1182–1191; English transl., *Siberian Math. J.* **43** (2002), 955–962.

[24] E. I. Khukhro, Groups with an automorphism of prime order that is almost regular in the sense of rank, *J. London Math. Soc.* **77** (2008), 130–148.

[25] E. I. Khukhro, Graded Lie rings with many commuting components and an application to 2-Frobenius groups, *Bull. London Math. Soc.* **40**, (2008), 907–912.

[26] E. I. Khukhro, Lie rings with a finite cyclic grading in which there are many commuting components, *Siberian Electron. Math. Rep.* (http://semr.math.nsc.ru) **6** (2009), 243–250 (Russian).

[27] E. I. Khukhro, Ant. A. Klyachko, N. Yu. Makarenko, Yu. B. Melnikova, Automorphism invariance and identities, *Bull. London Math. Soc.* **41** (2009), 804–816.

[28] E. I. Khukhro and N. Yu. Makarenko, Large characteristic subgroups satisfying multilinear commutator identities, *J. London Math. Soc.* **75**, no. 3 (2007), 635–646.

[29] E. I. Khukhro and N. Yu. Makarenko, Automorphically-invariant ideals satisfying multilinear identities, and group-theoretic applications, *J. Algebra* **320** (2008), 1723–1740.

[30] E. I. Khukhro, N. Yu. Makarenko, and P. Shumyatsky, Nilpotent ideals in graded Lie algebras and almost constant-free derivations, *Commun. Algebra* **36** (2008), 1869–1882.

[31] E. I. Khukhro and V. D. Mazurov, Finite groups with an automorphism of prime order whose centralizer has small rank, *J. Algebra* **301** (2006), 474–492.

[32] E. I. Khukhro and V. D. Mazurov, Automorphisms with centralizers of small rank, in *Groups St Andrews 2005, Vol. 2* (C. M. Campbell et al., eds.), London Math. Soc. Lecture Note Ser. **340** (CUP, Cambridge 2007), 564–585.

[33] E. I. Khukhro and P. V. Shumyatsky, On fixed points of automorphisms of Lie rings and locally finite groups, *Algebra Logika* **34** (1995), 706–723; English transl., *Algebra Logic* **34** (1995), 395–405.

[34] E. I. Khukhro and P. Shumyatsky, Nilpotency of finite groups with Frobenius groups of automorphisms, *submitted*, 2009.

[35] I. Kiming, Structure and derived length of finite p-groups possessing an automorphism of p-power order having exactly p fixed points, *Math. Scand.* **62** (1988), 153–172.

[36] L. G. Kovacs, Groups with regular automorphisms of order four, *Math. Z.* **75** (1961), 277–294.

[37] V. A. Kreknin, The solubility of Lie algebras with regular automorphisms of finite period, *Dokl. Akad. Nauk SSSR* **150** (1963), 467–469; English transl., *Math. USSR Doklady* **4** (1963), 683–685.

[38] V. A. Kreknin, Solvability of a Lie algebra containing a regular automorphism, *Sibirsk. Mat. Zh.* **8** (1967), 715–716; English transl., *Siberian Math. J.* **8** (1967), 536–537.

[39] V. A. Kreknin and A. I. Kostrikin, Lie algebras with regular automorphisms, *Dokl. Akad. Nauk SSSR* **149** (1963), 249–251; English transl., *Math. USSR Doklady* **4** (1963), 355–358.

[40] C. R. Leedham-Green, Pro-p-groups of finite coclass, *J. London Math. Soc.* **50** (1994), 43–48.

[41] C. R. Leedham-Green, The structure of finite p-groups, *J. London Math. Soc.* **50** (1994), 49–67.

[42] C. R. Leedham-Green and S. McKay, On p-groups of maximal class. I, *Quart. J. Math. Oxford Ser.* **27** (1976), 297–311; II, *ibid.* **29** (1978), 175–186; III, *ibid.* **29** (1978), 281–299.

[43] C. R. Leedham-Green, S. McKay, and W. Plesken, Space groups and groups of prime power order. V. A bound to the dimension of space groups with fixed coclass, *Proc. London Math. Soc.* **52** (1986), 73–94.

[44] C. R. Leedham-Green and M. F. Newman, Space groups and groups of prime power order. I, *Arch. Math. (Basel)* **35** (1980), 193–202.

[45] V. Linchenko, Identities of Lie algebras with actions of Hopf algebras, *Commun. Algebra* **25** (1997) 3179–3187; erratum, *ibid.* **31** (2003), 1045–1046.

[46] N. Yu. Makarenko, On almost regular automorphisms of prime order, *Sib. Matem. Zh.* **33**, no. 5 (1992), 206–208; English transl., *Sib. Math. J.* **33**, no. 5 (1992), 932–934.

[47] N. Yu. Makarenko, Finite 2-groups with automorphisms of order 4, *Algebra Logika* **40** (2001), 83–96; English transl., *Algebra Logic* **40** (2001), 47–54.

[48] N. Yu. Makarenko, A nilpotent ideal in the Lie rings with an automorphism of prime order, *Sibirsk. Mat. Zh.* **46** (2005), 1361–1374; English transl., *Siberian Math. J.* **46** (2005), 1097–1107.

[49] N. Yu. Makarenko, *Small centralizers in groups and Lie rings*, Diss. ... Doktor Fiz.-Mat. Nauk (Inst. Math., Novosibirsk 2006) (Russian).

[50] N. Yu. Makarenko, Graded Lie algebras with a few non-trivial components, *Sibirsk. Mat. Zh.* **48** (2007), 116–137; English transl., *Siberian Math. J.* **48** (2007), 95–111.

[51] N. Yu. Makarenko and E. I. Khukhro, Nilpotent groups admitting an almost regular

automorphism of order 4, *Algebra Logika* **35** (1996), 314–333; English transl., *Algebra Logic* **35** (1996), 176–187.

[52] N. Yu. Makarenko and E. I. Khukhro, Lie rings admitting an automorphism of order 4 with few fixed points. II, *Algebra Logika* **37** (1998), 144–166; English transl., *Algebra Logic* **37**, no. 2 (1998), 78–91.

[53] N. Yu. Makarenko and E. I. Khukhro, Almost solubility of Lie algebras with almost regular automorphisms, *J. Algebra* **277** (2004), 370–407.

[54] N. Yu. Makarenko and E. I. Khukhro, Finite groups with an almost regular automorphism of order four, *Algebra Logika* **45** (2006), 575–602; English transl., *Algebra Logic* **45** (2006), 326–343.

[55] N. Yu. Makarenko and P. Shumyatsky, *Frobenius groups as groups of automorphisms*, Preprint (Brazilia–Mulhouse 2009).

[56] V. D. Mazurov, Recognition of the finite simple groups $S_4(q)$ by their element orders, *Algebra Logika* **41** (2002), 166–198; English transl., *Algebra Logic* **41** (2002), 93–110.

[57] S. McKay, On the structure of a special class of p-groups, *Quart. J. Math. Oxford* **38** (1987), 489–502.

[58] Yu. Medvedev, Groups and Lie rings with almost regular automorphisms, *J. Algebra* **164** (1994), 877–885.

[59] Yu. Medvedev, p-Divided Lie rings and p-groups, *J. London Math. Soc.* **59** (1999), 787–798.

[60] M. R. Pettet, Automorphisms and Fitting factors of finite groups, *J. Algebra* **72** (1981), 404–412.

[61] P. Rowley, Finite groups admitting a fixed-point-free automorphism group, *J. Algebra* **174** (1995), 724–727.

[62] A. Shalev, Automorphisms of finite groups of bounded rank, *Israel J. Math.* **82** (1993), 395–404.

[63] A. Shalev, On almost fixed point free automorphisms, *J. Algebra* **157** (1993), 271–282.

[64] A. Shalev, The structure of finite p-groups: effective proof of the coclass conjectures, *Invent. Math.* **115** (1994), 315–345.

[65] A. Shalev and E. I. Zelmanov, Pro-p-groups of finite coclass, *Math. Proc. Cambridge Philos. Soc.* **111** (1992), 417–421.

[66] R. Shepherd, *p-Groups of maximal class*, Ph. D. Thesis (Univ. of Chicago 1971).

[67] P. Shumyatsky, Involutory automorphisms of finite groups and their centralizers, *Arch. Math. (Basel)* **71** (1998), 425–432.

[68] P. Shumyatsky, On locally finite groups and the centralizers of automorphisms, *Boll. Unione Mat. Italiana* **4** (2001), 731–736.

[69] P. Shumyatsky, Finite groups and the fixed points of coprime automorphisms, *Proc. Amer. Math. Soc.* **129** (2001), 3479–3484.

[70] V. P. Shunkov, On periodic groups with an almost regular involution, *Algebra Logika* **11** (1972), 470–493; English transl., *Algebra Logic* **11** (1973), 260–272.

[71] J. Thompson, Finite groups with fixed-point-free automorphisms of prime order, *Proc. Nat. Acad. Sci. U.S.A.* **45** (1959), 578–581.

[72] J. Thompson, Automorphisms of solvable groups, *J. Algebra* **1** (1964), 259–267.

[73] A. Turull, Fitting height of groups and of fixed points, *J. Algebra* **86** (1984), 555–566.

[74] D. J. Winter, On groups of automorphisms of Lie algebras, *J. Algebra* **8** (1968), 131–142.

PRONORMAL SUBGROUPS AND TRANSITIVITY OF SOME SUBGROUP PROPERTIES

LEONID A. KURDACHENKO*, JAVIER OTAL[†] and IGOR YA. SUBBOTIN[§]

*Department of Algebra, National University of Dnepropetrovsk, Dnipropetrovsk 10, Ukraine 49010
Email: lkurdachenko@i.ua

[†]Department of Mathematics, University of Zaragoza, 50009 Zaragoza, Spain
Email: otal@unizar.es

[§]Mathematics Department, National University, Los Angeles CA 90045, USA
Email: isubboti@nu.edu

Abstract

A subgroup H of a group G is called *pronormal* in G if for each element $g \in G$ the subgroups H and H^g are conjugate in $\langle H, H^g \rangle$. Pronormal subgroups have been introduced by P. Hall, and they play an important role in many studies dedicated to normal structure and Sylow theory of finite and infinite groups and in investigations of arrangement of subgroups in infinite linear groups over rings. Many interesting and important developments have been lately completed in this area by different authors. Thanks to these results, we can see that pronormal subgroups and some other types of subgroups related to them (such as contranormal, abnormal, polynormal, paranormal, permutable subgroups, and so on) are very closely connected to transitivity of some group properties (such as normality, permutability and other) and to (locally) nilpotency of a group. In the current survey, we try to reflect some important new results in this area.

1 Introduction

Recall that a subgroup H of a group G is called *pronormal* in G if for each element $g \in G$ the subgroups H and H^g are conjugate in $\langle H, H^g \rangle$. Pronormal subgroups have been introduced by P. Hall. Important examples of pronormal subgroups are the Sylow p-subgroups of finite groups, the Sylow π-subgroups of finite soluble groups, the Carter subgroups of finite soluble groups and many others. Finite groups with all pronormal subgroups have been described by T.A. Peng in [41, 42]. In infinite groups, the studies of pronormal subgroups and subgroups related to pronormal subgroups have been initiated by N.F. Kuzennyi and I.Ya. Subbotin [30, 31, 32, 33]. The importance of this research has been strongly justified by Z.I. Borevich and his students, who investigated the arrangement of subgroups in infinite linear groups over rings (see the survey [7]). They obtained promising results which stimulated further interest for pronormal subgroups and their influence on the structure of infinite groups (see surveys [34] and [19]). Thanks to these results,

Supported by Proyecto MTM2007-60994 of Dirección G. de Investigación (Spain)

we can see that pronormal subgroups and some other types of subgroups related to them (such as contranormal, abnormal, polynormal, paranormal, pemutable subgroups, and so on) are very closely connected to transitivity of some group properties (such as normality, permutability and other) and to (locally) nilpotency of a group. In the current survey, we try to reflect some important new results in this area.

2   Contranormal subgroups, descendant subgroups and some criteria of nilpotency

Let G be a group, H be a subgroup of G and X be a subset of G. Put

$$H^X = \langle\, h^x = x^{-1}hx \mid h \in H, x \in X \,\rangle.$$

In particular, H^G (the *normal closure* of H in G) is the smallest normal subgroup of G containing H. Following J.S. Rose [47], a subgroup H of a group G is called *contranormal* if $H^G = G$. As we can see from the definition, contranormal subgroups in some sense are a kind of antipodes to subnormal. Recall that a subgroup H of a group G is *abnormal* in G if $g \in \langle H, H^g\rangle$ for each element $g \in G$. Abnormal subgroups have been introduced by P. Hall in his paper [15], while the term "abnormal subgroup" belongs to R. Carter [9]. Abnormal subgroups are an important particular case of contranormal subgroups: abnormal subgroups are exactly the subgroups that are contranormal in each subgroup containing them. Abnormal subgroups are also a particular type of pronormal subgroups. All these subgroups and their generalizations have proved to be very useful in finite group theory.

While abnormal and contranormal subgroups are antipodes of normal subgroups, pronormal subgroups are a generalization of normality. Thus, every subgroup which is pronormal and subnormal is normal. Pronormal subgroups are connected to contranormal subgroups in the following way. If H is a pronormal subgroup of a group G and $H \leq K$, then its normalizer $N_K(H)$ in K is an abnormal and hence contranormal subgroup of K.

Starting from the normal closure of H, we can construct the *normal closure series of H in G*

$$H^G = H_0 \geq H_1 \geq \ldots H_\alpha \geq H_{\alpha+1} \geq \ldots H_\gamma$$

by the following rule: $H_{\alpha+1} = H^{H_\alpha}$ for every $\alpha < \gamma$, $H_\lambda = \bigcap_{\mu<\gamma} H_\mu$ for a limit ordinal λ. The term H_α of this series is called the αth *normal closure of H in G* and will be denoted by $H^{G,\alpha}$. The last term H_γ of this series is called *the lower normal closure of H in G* and will be denoted by $H^{G,\infty}$. Observe that every subgroup H is contranormal in its lower normal closure.

In finite groups, the subgroup $H^{G,\infty}$ is called *the subnormal closure of H in G*. The rationale for this is the following. In a finite group G, the normal closure series of every subgroup H is finite, and $H^{G,\infty}$ is the smallest subnormal subgroup of G containing H. The normal closure series play an important role in certain problems of group theory. Thus, the following two types of subgroups are connected to this series. A subgroup H *is called descendant in* G if H coincides with its lower normal closure $H^{G,\infty}$. An important particular case of descendant subgroups are subnormal subgroups. A subnormal subgroup is exactly a descending subgroup having finite normal closure series. These subgroups strongly affect on the structure of a group. For example, it is not hard to prove that if every subgroup of a locally (soluble-by-finite) group is descendant, then this group is locally nilpotent. If every subgroup of a group G is subnormal, then, by a remarkable result due to W. Möhres [38], G is soluble. Subnormal subgroups have been studied very deeply for a quite long period of time. We are not going to consider this topic here since it has been excellently presented in the survey of C. Casolo [10]. However, we need to admit that, with the exception of subnormal subgroups, we have no significant information regarding descendant subgroups.

If every subgroup of a group G is descendant, then G has no proper contranormal subgroups, and, in particular, no proper abnormal subgroups. On the other hand, if G is a locally nilpotent group, then G has proper abnormal subgroups (N.F. Kuzennyi and I.Ya. Subbotin [32]). In this connection, the following natural question arises: *For which groups does the absence of proper abnormal subgroups imply their locally nilpotence?* Obviously, all finite groups possess this property. In the articles [17, 18, 22, 23, 26], some classes of groups without proper abnormal subgroups have been considered and some criteria of locally nilpotency have been obtained. One can find a detailed description of this topic in [19]. The study of groups without contranormal subgroups is a logical continuation of this theme. We observe that every non-normal maximal subgroup of an arbitrary group is contranormal. Since a finite group whose maximal subgroups are normal is nilpotent, we come to the following criterion of nilpotency of finite groups in terms of contranormal subgroups: *A finite group G is nilpotent if and only if G has no proper contranormal subgroups.* However, in the general case, this property cannot serve as a characterization of contranormal subgroups. There exist non-nilpotent groups whose every subgroup is subnormal. The first such example has been constructed by H. Heineken and I.J. Mohamed [16]. It should be noted that in contrast to some other types of "anti-normal subgroups" (for example, abnormal subgroups), some locally nilpotent groups can contain contranormal subgroups. A simple example here is a group $G = K \rtimes \langle d \rangle$ where K is a Prüfer 2-group, d is an element of order 2 and $x^d = x^{-1}$ for each $x \in K$. This group is non-nilpotent but hypercentral, and $\langle d \rangle$ is a proper contranormal subgroup of G. This example confirms that the absence of contranormal subgroups cannot be a criterion for locally nilpotency. Nevertheless, for some classes of infinite groups the absense of contranormal subgroups implies nilpotency of a group. Some of these classes have been considered in the recent articles [20, 21]. We will present some results from there.

Theorem 2.1 ([20]) *Let G be group and H be a normal soluble-by-finite subgroup such that the factor-group G/H is nilpotent. Suppose that H satisfies the minimal condition on G-invariant subgroups (Min-G). If G has no proper contranormal subgroups, then G is nilpotent.*

We observe that an analog of 2.1 for the maximal condition on G-invariant subgroups (the condition Max-G) is not valid. In the paper [20], a corresponding contraexample has been constructed.

Let G be a group and let A be an infinite normal abelian subgroup of G. We say that A is a G-*quasifinite subgroup* if every proper G-invariant subgroup of A is finite. This means that either A includes a proper finite G-invariant subgroup B such that A/B is G-simple, or A is an union of all finite proper G-invariant subgroups.

Corollary 2.2 ([20]) *Let G be a polynilpotent group satisfying minimal condition for normal subgroups. If G has no proper contranormal subgroups, then G is nilpotent.*

Corollary 2.3 ([21]) *Let G be a group and H be a normal Chernikov subgroup. Suppose that G/H is nilpotent. If G has no proper contranormal subgroups, then G is nilpotent.*

We recall that a group G is said to be *minimax* if G has a finite subnormal series whose factors satisfy the condition Min (the minimal condition for all subgroups) or the condition Max (the maximal condition for all subgroups).

Theorem 2.4 ([20]) *Let G be group and H be a normal subgroup such that the factor-group G/H is nilpotent. Suppose that H has a finite series of G-invariant subgroups*

$$\langle 1 \rangle = C_0 \leq C_1 \leq \cdots \leq C_n = H$$

whose factors C_j/C_{j-1}, $1 \leq j \leq n$, satisfy one of the following conditions:
  (i) *C_j/C_{j-1} is finite;*
  (ii) *C_j/C_{j-1} is hyperabelian and minimax;*
  (iii) *C_j/C_{j-1} is hyperabelian and finitely generated;*
  (iv) *C_j/C_{j-1} is abelian and satisfies Min-G.*
  If G has no proper contranormal subgroups, then G is nilpotent.

Corollary 2.5 ([21]) *Let G be a group and let C be a normal subgroup of G such that G/C is nilpotent. Suppose that C is a hyperabelian finitely generated subgroup. If G has no proper contranormal subgroups, then G is nilpotent.*

In particular, if G is hyperabelian finitely generated group with no proper contranormal subgroups, then G is nilpotent.

Let G be a group and C be a normal subgroup of G. Then C is said to be a G-*minimax* if C has a finite series of G-invariant subgroups whose infinite factors are abelian and either satisfy Min-G or Max-G. Recall that a group G has *finite section rank* if every elementary abelian p-section of G is finite for all prime p.

Theorem 2.6 ([21]) *Suppose that the group G includes a normal G-minimax subgroup C such that G/C is a nilpotent group of finite section rank. If G has no proper contranormal subgroups, then G is nilpotent.*

Following A.I. Maltsev [35], we say that a group G is *a soluble A_3-group* if it has a finite series of normal subgroups whose factors are abelian and either are Chernikov or torsion-free groups of finite 0-rank. Generalizing this notion we say that a group G is *a generalized A_3-group* if G has a finite series of normal subgroups

$$\langle 1 \rangle = H_0 \leq H_1 \leq \cdots \leq H_n = G$$

every infinite factor H_j / H_{j-1} of which is abelian and satisfies one of the following conditions:

H_j / H_{j-1} is a torsion-free group of finite 0–rank;

H_j / H_{j-1} satisfies the condition Min–G;

H_j / H_{j-1} satisfies the condition Max–G.

Theorem 2.7 ([21]) *Let G be a generalized A_3–group. If G has no proper contranormal subgroups, then G is a nilpotent A_3–group.*

3 Generalized pronormality and transitivity of normality

Recall that a subgroup H of a group G is said to be *weakly pronormal* in G (or that H has *the Frattini property*) if H satisfies the following condition:

For any two subgroups K, L such that $H \leq K$ and K is normal in L we have $L \leq N_G(H)K$.

T.A. Peng in his article [42] characterized pronormal subgroups in finite soluble groups. He proved that a subgroup H of a finite soluble group G is pronormal if and only if H has the Frattini property.

Recall also that a group G is called *an N-group* or G is a group with the *normalizer condition* if $H < N_G(H)$ for each subgroup H. Note that the class of N-groups is a proper subclass of the class of all locally nilpotent groups.

In the article [18] the following generalization of the result of T.A. Peng.has been obtained.

Theorem 3.1 ([18]) *Let G be a group having an ascending series of normal subgroups whose factors are N-groups. Then a subgroup H of G is pronormal in G if and only if H is weakly pronormal in G.*

A group G is said to be *a T-group* if every subnormal subgroup of G is normal. A group G is said to be a $\bar{T}$-group, if every subgroup of G is a T-group. The structure of finite soluble T-groups has been described by W. Gaschütz [13]. In particular, he found that every finite soluble T-group is a $\bar{T}$-group. Recall that a finite $\bar{T}$-group is metabelian. Infinite soluble T-groups and $\bar{T}$-groups have been described by D.J.S. Robinson [43]. A locally soluble $\bar{T}$-group G has the following structure. If G is not periodic, then G is abelian. If G is periodic and L is the locally nilpotent residual of G, then G/L is a Dedekind group, $\pi(L) \cap \pi(G/L) = \varnothing$, $2 \notin \pi(L)$, and every subgroup of L is G-invariant. In particular, if $L \neq \langle 1 \rangle$, then $L = [L, G]$.

The following result provides us with a characterization of the $\bar{T}$-groups in terms of pronormal subgroups.

Theorem 3.2 ([33]) *Let G be a non-periodic locally soluble group. Then the following assertions are equivalent:*

(i) *G is a $\bar{T}$-group;*

(ii) *Every cyclic subgroup of G is pronormal in G; and*

(iii) *G is abelian.*

Recall that a group G is called *locally graded*, if every non-identity subgroup of G has a proper subgroup of finite index.

Theorem 3.3 ([33]) *Let G be a periodic locally graded group. Then the following assertion are equivalent:*

(i) *G is a soluble $\bar{T}$-group; and*

(ii) *Every cyclic subgroup of G is pronormal in G.*

A subgroup H of a group G is called *nearly pronormal* if $N_K(H)$ is contranormal in every subgroup K including H. In the paper [24], groups whose subgroups are nearly pronormal have been considered.

Theorem 3.4 ([24]) *Let G be a locally radical group.*

(i) *If every cyclic subgroup of G is nearly pronormal, then G is a $\bar{T}$-group.*

(ii) *If every subgroup of G is nearly pronormal, then every subgroup of G is pronormal in G.*

Infinite groups whose subgroups are pronormal have been considered in [30]. The authors completely described such infinite locally soluble non-periodic and infinite locally graded periodic groups. The main result of that paper is the following theorem.

Theorem 3.5 ([30]) *Let G be a group whose subgroups are pronormal, and L be the locally nilpotent residual of G.*

(i) *If G is periodic and locally graded, then G is a soluble $\bar{T}$-group, in which L is the complement for every Sylow $\pi(G/L)$-subgroup.*

(ii) *If G is non periodic and locally soluble, then G is abelian.*

Conversely, if G has a such structure, then every subgroup of G is pronormal in G.

In the paper [46], the second assertion (ii) of this theorem has been extended to non-periodic locally graded groups. It was proved that in this case such groups are also abelian, so no new group classes have been added to the above description.

J. Rose [48] introduced the notion of *a balanced chain* connecting a subgroup H to the group G, namely that is the chain of subgroups

$$H = H_0 \leq H_1 \leq \cdots \leq H_n = G$$

such that for each j, $0 \leq j \leq n-1$, either H_j is normal in H_{j+1}, or H_j is abnormal in H_{j+1}. In a finite group, every subgroup can be connected to the group by some balanced chain. It is natural to consider the case when these balanced chains are short, i.e., their lengths are bounded by a small number. If these lengths are

≤ 1, then every subgroup is either normal or abnormal in the group. Such finite groups were studied in [12]. The infinite groups of this kind and some of their generalizations were described in [50] and [11]. Observe that in the groups in which the normalizer of any subgroup is abnormal, and in the groups whose every subgroup is abnormal in its normal closure, the mentioned above lengths are ≤ 2. The groups with these properties have been considered in [25].

Theorem 3.6 ([25]) *Let G be radical group. Then G is a $\bar{T}$-group, if and only if every cyclic subgroup of G is abnormal in its normal closure.*

Theorem 3.7 ([25]) *Let G be periodic soluble group. Then G is a $\bar{T}$-group, if and only if its locally nilpotent residual L is abelian and the normalizer of each cyclic subgroup of G is abnormal in G.*

The following theorem is a new interesting and useful characterization of groups with pronormal subgroups.

Theorem 3.8 ([25]) *Let G be periodic soluble group. Then every subgroup of G is pronormal if and only if its locally nilpotent residual L is abelian and the normalizer of every subgroup of G is abnormal in G.*

The article [25] includes an example showing that the last theorem is not valid for the non-periodic case.

Observe further the following interesting property of pronormal subgroups already recalled above:

> Let G be a group and H, K be subgroups of G such that $H \leq K$. If H is a subnormal and pronormal subgroup in K, then H is normal in K.

In connection with this, the following question naturally arises: *In what cases does this property characterize pronormal subgroups?*

We say that a subgroup H of a group G *is transitively normal* if H is normal in every subgroup $K \geq H$ in which H is subnormal [27]. In [40], these subgroups have been introduced under a different name, namely, a subgroup H of a group G is said to satisfy *the subnormalizer condition* in G if for every subgroup K such that H is normal in K we have $N_G(K) \leq N_G(H)$. It is not hard to see that every transitively normal subgroup satisfies the subnormalizer condition and, conversely, every subgroup satisfying the subnormalizer condition is transitively normal.

We say that a subgroup H of a group G is *strong transitively normal* if HA/A is transitively normal for every normal subgroup A of a group G [27]. Since the homomorphic image of pronormal subgroup is pronormal, we can conclude that every pronormal subgroup is a strong transitively normal subgroup. In the paper [27] the restrictions under which strong transitively normal and transitively normal subgroups are pronormal have been established. We will present them here together with some useful corollaries.

Theorem 3.9 ([27]) *Let G be a group and H be a hypercentral subgroup of G. Suppose that G has a normal soluble subgroup R such that G/R is hypercentral. If*

H is strong transitively normal in G and R satisfies Min-H, then H is a pronormal subgroup of G.

Corollary 3.10 ([27]) *Let G be a group and H be a hypercentral subgroup of G. Suppose that G has a normal soluble Chernikov subgroup R such that G/R is hypercentral. If H is strong transitively normal in G, then H is a pronormal subgroup of G. In particular, if G is a soluble Chernikov group and H is a hypercentral strong transitively normal subgroup of G, then H is pronormal in G.*

In particular, the result due to T.A. Peng [42] follows from Corollary 13.1.

A subgroup H is said to be *polynormal* in a group G, if H is contranormal in H^S for every subgroup S including H (M.S. Ba and Z.I. Borevich [7]). We observe that every pronormal subgroup is polynormal [7, section 8, Theorem 2].

Corollary 3.11 ([27]) *Let G be a group and H be a hypercentral subgroup of G. Suppose that G has a normal soluble subgroup R such that G/R is hypercentral. If H is a polynormal subgroup of G and R satisfies Min-H (in particular, if R is Chernikov), then H is pronormal in G.*

Corollary 3.12 ([40]) *Let G be a soluble finite group and H be a nilpotent subgroup of G. If H is a polynormal subgroup of G, then H is a pronormal subgroup of G.*

A subgroup H is said to be *paranormal* in a group G if H is contranormal in $\langle H, H^g \rangle$ for all elements $g \in G$ (M.S. Ba and Z.I. Borevich [7]). Every pronormal subgroup is paranormal [7, section 8, Theorem 2], and every paranormal subgroup is polynormal [7, section 8, Proposition 1]. Thus we have

Corollary 3.13 ([27]) *Let G be a group and H be a hypercentral subgroup of G. Suppose that G has a normal soluble subgroup R such that G/R is hypercentral. If H is a paranormal subgroup of G and R satisfies Min-H (in particular, if R is Chernikov), then H is pronormal in G.*

Corollary 3.14 ([27]) *Let G be a soluble finite group and H be a nilpotent subgroup of G. If H is a paranormal subgroup of G, then H is a pronormal subgroup of G.*

Theorem 3.15 ([27]) *Let G be a group and H be a hypercentral subgroup of G. Suppose that G has a normal nilpotent subgroup R such that G/R is hypercentral. If H is transitively normal in G and R satisfies Min-H (in particular, if R is Chernikov), then H is a pronormal subgroup of G.*

Corollary 3.16 ([27]) *Let G be a nilpotent-by-hypercentral Chernikov group and H be a hypercentral subgroup of G. If H is transitively normal in G, then H is a pronormal subgroup of G.*

Corollary 3.17 ([42]) *Let G be a nilpotent-by-abelian finite group and H be a nilpotent subgroup of G. If H is transitively normal in G, then H is a pronormal subgroup of G.*

A subgroup H of a group G is called *weakly normal* if $H^g \leq N_G(H)$ implies that $g \in N_G(H)$ (K.H. Müller [39]). We note that every pronormal subgroup is weakly normal [2], and every weakly normal subgroup satisfies the subnormalizer condition [2], and hence it is transitively normal in G. Thus we have

Corollary 3.18 ([27]) *Let G be a group and H be a hypercentral subgroup of G. Suppose that G has a normal nilpotent subgroup R such that G/R is hypercentral. If H is weakly normal in G and R satisfies Min-H (in particular, if R is Chernikov), then H is a pronormal subgroup of G.*

A subgroup H of a group G is called *an $\mathfrak{H}$-subgroup* if $N_G(H) \cap H^g \leq H$ for all elements $g \in G$ [8]. Note that every $\mathfrak{H}$-subgroup is transitively normal [8]. Therefore we obtain

Corollary 3.19 ([27]) *Let G be a group and H be a hypercentral subgroup of G. Suppose that G has a normal nilpotent subgroup R such that G/R is hypercentral. If H is an $\mathfrak{H}$-subgroup of G and R satisfies Min-H (in particular, if R is Chernikov), then H is a pronormal subgroup of G.*

Some properties of transitively normal subgroups (under another name) in FC-groups have been considered in the paper [14]. In particular, this paper contains the following important result.

Theorem 3.20 ([14]) *Let G be an FC-group and H be a transitively normal subgroup of G. If H is a p-subgroup for some prime p, then H is a pronormal subgroup of G.*

4   Pronormality and transitivity of permutability

As we have seen above, pronormality is strongly connected to transitivity of normality. It also play a key role in the study of transitivity of other important subgroups properties. In this passing, it is worthy to mention that groups with transitivity of pronormality, abnormality and other connected properties have been studied by L.A. Kurdachenko and I.Ya. Subbotin (see [28], [29], and [19]). Here we focus on transitivity of another natural important group property-permutability. A subgroup H of a group G is said to be *permutable* in G (or quasinormal in G), if $HK = KH$ for every subgroup K of G. This concept arises as a generalization of a normal subgroup. The study of the properties of the permutable subgroups began a rather long time ago, and was reflected in the book [45]. According to a well-known theorem due to E. Stonehewer [49], permutable subgroups are always ascendant. Therefore it is natural to consider the opposite case: the groups whose ascendant subgroups are permutable. A group G is said to be *an AP-group* if every ascendant subgroup of G is permutable in G. These groups are very close to the groups in which the relation "to be a permutable subgroup" is transitive. A group G is said to *be a PT-group* if permutability is a transitive relation in G, that is, if K is a permutable subgroup of H and H is a permutable subgroup of G, then K is a permutable subgroup of G. The description of finite soluble PT-groups has been obtained by G. Zacher [51]. It looks close to the description of finite soluble T-groups due to W. Gaschütz [13]. Namely, if G is finite soluble group and L is a nilpotent residual of G, then every subgroup of G/L is permutable, $\pi(L) \cap \pi(G/L) = \varnothing$, $2 \notin \pi(L)$ and every subgroup of L is G-invariant. Some properties of finite soluble PT-groups have been considered in several articles (see, for example, [1, 6, 3]). The soluble infinite PT-groups have been described by F. Menegazzo [36, 37].

Obviously, a finite group G is a PT-group if and only if every subnormal subgroup of G is permutable. Therefore, for infinite groups the following natural question arises: *Do the classes of infinite PT-groups and AP-groups coincide?* In [44, Lemma 4], it is claimed that in an arbitrary PT-group every ascendant subgroup is permutable. The following simple counterexample shows that this statement is incorrect. Let $G = A \rtimes \langle b \rangle$ be a semidirect product of a Prüfer 2-group A and a group $\langle b \rangle$ of order 2 that acts on A by $a^b = a^{-1}$ for each $a \in A$ (an infinite dihedral group). Clearly, G is hypercentral, and in particular, every subgroup of G is ascendant. However, it is not hard to see, that the subgroup $\langle b \rangle$ is not permutable. It means that for infinite groups the classes of AP-groups and PT-groups do not coincide. The paper [4] initiated the study of infinite AP-groups. Consider the main results of this work.

Theorem 4.1 ([4]) *Let G be a radical hyperfinite AP-group. Then the following assertions hold:*

(i) *G is metabelian;*

(ii) *if R is the locally nilpotent radical of G, then $R = L \times Z$, where L is the locally nilpotent residual of G and Z is the upper hypercenter of G;*

(iii) *$\pi(L) \cap \pi(G/L) = \varnothing$, $2 \notin \pi(L)$;*

(iv) *L is abelian and every subgroup of L is G-invariant; and*

(v) *every subgroup of G/L is permutable (in particular, G/L is nilpotent).*
Moreover, if the factor-group G/L is countable, then G splits over L.

On the other hand, if G is a periodic group having a normal abelian subgroup L that satisfies the conditions (iii)–(v), then G is an AP-group.

Note that this theorem describes a much wider class of groups. The following result justifies this.

Theorem 4.2 ([4]) *Let G be a periodic AP-group. If G is a hyper-N-group, then G is hyperfinite. In particular, G is a hypercyclic metabelian AP-group.*

Corollary 4.3 ([4]) *Let G be a periodic AP-group. If G is a hyper-Gruenberg group, then G is a hypercyclic metabelian AP-group.*
In particular, if G is a countable radical group, then G is a hypercyclic metabelian AP-group.

Corollary 4.4 ([4]) *Let G be a periodic AP-group. If G is residually soluble, then G is a hypercyclic metabelian AP-group.*

Let G be a group and let p be a prime. We say that G *belongs to the class* $\mathfrak{B}_p$ if each Sylow p-subgroup P of G satisfies the following condition: (i) every subgroup of P is permutable in P; (ii) every normal subgroup of P is pronormal in G. The following result from [5] shows the role of pronormal subgroups in AP-groups.

Theorem 4.5 ([4]) *Let G be a periodic locally soluble group. If G belongs to the class $\mathfrak{B}_p$ for all primes p, then G is a hypercyclic AP-group. Moreover, if L is the locally nilpotent residual, then L has a complement in G.*

References

[1] M.J. Alejandre, A. Ballester-Bolinches and M.C. Pedraza-Aguilera, Finite soluble groups with permutable subnormal subgroups, *J. Algebra* **240** (2001), 705–722.

[2] A. Ballester-Bolinches and R. Esteban-Romero, On finite $\mathfrak{F}$-groups, *J. Austral. Math. Soc.* **75** (2003), 1–11.

[3] A. Ballester-Bolinches, R. Esteban-Romero and M.C. Pedraza-Aguilera, On finite groups in which subnormal subgroups satisfy certain permutability conditions, in *Advances in Algebra*, (World Sci. Publ., River Edge, New York, 2003), 38–45

[4] A. Ballester-Bolinches, L.A. Kurdachenko, J. Otal and T. Pedraza, Infinite groups with many permutable subgroups, *Rev. Mat. Iberoam.* **24** (2008), 745–764.

[5] A. Ballester-Bolinches, L.A. Kurdachenko, J. Otal and T. Pedraza, Local characterizations of infinite groups whose ascendant subgroups are permutable, *Forum Math.*, to appear.

[6] J.C. Beidleman, B. Brewster and D.J.S. Robinson, Criteria for permutability to be transitive in finite groups, *J. Algebra* **222** (1999), 400–412.

[7] M.S. Ba and Z.I. Borevich, Arrangements of intermediate subgroups (Russian), in *Rings and linear groups*, Kuban. Gos. Univ (Krasnodar 1988), 14–41.

[8] M. Bianchi, A.G.B. Mauri, M. Herzog and L. Verardi, On finite solvable groups in which normality is a transitive relation, *J. Group Theory* **3** (2000), 147–156.

[9] R.W. Carter, Nilpotent self-normalizing subgroups of soluble groups, *Math. Z.* **75** (1961), 136–139.

[10] C. Casolo, Groups with all subgroups subnormal, in *Conference Advances in Group Theory and Applications 2007* (Otranto 2007), to appear.

[11] M. de Falco, L.A. Kurdachenko and I.Ya. Subbotin, Groups with only abnormal and subnormal subgroups, *Atti Sem. Mat. Fis. Univ. Modena* **46** (1998), 435–442.

[12] A. Fattahi, Groups with only normal and abnormal subgroups, *J. Algebra* **28** (1974), 15–19.

[13] W. Gaschütz, Gruppen in denen das Normalteilersein transitiv ist, *J. Reine Angew. Math.* **198** (1957), 87–92.

[14] F. De Giovanni and G. Vincenzi, Pseudonormal subgroups of groups *Ricerche Mat.* **52** (2003), 91–101.

[15] P. Hall, On the system normalizers of a soluble groups, *Proc. London Math. Soc.* **43** (1937), 507–528.

[16] H. Heineken and I.J. Mohamed, Groups with normalizer condition, *Math. Ann.* **198** (1972), 178–187.

[17] L.A. Kurdachenko, J. Otal and I.Ya. Subbotin, On some criteria of nilpotency, *Comm. Algebra* **30** (2002), 3755–3776.

[18] L.A. Kurdachenko, J. Otal J. and I.Ya. Subbotin, Abnormal, pronormal, contranormal and Carter subgroups in some generalized minimax groups, *Comm. Algebra* **33** (2005), 4595–4616.

[19] L.A. Kurdachenko, J. Otal and I.Ya. Subbotin, On properties of abnormal and pronormal subgroups in some infinite groups, in *Groups St Andrews 2005, Vol. 2*, London Mathematical Society Lecture Note Series **339** (Cambridge Univ. Press 2007), 597–604.

[20] L.A. Kurdachenko, J. Otal and I.Ya. Subbotin, Criteria of nilpotency and influence of contranormal subgroups on the structure of infinite groups, *Turkish J. Math.*, to appear.

[21] L.A. Kurdachenko, J. Otal and I.Ya. Subbotin, On influence of contranormal subgroups on the structure of infinite groups, *Comm. Algebra*, to appear.

[22] L.A. Kurdachenko, J. Otal, A. Russo and G. Vincenzi, Abnormal subgroups and Carter subgroups in some classes of infinite groups, *J. Algebra* **297** (2006), 273–291.

[23] L.A. Kurdachenko, A. Russo and G. Vincenzi, Groups without the proper abnormal subgroups, *J. Group Theory* **9** (2006), 507–518.

[24] L.A. Kurdachenko, A. Russo and G. Vincenzi, On some groups all subgroups of which are near to pronormal, *Ukrainian Math. J.* **59** (2007), 1332–1339.

[25] L.A. Kurdachenko, A. Russo, I.Ya. Subbotin and G. Vincenzi, Infinite groups with short balanced chains of subgroups, *J. Algebra* **319** (2008), 3901–3917.

[26] L.A. Kurdachenko and I.Ya. Subbotin, Pronormality, contranormality and generalized nilpotency in infinite groups, *Publ. Mat.* **47** (2003), 389–414.

[27] L.A. Kurdachenko and I.Ya. Subbotin, Transitivity of normality and pronormal subgroups, in *AMS Special session on infinite groups* (Bard College 2005), *Contemporary Mathematics* **421** (2006), 201–212.

[28] L.A. Kurdachenko and I.Ya. Subbotin, On transitivity of pronormality, *Comment. Matemat. Univ. Caroline* **43** (2002), 583–594.

[29] L.A. Kurdachenko and I.Ya. Subbotin, On U-normal subgroups, *Southeast Asian Bull. Math.* **26** (2002), 789–801

[30] N.F. Kuzennyi and I.Ya. Subbotin, The groups whose subgroups are pronormal, *Ukrainian Math. J.* **39** (1987), 325–329.

[31] N.F. Kuzennyi and I.Ya. Subbotin, Locally soluble groups whose infinite subgroups are pronormal, *Izvestiya VUZ Math. J.* **11** (1987), 77–79.

[32] N.F. Kuzennyi and I.Ya. Subbotin, New characterization of locally nilpotent IH-groups, *Ukrainian Math. J.* **40** (1988), 274–277.

[33] N.F. Kuzennyi and I.Ya. Subbotin, The groups with pronormal primary subgroups, *Ukrainian Math. J.* **41** (1989), 286–289.

[34] N.F. Kuzennyi and I.Ya. Subbotin, On groups with fan subgroups, in *Proc. Int. Conference in Algebra, Part 1*, (Novosibirsk 1989), *Contemp. Math.* **131** (1992), 383–388.

[35] A.I. Maltsev, On certain classes of infinite soluble groups, *Mat. Sb.* **28** (1951), 567–588. English translation *Amer. Math. Soc. Translations* **2** (1956), 1–21.

[36] F. Menegazzo, Gruppi nei quail la relazione di guasi-normalita e transitiva, *Rend. Semin. Univ. Padova* **40** (1968), 347–361.

[37] F. Menegazzo, Gruppi nei quail la relazione di guasi-normalita e transitiva, *Rend. Semin. Univ. Padova* **42** (1969), 389–399.

[38] W. Möhres, Auflösbarkeit von Gruppen deren Untergruppen alle subnormal sind, *Arch. Math.* **54** (1990), 232–235.

[39] K.H. Müller, Schwachnormale Untergruppen: Eine gemeinsame Verallgemeinerung der normalen und normalisatorgleichen Untergruppen, *Rend. Semin. Mat. Univ. Padova*

36 (1966), 129–157.

[40] V.I. Mysovskikh, Subnormalizers and properties of embedding of subgroups in finite groups, *Zapiski nauchnyh semin. POMI* **265** (1999), 258–280.

[41] T.A. Peng, Finite groups with pronormal subgroups, *Proc. Amer. Math. Soc.* **20** (1969), 232–234.

[42] T.A. Peng, Pronormality in finite groups, *J. London Math. Soc.* **3** (1971), 301–306.

[43] D.J.S. Robinson, Groups in which normality is a transitive relation, *Proc. Cambridge Philos. Soc.* **60** (1964), 21–38.

[44] D.J.S. Robinson, Minimality and Sylow permutability in locally finite groups, *Ukrainian Math. J.* **54** (2002), 1038–1049.

[45] R. Schmidt, *Subgroup lattices of groups* (Walter de Gruyter, Berlin 1994).

[46] D.J.S. Robinson, A. Russo and G. Vincenzi, On groups which contain no HNN-extensions, *Internat. J. Algebra Comput.* **17** (2007), 1377–1387.

[47] J.S. Rose, Nilpotent subgroups of finite soluble groups, *Math. Z.* **106** (1968), 97–112.

[48] J.S. Rose, Abnormal depth and hypereccentric length in finite soluble groups, *Math. Z.* **90** (1965), 29–49.

[49] S.E. Stonehewer, Permutable subgroups of infinite groups, *Math. Z.* **125** (1972), 1–16.

[50] I.Ya. Subbotin, Groups with alternatively normal subgroups, *Izvestiya. VUZ Mat.* **3** (1992), 86–88.

[51] G. Zacher, I gruppi risolubiliniti in cui i sottogruppi di composizione coincidono con i sottogruppi quasi-normali, *Atti Accad. Naz. Lincei Rend. Cl. Sci. Fis. Mat. Natur.* **37** (1964), 150–154.

ON ENGEL AND POSITIVE LAWS

O. MACEDOŃSKA and W. TOMASZEWSKI

Institute of Mathematics, Silesian University of Technology, Gliwice 44-100, Poland
Email: O.Macedonska@polsl.pl, W.Tomaszewski@polsl.pl

Abstract

Engel laws and positive laws have attracted the attention of many authors. These laws have many common properties, e.g. each finitely generated group satisfying any of these laws has a finitely generated commutator subgroup. There are still many open problems concerning these laws.
MSC-2000: 20E10, 20F45, keywords: positive laws, Engel laws.

1 Notation

Let $F = \langle x, y \rangle$ be a free group of rank 2. We denote $x^{y^i} = y^{-i} x y^i$, $[x, y] = x^{-1} y^{-1} xy$, $[x, {}_0 y] = x$ and $[x, {}_{i+1} y] = [[x, {}_i y], y]$. The law $[x, {}_n y] \equiv 1$ is the n-Engel law. We introduce the following subgroups.

$$E_n = \langle\, [x, {}_i y],\ 0 \le i \le n\,\rangle, \qquad E = \langle\, [x, {}_i y],\ 0 \le i\,\rangle.$$

Let $\mathfrak{A}$ be the variety of all abelian groups, and $\mathfrak{A}_p$ the variety of all abelian groups of exponent p. By $\mathfrak{N}_c$ we denote the variety of all nilpotent groups of nilpotency class c, and by $\mathfrak{S}_d$ the variety of all soluble groups of solubility class d. By $\mathfrak{B}_e$ we denote so called restricted Burnside variety of exponent e, that is the variety generated by all finite groups of exponent e. It follows from Zelmanov positive solution of the Restricted Burnside Problem that all groups in $\mathfrak{B}_e$ are locally finite of exponent dividing e.

For the text below we have to recall the following

Definition A group G is called **locally graded** if every nontrivial finitely generated subgroup of G has a proper subgroup of finite index.

The class of locally graded groups was introduced in 1970 by S. N. Černikov [8] in order to avoid groups such as infinite Burnside groups or Ol'shanskii–Tarski monsters. This class contains all *soluble* groups, *locally finite* groups, *residually finite* groups. It is closed under taking *subgroups* and *extensions*. It is also closed under taking groups which are *locally-* or *residually-* in this class.

2 Positive laws

Positive laws are the laws of the form $u(x_1, x_2, \ldots, x_n) \equiv v(x_1, x_2, \ldots, x_n)$, where u, v are distinct words in the free group $\langle x_1, x_2, \ldots \rangle$, written without negative powers of the variables $x_1, x_2, \ldots, x_n$. The law is cancelled if u and v have different first (and last) letters. The degree of a cancelled law is the length of the longer word u or v.

Each positive law implies a binary positive law $u(x, y) = v(x, y)$ if substitute $x_i \rightarrow xy^i$.

It was shown by J. & T. Lewins [18], that if a group satisfies a positive law, then the variety it generates has a basis consisting of positive laws.

In 1953, A. I. Mal'tsev [21] and, independently in 1963, B. H. Neumann and T. Taylor [27] proved that nilpotency can be defined by a positive law. Indeed, let $P \equiv Q$ be a positive law defining nilpotency of class $n - 1$, and z be a variable which does not occur in P, Q. Let G satisfy the law $PzQ \equiv QzP$. This law implies $PQ \equiv QP$ and then $PQ^{-1}z \equiv zPQ^{-1}$. Hence the quotient of G by its center satisfies the law $P \equiv Q$, and is by assumption nilpotent of class $n - 1$. So G is nilpotent of class n.

Let $P_1 \equiv Q_1$ be the abelian law $xy \equiv yx$. Let $P_2 = P_1 z_1 Q_1$, $Q_2 = Q_1 z_1 P_1$ and inductively $P_n = P_{n-1} z_{n-1} Q_{n-1}$, and $Q_n = Q_{n-1} z_{n-1} P_{n-1}$. Then the positive law $P_n \equiv Q_n$ defines the variety $\mathfrak{N}_n$. Note that this law has $n + 1$ variables $x, y, z_1, \ldots, z_{n-1}$. If we put 1 for each z_i we get a binary positive law

$$P_n(x, y, 1) \equiv Q_n(x, y, 1).$$

It follows that groups which are nilpotent-by-(finite exponent), in particular virtually nilpotent groups (i.e., nilpotent-by-finite groups) satisfy positive laws.

The question whether a finitely generated group satisfying a positive law must be nilpotent-by-finite has a negative answer. For example a free Burnside group $B(r, n)$, $r > 1$, satisfying the law $x^n \equiv 1$, or a free finitely generated group satisfying the law $xy^n = y^n x$, are not nilpotent-by-finite for n sufficiently large by results of Novikov and Adian (see [1]).

The question whether a finitely generated group satisfying a positive law must be nilpotent-by-(finite exponent) was open for more then forty years. A counterexample was found in 1996 by Ol'shankii and Storozhev [25]. Their groups are not even (locally soluble)-by-(finite exponent).

However the affirmative answer was found for many types of groups. In 1997 it was done for so-called class $\mathcal{C}$ ([5], Theorem B) defined inductively. The class $\mathcal{C}$, as it was shown later [4], consists of groups which locally are residually SB-groups (groups lying in finite products of varieties $\mathfrak{S}_d$ and $\mathfrak{B}_e$ for varying d, e). This result, combined with results of Kim and Rhemtulla ([15], Theorem A), saying that a finitely generated locally graded group G satisfying a positive law must be polycyclic-by-finite (hence in the class $\mathcal{C}$) implies the affirmative answer for locally graded groups. So since 1997 it is known that

> a locally graded group G satisfying a positive law of degree n, must be in the product variety $\mathfrak{N}_c \mathfrak{B}_e$, where c and e depend on n only.

3   Three Questions on Engel laws

In 1936 M. Zorn proved that every finite n-Engel group is nilpotent. This is not true in general. An n-Engel law for $n > 2$ does not imply nilpotency (see [24], 34.62). In

1971 S. Bachmuth and H. Y. Mochizuki [2] constructed a non-soluble locally finite 3-Engel group of exponent 5, every n-generator subgroup of which is nilpotent of class at most $2n - 1$. The local nilpotence of 4-Engel groups of exponent 5 was shown in 1997 by M. Vaughan-Lee [35].

The following question is open in general.

Q1: Is every n-Engel group locally nilpotent? (In other words, is every finitely generated n-Engel group nilpotent?)

Question Q1 is approached in two main ways: one is to examine n-Engel groups for different n. In that case an affirmative answer has been found

- 1942: for $n = 2$ — F. W. Levi [17],
- 1961: for $n = 3$ — H. Heineken [14],
- 2005: for $n = 4$ — G. Havas and M. R. Vaughan-Lee [13] (see also [32] for a simplification of a part of the proof of Havas and Vaughan-Lee).

The second approach is to investigate the problem in certain classes of groups. It has been shown that *n-Engel groups are locally nilpotent if additionally they* are

- 1953: soluble groups — K. W. Gruenberg [12],
- 1957: groups with the maximal condition — R. Baer [3],
- 1991: residually finite groups — J. S. Wilson [36],
- 1992: profinite groups — J. S. Wilson and E. I. Zelmanov [37],
- 1994: locally graded groups — Y. Kim and A. H. Rhemtulla [15],
- 1998: groups in the class $\mathcal{C}$ — R. Burns and Yu. Medvedev [6]. They proved that all n-Engel groups in the class $\mathcal{C}$ are contained in the product varieties $\mathfrak{N}_c \mathfrak{B}_e \cap \mathfrak{B}_e \mathfrak{N}_c$, where c, e depend on n only.
- 2003: compact groups — Yu. Medvedev [22].

Q2: Does there exist a finitely generated infinite simple n-Engel group?

The Questions **Q1, Q2** are equivalent in a sense that one has an affirmative answer if and only if the other has a negative answer.

Proposition 3.1 *There exists a non-(locally nilpotent) n-Engel group if and only if there exists a finitely generated infinite simple n-Engel group.*

Proof If there exists an n-Engel group G which is not locally nilpotent then by [15], G is not locally graded. Thus G must contain a finitely generated subgroup H, which has no proper subgroup of finite index. Since H is finitely generated, by Zorn's Lemma it has a maximal proper normal subgroup N. Then N is of infinite index and the factor H/N is a finitely generated infinite simple n-Engel group.

Conversely, if there exists a finitely generated infinite simple n-Engel group then it is not nilpotent, since the only finitely generated nilpotent simple groups are cyclic of prime orders. $\qquad\square$

The following question was posed by A. I. Shirshov in 1963 (Problem 2.82, The Kourovka Notebook [34]) and still is open.

Q3: Are n-Engel varieties defined by positive laws?

An affirmative answer has been given for 2- and 3-Engel groups by A. I. Shirshov [30] and for 4-Engel groups by G. Traustason [31].

4   Observation

It is not known whether n-Engel groups (for $n > 4$) satisfy positive laws, however the Engel laws and positive laws have some common properties:
First it was shown for finitely generated **residually finite groups** G:

- 1991: If G satisfies **an Engel law**, then G is nilpotent (Wilson [36]),
- 1993: If G satisfies **a positive law**, then G is virtually nilpotent (Semple and Shalev [29]).

Later for finitely generated **locally graded groups** G:

- 1994: If G satisfies **an Engel law**, then G is virtually nilpotent (Kim and Rhemtulla [15]),
- 1997: If G satisfies **a positive law**, then G is virtually nilpotent (see page 462 and [7] Corollary 1).

The following question was posed by R. Burns:

> *What do the Engel laws and positive laws have in common that forces finitely generated locally graded groups satisfying them to be nilpotent-by-finite?*

The answer is that these laws have the same Engel construction.

5   Engel construction of the laws

Let $u(x, y)$ be a word, and S be a subset in the free group $F = \langle x, y \rangle$.

Definition 5.1 We say that a law $w(x, y) \equiv 1$ has construction

$$u(x, y) \mathrel{\widetilde{\in}} S$$

if it is equivalent to a law $u(x, y) \equiv s(x, y)$ for some word $s(x, y) \in S$.

The laws with the same construction have similar properties. For example
- Every law of the form $[x, y] \equiv x^p$ for some prime p has construction $[x, y] \mathrel{\widetilde{\in}} \{x^p, \, p \in \mathbb{P}\}$. These laws define abelian varieties $\mathfrak{A}_p$.
- The laws with construction $[x, y] \mathrel{\widetilde{\in}} F''$ define varieties of groups with perfect commutator subgroups (i.e., $G' = G''$).

Definition 5.2 The **general Engel construction** is a construction of the form

$$x^{k_0}[x,y]^{k_1}[x,{}_2y]^{k_2}\ldots[x,{}_ny]^{k_n} \widetilde{\in} E',$$

where $E = \langle\, [x,{}_iy],\ 0 \leq i\,\rangle$, $n \in \mathbb{N}$, $k_i \in \mathbb{Z}$.

To show that every law is equivalent to a law which has the general Engel construction, we need the following technical results.

Lemma 5.3 *For $k \geq 1$ the following inclusions hold*

$$[x,y^{k+1}] \in \langle\, [x,{}_iy],\ 1 \leq i \leq k\,\rangle \cdot [x,{}_{k+1}y], \tag{1}$$

$$[x,{}_{k+1}y] \in \langle\, [x,y^i],\ 1 \leq i \leq k\,\rangle \cdot [x,y^{k+1}]. \tag{2}$$

Proof From the commutator identity $[a,bc] = [a,c][a,b][a,b,c]$ it follows

$$[x,y^{k+1}] = [x,y^k][x,y][[x,y],y^k] \tag{3}$$

$$[x,y,y^k] = [x,y]^{-1}[x,y^k]^{-1}[x,y^{k+1}] \tag{4}$$

For $k = 1$ in (1) we have by (3): $[x,y^2] = [x,y]^2[x,y,y] \in \langle [x,y]\rangle \cdot [x,{}_2y]$. Now we assume that

$$[x,y^k] \in \langle\, [x,{}_iy],\ 1 \leq i \leq (k-1)\,\rangle \cdot [x,{}_ky].$$

If replace $x \to [x,y]$, we obtain the following consequence of the assumption

$$[[x,y],y^k] \in \langle\, [[x,y],{}_iy],\ 1 \leq i \leq (k-1)\,\rangle\cdot[[x,y],{}_ky] = \langle\, [x,{}_iy],\ 1 < i \leq k\,\rangle\cdot[x,{}_{k+1}y].$$

Now we apply the assumption and its consequence to (3) to get required (1),

$$[x,y^{k+1}] \in \langle\, [x,{}_iy],\ 1 \leq i \leq k\,\rangle \cdot [x,{}_{k+1}y].$$

For $k = 1$ in (2) we have by (4): $[x,y,y] \in \langle [x,y]\rangle \cdot [x,y^2]$. Now assume

$$[x,{}_ky] \in \langle\, [x,y^i],\ 1 \leq i \leq (k-1)\,\rangle \cdot [x,y^k],$$

then if replace here $x \to [x,y]$ and write $[[x,y],{}_ky]$ as $[x,{}_{k+1}y]$, we have

$$[x,{}_{k+1}y] \in \langle\, [[x,y],y^i],\ 1 \leq i \leq (k-1)\,\rangle \cdot [[x,y],y^k]$$
$$\overset{(4)}{\subseteq} \langle\, [x,y]^{-1}[x,y^i]^{-1}[x,y^{i+1}],\ 1 \leq i \leq (k-1)\,\rangle \cdot [x,y]^{-1}[x,y^k]^{-1}[x,y^{k+1}]$$
$$\subseteq \langle\, [x,y^i],\ 1 \leq i \leq k\,\rangle \cdot [x,y^{k+1}].$$

$\square$

Corollary 5.4 *The following subgroups coincide for every n, $(n \geq 0)$.*

$$E_n := \langle\, [x,{}_iy],\ 0 \leq i \leq n\,\rangle = \langle\, x^{y^i},\ 0 \leq i \leq n\,\rangle.$$

Proof By (1) and (2) we have the following equality

$$\langle\, [x, {}_iy], 0 \leq i \leq n\,\rangle = \langle\, x, [x, y^i], 0 \leq i \leq n\,\rangle.$$

Now it suffices to show that $\langle\, x, [x, y^i], 1 \leq i \leq n\,\rangle = \langle\, x^{y^i}, 0 \leq i \leq n\,\rangle$. Indeed, the normal closure of x, denoted by $\langle x\rangle^F$, is freely generated by all conjugates x^{y^i}, $i \in \mathbb{Z}$ (see [20], p.138). Hence in the subgroup $\langle\, x^{y^i}, 0 \leq i \leq n\,\rangle$ we can replace the free generators x^{y^i}, $i \neq 0$, by $x^{-1}x^{y^i} = [x, y^i]$, which gives the required equality. $\square$

We can now prove that every binary law has the general Engel construction.

Theorem 5.5 *Every binary law is equivalent to a law which has construction*

$$x^{k_0}[x, y]^{k_1}[x, {}_2y]^{k_2}\ldots[x, {}_ny]^{k_n} \,\widetilde{\in}\, E',$$

where $E = \langle\, [x, {}_iy], 0 \leq i\,\rangle$, $n \in \mathbb{N}$, $k_i \in \mathbb{Z}$.

Proof By ([24], 12.12) every word is equivalent to a pair of words, one of the form x^k, $k \geq 0$, the other a commutator word. Thus every binary law is equivalent to a law $x^k w(x, y) \equiv 1$, where $w \in F'$. Since $F' \subseteq \langle x\rangle^F$, w is a product of some x^{y^i} with say, $i \geq -m$. Conjugation by y^m gives us the equivalent law $w \equiv 1$ where $w \in \langle\, x^{y^i}, 0 \leq i\,\rangle$. Since by Corollary 5.4, $E := \langle\, [x, {}_iy], 0 \leq i\,\rangle = \langle\, x^{y^i}, 0 \leq i\,\rangle$, we have

$$w \in \langle\, [x, {}_iy], 0 \leq i\,\rangle = E.$$

Hence every commutator law is equivalent to a law $w \equiv 1$, where the word w is a product of commutators $[x, {}_iy]$, $0 \leq i$. By ordering these factors *modulo* E', we get the law which has construction

$$[x, y]^{k_1}[x, {}_2y]^{k_2}\ldots[x, {}_ny]^{k_n} \,\widetilde{\in}\, E', \quad k_i \in \mathbb{Z}.$$

Now we add x^{k_0} if the initial law defines a variety of exponent k_0. $\square$

The parameters n, k_i and a subset $S \subseteq E'$ define specific Engel constructions, responsible for some properties of groups. However the same property can be defined by different equivalent constructions.

6 Engel construction and properties of $\mathfrak{R}$-laws

In 1968 Milnor ([23], Lemma 3) proved that if for all elements g, h in a finitely generated group G the subgroup $\langle g^{h^i}, i \in \mathbb{N}\rangle$ is finitely generated, and A is an abelian normal subgroup in G so that G/A is cyclic then A is finitely generated.

The property that *for all elements g, h in a finitely generated group G, the subgroup $\langle g^{h^i}, i \in \mathbb{N}\rangle$ is finitely generated* was considered by many authors. It is called *the Milnor property* in F. Point [26]. The groups with this property are called *restrained* (Kim and Rhemtulla [15]). We shall call this property *the restraining property* and varieties consisting of groups with this property *the restrained varieties*.

Definition 6.1 A law is an $\mathfrak{R}$-law (*restraining law*) if for all elements g, h in a group G satisfying this law, the subgroup $\langle g^{h^i}, i \in \mathbb{N} \rangle$ is finitely generated.

By another words, a law is an $\mathfrak{R}$-law if it provides the restraining property.

In 1976 Rosset [28] noticed that in the mentioned above Milnor's proof the assumption that A is abelian can be dropped. He proved that *if for all elements g, h in a group G the subgroup $\langle g^{h^i}, i \in \mathbb{N} \rangle$ is finitely generated* then
  (i) the commutator subgroup G' is finitely generated and
  (ii) if G/N is cyclic then N is finitely generated.
In view of (i) we obtain another equivalent definition of the $\mathfrak{R}$-laws [19].

Definition 6.2 A law $w \equiv 1$ is an $\mathfrak{R}$-law if every finitely generated group G satisfying this law has its commutator subgroup G' finitely generated.

The following observation shows more clearly how restrained varieties in general are related to Engel varieties.

Proposition 6.3 *A law is an $\mathfrak{R}$-law if and only if it implies a law with the following Engel construction*

$$[x, {}_ny] \;\widetilde{\in}\; \langle x, [x,y], [x, {}_2y], \dots, [x, {}_{n-1}y] \rangle. \tag{5}$$

Proof A variety is restrained if and only if for some positive integer n it satisfies a law with construction

$$x^{y^n} \;\widetilde{\in}\; \langle x, x^y, x^{y^2}, \dots, x^{y^{n-1}} \rangle, \tag{6}$$

which in view of (1) is equivalent to a law with construction (5). □

We summarize that a law $w \equiv 1$ is an $\mathfrak{R}$-law if one of the following holds
  • each group satisfying $w \equiv 1$ is restrained,
  • each finitely generated group satisfying $w \equiv 1$ has G' finitely generated,
  • the law $w \equiv 1$ implies a law with the Engel construction $[x, {}_ny] \;\widetilde{\in}\; E_{n-1}$.

Corollary 6.4 *All Engel laws and all positive laws are $\mathfrak{R}$-laws.*

Proof Each Engel law has the required Engel construction. In [28] Rosset proved that a finitely generated group of subexponential growth is restrained. He used only fact that such a group does not contain a free subsemigroup (by growth reason). Since groups satisfying positive laws also do not contain free subsemigroups the same proof implies that groups satisfying positive laws are restrained (see also, e.g., [12], [16], [26], [5] p.520). So we conclude that positive laws are the $\mathfrak{R}$-laws. □

We need the following

Proposition 6.5 *Every finitely generated residually finite group G satisfying an $\mathfrak{R}$-law is nilpotent-by-finite, $G \in \mathfrak{N}_c\mathfrak{B}_e$, where c, e depend on the law only.*

Proof By result of G. Endimioni [9] every finitely generated residually finite group G in a variety is nilpotent-by-finite if and only if the variety does not contain a group C_p wr C for all primes p. Since the groups C_p wr C have infinitely generated commutator subgroups they do not satisfy an $\mathfrak{R}$-law and hence the first statement follows. It is also shown in [9] that c and e depend only on the variety and hence on the $\mathfrak{R}$-law (see also [6], Theorem A). □

A result of a similar kind as the above by Endimioni has been obtained by G. Traustason [33]. Namely, the property that every finitely generated residually nilpotent group in the variety is nilpotent holds if and only if the variety contains neither $C_p wr C$ nor $C wr C_p$. It is interesting to note that the Engel varieties satisfy but the varieties with positive laws do not in general satisfy this property.

Lemma 6.6 *Every finitely generated group G satisfying an $\mathfrak{R}$-law has its finite residual R (intersection of all subgroups of finite index in G) finitely generated.*

Proof The group G/R is residually finite and by assumption satisfies an $\mathfrak{R}$-law, hence by Proposition 6.5, G/R contains a nilpotent subgroup H/R of finite index. Now, since $|G : H| = |(G/R) : (H/R)| < \infty$ and G is finitely generated, both H and H/R are finitely generated. Being a finitely generated nilpotent group, H/R is polycyclic (see [24], 31.12). So there is a finite subnormal series with cyclic factors $H = N_0 \triangleright N_1 \triangleright \cdots \triangleright N_m = R$. Then by means of m successive applications of Rosset's result mentioned as (ii) on the page 467, we conclude that R is finitely generated. □

Corollary 6.4 and the following Theorem answer the question: *What do the Engel laws and positive laws have in common that forces finitely generated locally graded groups satisfying them to be nilpotent-by-finite?*

Theorem 6.7 (cf. [19]) *Let G be a finitely generated group satisfying an $\mathfrak{R}$-law. Then either G is nilpotent-by-finite or its finite residual has a finitely generated, infinite simple quotient.*

Proof By Lemma 6.6, finite residual R of G is finitely generated. Let R contain a proper subgroup T say, of finite index. By a result of M. Hall, R has only finitely many subgroups of index $k = |R : T|$ and their intersection K is characteristic of finite index in R. So K is normal in G and $K \subseteq T \subsetneq R$. Since R/K is finite and G/R is nilpotent-by-finite, the isomorphism $(G/K)/(R/K) \cong G/R$ implies that G/K is finite-by-(nilpotent-by-finite). Since finite-by-nilpotent group is nilpotent-by-finite, whence G/K is nilpotent-by-finite and then residually finite. Then $R \subseteq K$, which contradicts to $K \subseteq T \subsetneq R$ unless $R = 1$ in which case G is nilpotent-by-finite by Proposition 6.5.

If R contains no proper subgroup of finite index then by Zorn Lemma it has a maximal proper normal subgroup N of infinite index. The quotient R/N is the finitely generated infinite simple group satisfying the $\mathfrak{R}$-law. □

Moreover, in view of Proposition 6.5 we obtain the following.

Corollary 6.8 *For every $\mathfrak{R}$-law there exist positive integers c and e depending only on the law, such that every locally graded group satisfying this law lies in the product variety $\mathfrak{N}_c\mathfrak{B}_e$.*

Note There are non-(locally graded), non-(virtually nilpotent) finitely generated groups satisfying $\mathfrak{R}$-laws. For example free Burnside groups $B(r,n)$ for $r > 1$ and sufficiently large n, the groups satisfying the law $xy^n = y^nx$ also for n sufficiently large. Another example was given by Ol'shanskii and Storozhev in [25]. The $\mathfrak{R}$-laws in these examples are positive laws.

In [19] a problem was formulated whether there are $\mathfrak{R}$-laws that imply neither positive nor Engel law? Note that every law $[x, {}_ny] \equiv [x, {}_my]$ for $n > m$ has Engel construction (5), so it is an $\mathfrak{R}$-law. Every finite group satisfies such a law, hence the law need not be an Engel law. We do not know whether it is then a positive law.

7 Special kind of $\mathfrak{R}$-laws, called L_n

We denote by L_n the laws of the form $[x, y] \equiv [x, {}_ny]$, where $n > 1$. These laws are the $\mathfrak{R}$-laws because they have construction of the form (5) if we write them as $[x, {}_ny] \equiv [x, y]$.

Proposition 7.1 (cf. [10]) *Every metabelian group and every finite group satisfying the law L_n is abelian.*

Proof If substitute $[y, {}_{n-1}x]$ for y in L_n, we get $[x, [y, {}_{n-1}x]] \equiv [x, {}_n[y, {}_{n-1}x]]$. By taking inverse and interchanging $x \rightleftarrows y$ we obtain a law with construction $[x, {}_ny] \mathrel{\widetilde{\in}} F''$ which in view of L_n, implies a law with construction $[x, y] \mathrel{\widetilde{\in}} F''$. Clearly, each metabelian group satisfying a law with such construction is abelian.

Now, if there exists a finite non-abelian group satisfying the law L_n, then there exists such a group G of the smallest order. By O. Schmidt, a finite group G, all whose proper subgroups are abelian, is metabelian. Hence by the above, G must be abelian, which is a contradiction. $\square$

So every law of the form $[x, y] \equiv [x, {}_ny]$, $n > 1$, is either abelian or pseudo-abelian (see [24], Problem 5). It was conjectured in 1966 by N. Gupta [10] that each such a law is abelian. The proofs were given only for $n = 2$ and $n = 3$. We have looked for a shorter proof and its possible extension for $n > 3$. Our proofs are based on the following

Observation *Let a group G satisfy the law L_n. If an element $b \in G$ is conjugate to its inverse then $b^2 = 1$.*

Proof Let $b^{-1} = b^a$ say. Then $b^2 = (b^a)^{-1}b = [a, b] \equiv [a, {}_nb] = [b^2, {}_{n-1}b] = 1$. $\square$

In both cases for L_2 and L_3 we could deduce the law $[x, y]^2 \equiv 1$, which leads to the abelian law. To do this we used the following

Proposition 7.2 *In the free group F the word $[x, {}_ny^{-1}]$ is conjugate to $[x, {}_ny]^{(-1)^n}$ for $n = 1, 2, 3$.*

Proof It suffices to check the following commutator identities which are the equalities in the free group F. Namely (i) $[x, y^{-1}] = [x, y]^{-y^{-1}}$ which is clear,

(ii) $[x, {}_2y^{-1}] = [x, {}_2y]^{y^{-1}[x,y]^{-1}y^{-1}}$,   (iii) $[x, {}_3y^{-1}] = [x, {}_3y]^{-y^{-1}[x,y]^{-1}y^{-2}}$.

We prove here only (iii) for which we use the commutator identities (i) and

(iv) $[a^{-1}, b] = [a, b]^{-a^{-1}}$,   (v) $a^b = a[a, b]$,   (vi) $[a, bc] = [a, c]\,[a, b]^c$.

Namely,

$$
\begin{aligned}
[x, {}_3y^{-1}] &\overset{(i)}{=} [[[x, y]^{-y^{-1}}, y]^{-y^{-1}}, y]^{-y^{-1}} \\
&= [[[x, y]^{-1}, y]^{-1}, y]^{-y^{-3}} \\
&\overset{(iv)}{=} [[x, {}_2y]^{[x,y]^{-1}}, y]^{-y^{-3}} \\
&= [[x, {}_2y], y^{[x,y]}]^{-[x,y]^{-1}y^{-3}} \\
&\overset{(v)}{=} \left[[x, {}_2y], y \cdot [x, {}_2y]^{-1}]\right]^{-[x,y]^{-1}y^{-3}} \\
&\overset{(vi)}{=} [x, {}_3y]^{-[x,{}_2y]^{-1}[x,y]^{-1}y^{-3}} \\
&= [x, {}_3y]^{-y^{-1}[x,y]^{-1}y^{-2}}.
\end{aligned}
$$

$\square$

However Proposition 7.2 is not valid for $n = 4$. We conjecture that the laws L_n need not be abelian for $n > 3$.

Acknowledgement: The authors are very grateful to the referee for valuable remarks.

References

[1] S. I. Adian, *The problem of Burnside and identities in groups*, Nauka, Moscow, 1975 (Russian), (see also, trans. J. Lennox and J. Wiegold, Ergebnisse der Mathematik und ihrer Grenzgebiete **92**, Springer-Verlag, Berlin, 1979).

[2] S. Bachmuth and H. Y. Mochizuki, Third Engel groups and the Macdonald–Neumann conjeture, *Bull. Austral. Math. Soc.* **5** (1971), 379–386.

[3] R. Baer, Engelsche Elemente Noetherscher Gruppen, *Math. Ann.* **133** (1957), 256–270.

[4] B. Bajorska, On the smallest locally and residually closed class of groups, containing all finite and all soluble groups, *Publ. Math. Debrecen* **64** (2006), no. 4, 423–431.

[5] R. G. Burns, O. Macedońska and Yu. Medvedev, Groups satisfying semigroup laws and nilpotent-by-Burnside varieties, *J. Algebra* **195** (1997), 510–525.

[6] R. G. Burns and Yu. Medvedev, A note on Engel groups and local nilpotence, *J. Austral. Math. Soc. (Series A)* **64** (1998), 92–100.

[7] R. G. Burns, Yu. Medvedev, Group laws implying virtual nilpotence, *J. Austral. Math. Soc.* **74** (2003), 295–312.

[8] S. N. Černikov, Infinite nonabelian groups with an invariance condition for infinite nonabelian subgroups, *Dokl. Akad. Nauk SSSR* **194** (1970), 1280–1283.

[9] G. Endimioni, Bounds for nilpotent-by-finite groups in certain varieties of groups, *J. Austral. Math. Soc.* **73** (2002), 393–404.

[10] N. D. Gupta, Some group-laws equivalent to the commutative law, *Arch. Math. (Basel)* **17** (1966), 97–102.

[11] J. R. J. Groves, Varieties of soluble groups and a dichotomy of P. Hall, *Bull. Austral. Math. Soc.* **5** (1971), 391–410.

[12] K. W. Gruenberg, Two Theorems on Engel Groups, *Proc. Cambridge Philos. Soc.* **49** (1953), 377–380.

[13] G. Havas and M. R. Vaughan-Lee, 4-Engel groups are locally nilpotent, *Internat. J. Algebra and Comput.* **15** (2005), no. 4, 649–682.

[14] H. Heineken, Engelsche Elemente der Lange drei, *Illinois J. Math.* **5** (1961), 681–707.

[15] Y. Kim and A. H. Rhemtulla, On locally graded groups, *Proceedings of the Third International Conference on Group Theory, Pusan, Korea 1994* (Springer-Verlag, Berlin—Heidelberg—New York, 1995), 189–197.

[16] Y. K. Kim and A. H. Rhemtulla, Weak maximality condition and polycyclic groups, *Proc. Amer. Math. Soc.* **123** (1995), 711–714.

[17] F. W. Levi Groups in which the commutator operation satisfies certain algebraic conditions, *J. Indian Math. Soc. (N.S.)* **6** (1942), 87–97.

[18] J. Lewin and T. Lewin Semigroup laws in varieties of soluble groups, *Proc. Cambridge Philos. Soc.* **65** (1969), 1–9.

[19] Olga Macedonska, What do the Engel laws and positive laws have in common, *Fundamental and Applied Mathematics* **14** (2008), no. 7, 175–183 (Russian).

[20] W. Magnus, A. Karrass, D. Solitar, *Combinatorial Group Theory*, 2nd ed., Dover Publications, New York 1976.

[21] A. I. Mal'tsev, Nilpotent semigroups, *Ivanov. Gos. Ped. Inst. Uc. Zap.* **4** (1953), 107–111 (Russian).

[22] Yu. Medvedev On compact Engel groups, *Israel J. Math.* **135** (2003), no. 1, 147–156.

[23] J. Milnor, Growth of finitely generated solvable groups, *J. Diff. Geom.* **2** (1968), 447–449.

[24] H. Neumann, *Varieties of groups*, Springer-Verlag, Berlin, Heidelberg, New York 1967.

[25] A. Yu. Ol'shanskii and A. Storozhev, A group variety defined by a semigroup law, *J. Austral. Math. Soc. (Series A)* **60** (1996), 255–259.

[26] F. Point, Milnor identities, *Comm. Algebra* **24** (1996), no. 12, 3725–3744.

[27] B. H. Neumann and T. Taylor, Subsemigroups of nilpotent groups, *Proc. Roy. Soc. (Series A)* **274** (1963), 1–4.

[28] S. Rosset, A property of groups of non-exponential growth, *Proc. Amer. Math. Soc.* **54** (1976), 24–26.

[29] J. F. Semple and A.Shalev, Combinatorial conditions in residually finite groups, I, *J. Algebra* **157** (1993), 43–50.

[30] A. I. Shirshov, On certain near-Engel groups, *Algebra i Logika* **2** (1963), no. 1, 5–18 (Russian).

[31] G. Traustason, Semigroup identities in 4-Engel groups, *J. Group Theory* **2** (1999), 39–46.

[32] G. Traustason, A note on the local nilpotence of 4-Engel groups, *Internat. J. Algebra Comput.* **15** (2005), no. 4, 757–764.

[33] G. Traustason, Milnor groups and (virtual) nilpotence, *J. Group Theory* **8** (2005), 203–221.

[34] *Unsolved problems in group theory: The Kourovka Notebook*, Fourteenth edition, Russian Academy of Sciences, Siberian Division, Institute of Mathematics, Novosibirsk

1999.

[35] M. Vaughan-Lee Engel-4 groups of exponent 5, *Proc. London Math. Soc.* **74** (1997), no. 3, 306–334.

[36] J. S. Wilson, Two-generator conditions for residually finite groups, *Bull. London Math. Soc.* **23** (1991), 239–248.

[37] J. S. Wilson and E. I. Zelmanov, Identities for Lie algebras of pro-p groups, *J. Pure Appl. Algebra* **81** (1992), 103–109.

MAXIMAL SUBGROUPS OF ODD INDEX IN FINITE GROUPS WITH SIMPLE CLASSICAL SOCLE

N. V. MASLOVA

Institute of Mathematics and Mechanics of UB RAS, Ekaterinburg, 620219, Russia
Email: butterson@mail.ru

Abstract

We discuss the completion of the classification of maximal subgroups of odd index in finite groups with simple classical socle.

1 Introduction

The subgroup of a finite group G generated by the set of all its minimal non-trivial normal subgroups is called the socle of G and is denoted by $\mathrm{Soc}(G)$. A finite group is almost simple if its socle is a nonabelian simple group. It is well known that a finite group G is almost simple if and only if there exists a nonabelian finite simple group L such that $L \cong \mathrm{Inn}(L) \trianglelefteq G \leq \mathrm{Aut}(L)$. In this case $\mathrm{Inn}(L) = \mathrm{Soc}(G)$. One of the greatest results in the theory of finite permutation groups was obtained by Liebeck and Saxl [7] and independently by Kantor [3]. They gave the classification of finite primitive permutation groups of odd degree. In particular, for each finite group G whose socle is a simple classical group they specified types of subgroups which can be maximal subgroups of odd index in G. However, not every subgroup of these types is a maximal subgroup of odd index in G. Thus, the classification of maximal subgroups of odd index in finite groups with a simple classical socle is not complete. In this paper, we discuss the completion of the classification of maximal subgroups of odd index in finite groups with simple classical socle.

2 Results by Liebeck, Saxl and Kantor

We use basically the standard terminology and notation (see, for example, [2, 6]). Let q be a power of a prime and let $L = L(q)$ be one of the finite simple classical groups $PSL_n(q)$, $PSU_n(q)$, $PSp_n(q)$, where n is even, $P\Omega_n(q)$, where nq is odd, and $P\Omega_n^\varepsilon(q)$, where n is even and $\varepsilon \in \{+, -\}$. Let V be the natural n-dimensional vector space over the field F associated with L, where $F = F_{q^2}$ if L is unitary and $F = F_q$ otherwise. Let $L \cong \mathrm{Inn}(L) \trianglelefteq G \leq \mathrm{Aut}(L)$.

Liebeck, Saxl and Kantor proved the following.

If H is a maximal subgroup of odd index in G, then either q is even and $L \cap H$ is a parabolic subgroup of L or q is odd, $L = L(q)$ and one of the following holds:

(1) $H = N_G(L(q_0))$, where $q = q_0^s$ and s is an odd prime;

The author is supported by RFBR grant nos. 07-01-00148 and 10-01-00324, UB RAS grant for young scientists.

(2) H is the stabilizer of a non-singular subspace (any subspace for $L = PSL_n(q)$);

(3) $L \cap H$ is the stabilizer of an orthogonal decomposition $V = \bigoplus V_i$ with all V_i isometric (any decomposition $V = \bigoplus V_i$ with $\dim V_i$ constant for $L = PSL_n(q)$);

(4) $L = PSL_n(q)$, H is the stabilizer of a pair $\{U, W\}$ of subspaces of complementary dimensions with $U \leq W$ or $U \oplus W = V$, and G contains an automorphism of L interchanging U and W;

(5) $L = PSL_2(q)$ and $L \cap H$ is A_4, S_4, A_5, $PGL_2(q^{1/2})$ or dihedral (note that this case is omitted in [3]);

(6) $L = PSU_3(5)$ and $L \cap H \cong M_{10}$;

(7) $L \cap H$ is $\Omega_7(2)$ or $\Omega_8^+(2)$ and L is $P\Omega_7(q)$ or $P\Omega_8^+(q)$, respectively, q is prime and $q \equiv \pm 3 \pmod 8$;

(8) $L = P\Omega_8^+(q)$, q is prime and $q \equiv \pm 3 \pmod 8$, G contains a triality automorphism of L and $L \cap H$ is $2^3.2^6.PSL_3(2)$.

The proof of the results [3, 7] uses the Classification of Finite Simple Groups.

If q is even or q is odd and (1) holds, then the subgroup H is always a subgroup of odd index in G (see [3, 7]). In (2)–(8) the oddness of the index $|G : H|$ depends on some additional conditions. Note that if G is a finite group, $L \trianglelefteq G$ and H is a maximal subgroup of G not containing L, then $G = LH$ and consequently $|G : H| = |L : L \cap H|$. So it is important to find the conditions for the subgroups of the group G obtained as $L \cap H$, where the subgroups H satisfy (2)–(8), to be subgroups of odd index in L. In (6)–(8) we can prove the oddness of this index by direct calculation. In other cases the oddness of the index of $L \cap H$ in L essentially depends on the values of several parameters, including n and q.

3   Main results

Using the classification of finite simple groups, Aschbacher [1] described a large family of natural geometrically defined subgroups of finite simple classical groups. He divided this family into eight classes C_i ($1 \leq i \leq 8$), now called Ashbacher's classes. Subgroups P obtained as $P = L \cap H$, where H satisfies (1), (2), (3) or (4) (see section 2), are contained in Ashbacher's classes C_5, C_1, C_2 or C_1 of L, respectively. The group-theoretical structure of subgroups of finite simple classical groups contained in Ashbacher's classes was described by Kleidman and Liebeck [6]. We will use that result below.

If $L = P\Omega_n^\varepsilon(q)$, where n is even, then $\varepsilon \in \{+, -\}$ is called the sign of the group L and of the associated vector space V and is denoted by $\mathrm{sgn}(V)$. For every non-singular subspace $U \leq V$, where $m = \dim U$ is even, $v = \mathrm{sgn}(U)$ is defined (see [6, Ch. 2]).

Define

$$D(U) = D_m^v(q) = \begin{cases} 1 & \text{if } v = + \text{ and } (q-1)m/4 \text{ is even,} \\ 1 & \text{if } v = - \text{ and } (q-1)m/4 \text{ is odd,} \\ -1 & \text{otherwise.} \end{cases}$$

If $V = U \oplus W$ is an orthogonal decomposition, where U and W are non-singular subspaces of even dimensions, then $D(V) = D(U) \cdot D(W)$ and $\mathrm{sgn}(V)1 = \mathrm{sgn}(U)1 \cdot \mathrm{sgn}(W)1$ (see [6, Proposition 2.5.11]).

Let $\mathcal{M}$ be the set of all sequences $(x_0, x_1, \ldots, x_n, \ldots)$, where $x_i \in \{0,1\}$ for all i. Define an order $\geq$ on $\mathcal{M}$ as follows: $1 \geq 0$ and $u = (u_0, u_1, \ldots, u_n, \ldots) \geq v = (v_0, v_1, \ldots, v_n, \ldots)$ if and only if $u_i \geq v_i$ for all i.

Let $\psi : \mathbb{N} \to \mathcal{M}$ be the function mapping a natural number s to the sequence $(s_0, s_1, \ldots, s_k, \ldots) \in \mathcal{M}$, where $s = s_0 + s_1 \cdot 2 + \ldots + s_k \cdot 2^k$ is the binary-coded number s and $s_n = 0$ for $n > k$.

We prove the following theorem.

Theorem 3.1 *Let q be odd and let L be one of the finite simple classical groups $PSL_n(q)$, where $n \geq 2$, $PSU_n(q)$, where $n \geq 3$, $PSp_n(q)$, where $n \geq 4$ and n is even, $P\Omega_n(q)$, where $n \geq 7$ and n is odd, $P\Omega_n^{\pm}(q)$, where $n \geq 8$ and n is even. Let V be the natural vector space associated with L and let P be a subgroup of L. Then the following statements hold:*

(1) *If $L = PSL_n(q)$, P is the stabilizer of a subspace $U \leq V$ and $\dim U = m$, then $|L : P|$ is odd if and only if $\psi(n) \geq \psi(m)$;*

(2) *If $L = PSL_n(q)$ and P is the stabilizer of a pair $\{U, W\}$ of subspaces of complementary dimensions $m < n/2$ and $n - m$, respectively, with $U \leq W$, then $|L : P|$ is always even;*

(3) *If $L = PSL_n(q)$ and P is the stabilizer of a pair $\{U, W\}$ of subspaces of complementary dimensions $m < n/2$ and $n - m$, respectively, with $U \oplus W = V$, then $|L : P|$ is odd if and only if $\psi(n) \geq \psi(m)$;*

(4) *If $L = PSL_n(q)$ and P is the stabilizer of a decomposition $V = \bigoplus V_i$ with $\dim V_i = m$, then $|L : P|$ is odd if and only if either $m = 2^w \geq 2$, where w is an integer or $m = 1$ and $q \equiv 1 \pmod 4$;*

(5) *If $L = PSU_n(q)$ or $L = PSp_n(q)$, P is the stabilizer of a non-singular subspace $U \leq V$ and $\dim U = m$, then $|L : P|$ is odd if and only if $\psi(n) \geq \psi(m)$;*

(6) *If $L = PSU_n(q)$ and P is the stabilizer of an orthogonal decomposition $V = \bigoplus V_i$ with all V_i isometric and $\dim V_i = m$, then $|L : P|$ is odd if and only if either $m = 2^w \geq 2$, where w is an integer or $m = 1$ and $q \equiv 3 \pmod 4$;*

(7) *If $L = PSp_n(q)$ and P is the stabilizer of an orthogonal decomposition $V = \bigoplus V_i$ with all V_i isometric and $\dim V_i = m$, then $|L : P|$ is odd if and only if $m = 2^w \geq 2$, where w is an integer;*

(8) *If $L = P\Omega_n(q)$ and P is the stabilizer of a non-singular subspace $U \leq V$ and $\dim U = m$ is even, then $|L : P|$ is odd if and only if $D(U) = 1$ and $\psi(n) \geq \psi(m)$;*

(9) *If $L = P\Omega_n(q)$ and P is the stabilizer of an orthogonal decomposition $V = \bigoplus V_i$ with all V_i isometric and $\dim V_i > 1$, then $|L : P|$ is always even;*

(10) *If $L = P\Omega_n(q)$, where q is prime and P is the stabilizer of an orthogonal decomposition $V = \bigoplus V_i$ with all V_i isometric and $\dim V_i = 1$, then $|L : P|$ is odd if and only if $q \equiv \pm 3 \pmod 8$;*

(11) If $L = P\Omega_n^\varepsilon(q)$, P is the stabilizer of a non-singular subspace $U \le V$ and $\dim U = m$, then $|L : P|$ is odd if and only if either m is odd, $D(V) = -1$ and $\psi(n - 2) \ge \psi(m - 1)$ or m is even and either $D(U) = D(V) = -1$ and $\psi(n - 2) \ge \psi(m - 2)$ or $D(U) = D(V) = 1$ and $\psi(n) \ge \psi(m)$;

(12) If $L = P\Omega_n^\varepsilon(q)$ and P is the stabilizer of an orthogonal decomposition $V = \bigoplus V_i$ with all V_i isometric and $\dim V_i > 1$, then $|L : P|$ is odd if and only if $m = 2^w \ge 2$, where w is an integer and $D(V) = D(V_i) = 1$;

(13) If $L = P\Omega_n^\varepsilon(q)$, where q is prime and P is the stabilizer of an orthogonal decomposition $V = \bigoplus V_i$ with all V_i isometric and $\dim V_i = 1$, then $|L : P|$ is odd if and only if $q \equiv \pm 3 \pmod 8$.

The proof of clauses (1) and (4)–(13) is given in [8], and (2)–(3) are proved in [9]. Our proof uses the description of the group-theoretical structure of subgroups from Ashbacher's classes of finite simple classical groups obtained in [6] and a number-theoretical result closely related to [10, Lemma 2].

Let G be a finite group, let $L = \mathrm{Soc}(G)$ be a simple group, and let P be a subgroup of L. If H is a maximal subgroup of G such that $1 \ne P = L \cap H < L$, then $H = N_G(P)$. Note that the condition that q is prime in clauses (10), (13) of Theorem 3.1 is natural because otherwise the subgroup $N_G(P)$ cannot be maximal in G for any group G such that $L = \mathrm{Soc}(G)$ (see [6, § 4.2]).

If the characteristic of the field is even, the maximal subgroups of odd index of a finite simple classical group are precisely the parabolic maximal subgroups.

If q is odd, descriptions [3, 7] and Theorem 3.1 imply the following classification of the maximal subgroups of odd index of finite simple classical groups over a field of odd characteristic. This classification was obtained by the author in [8].

Theorem 3.2 Let q be odd and let $G = G(q)$ be one of the finite simple classical groups $PSL_n(q)$, where $n \ge 2$, $PSU_n(q)$, where $n \ge 3$, $PSp_n(q)$, where $n \ge 4$ and n is even, $P\Omega_n(q)$, where $n \ge 7$ and n is odd, $P\Omega_n^\pm(q)$, where $n \ge 8$ and n is even. Let V be the natural vector space associated with L. If H is a maximal subgroup of odd index in G then one of the following statements holds:

(1) $H = N_G(G(q_0))$, where $q = q_0^s$ and s is an odd prime;

(2) $G = PSL_n(q)$, H is the stabilizer of a subspace $U \le V$, where $\dim U = m \le n - 1$ and $\psi(n) \ge \psi(m)$;

(3) $G = PSU_n(q)$ or $G = PSp_n(q)$, H is the stabilizer of a non-singular subspace $U \le V$, $\dim U = m$ and $\psi(n) \ge \psi(m)$;

(4) $G = P\Omega_n(q)$, H is the stabilizer of a non-singular subspace $U \le V$, $\dim U = m$ is even, $D(U) = 1$ and $\psi(n) \ge \psi(m)$;

(5) $G = P\Omega_n^\varepsilon(q)$, H is the stabilizer of a non-singular subspace $U \le V$, $\dim U = m$, and either m is odd, $D(V) = -1$ and $\psi(n-2) \ge \psi(m-1)$ or m is even and either $D(U) = D(V) = -1$ and $\psi(n - 2) \ge \psi(m - 2)$ or $D(U) = D(V) = 1$ and $\psi(n) \ge \psi(m)$;

(6) $G = PSL_n(q)$, H is the stabilizer of a decomposition $V = \bigoplus V_i$ with $\dim V_i = m$, and either $m = 2^w \ge 2$, where w is an integer or $m = 1$ and $q \equiv 1$

(mod 4);

(7) $G = PSU_n(q)$, H is the stabilizer of an orthogonal decomposition $V = \bigoplus V_i$ with all V_i isometric, $\dim V_i = m$ and either $m = 2^w \geq 2$, where w is an integer or $m = 1$ and $q \equiv 3$ (mod 4);

(8) $G = PSp_n(q)$, H is the stabilizer of an orthogonal decomposition $V = \bigoplus V_i$ with all V_i isometric, $\dim V_i = m$ and $m = 2^w \geq 2$, where w is an integer;

(9) $G = P\Omega_n(q)$, H is the stabilizer of an orthogonal decomposition $V = \bigoplus V_i$ with all V_i isometric, $\dim V_i = 1$, q is prime and $q \equiv \pm 3$ (mod 8);

(10) $G = P\Omega_n^\varepsilon(q)$, H is the stabilizer of an orthogonal decomposition $V = \bigoplus V_i$ with all V_i isometric, $\dim V_i = m$ and either $m = 1$, q is prime and $q \equiv \pm 3$ (mod 8) or $m = 2^w \geq 2$, where w is an integer and $D(V) = D(V_i) = 1$;

(11) $G = PSL_2(q)$ and H is $PGL_2(q_0)$, where $q = q_0^2$;

(12) $G = PSL_2(q)$ and H is A_4, where $q \equiv \pm 3$ (mod 8);

(13) $G = PSL_2(q)$ and H is S_4, where $q \equiv \pm 7$ (mod 16);

(14) $G = PSL_2(q)$ and H is A_5, where $q \equiv \pm 3$ (mod 8);

(15) $G = PSL_2(q)$ and H is D_{q+1}, where $q \equiv 3$ (mod 4);

(16) $G = PSU_3(5)$ and H is M_{10};

(17) $G = PSL_4(q)$ and H is $2^4.A_6$, where q is prime and $q \equiv 5$ (mod 8);

(18) $G = PSL_4(q)$ and H is $PSp_4(q).2$, where $q \equiv 3$ (mod 4);

(19) $G = PSU_4(q)$ and H is $2^4.A_6$, q is prime and $q \equiv 3$ (mod 8);

(20) $G = PSU_4(q)$ and H is $PSp_4(q).2$, where $q \equiv 1$ (mod 4);

(21) $G = PSp_4(q)$ and H is $2^4.A_5$, where q is prime and $q \equiv 3$ (mod 8);

(22) $G = P\Omega_7(q)$ and H is $\Omega_7(2)$, where q is prime and $q \equiv \pm 3$ (mod 8);

(23) $G = P\Omega_8^+(q)$ and H is $\Omega_8^+(2)$, where q is prime and $q \equiv \pm 3$ (mod 8).

Note that the results in [3, 7] were obtained for the groups $P\Omega_6^+(q)$, $P\Omega_6^-(q)$ and $P\Omega_5(q)$ but not for their isomorphic copies $PSL_4(q)$, $PSU_4(q)$ and $PSp_4(q)$, respectively. These copies are examined in our paper [8]. Using the description of all maximal subgroups of finite simple classical groups of small degree, given in [4], and directly calculating indices of all maximal subgroups, we obtain that, except the cases stated in [3, 7], exceptions (17)–(21) appear. These exceptions are subgroups from $C_1 \cup C_2$ in the natural representations of the orthogonal groups.

The maximal subgroups of finite groups with simple classical socle of degree not greater than 12 are known (see [6, Theorem 1.2.2] and [4, 5]). The maximal subgroups from Ashbacher's classes of finite simple classical groups of degree greater than 12 are described by Kleidman and Liebeck [6, Main theorem]. In this proof the classification of finite simple groups is used essentially.

Recall that if G is a finite group, $L = \mathrm{Soc}(G)$ is a simple group and H is a maximal subgroup in G such that $1 \neq P = L \cap H < L$, then $H = N_G(P)$.

If in this case P is a maximal subgroup of L, then H is a maximal subgroup of G if and only if $G = N_G(P)L$, i.e., each subgroup of L conjugate to P in G

is conjugate to P in L. Conjugation within Ashbacher's classes is described in [6, § 3.2].

In the other case when P is a non-maximal subgroup of L, H is a maximal subgroup of G if and only if $G = N_G(P)L$ and for each subgroup Y such that $P < Y < L$ the statement $N_G(P) \not< N_G(Y)$ is true. The overgroups of non-maximal subgroups from Ashbacher's classes of finite simple groups of degree greater than 12 are described in [6, § 3.3, Tables 3.5.H, 3.5.I]. The conditions under which the statement $N_G(P) \not\leq N_G(Y)$ holds for each pair of subgroups $P < Y$ is true are also stated in [6].

Acknowledgement   The author wishes to thank her supervisor Prof. A. S. Kondratiev for setting of the problem and attention to work.

References

[1] M. Aschbacher, On the maximal subgroups of the finite classical groups, *Invent. Math.* **76** (1984), 469–514.

[2] J. H. Conway, R. T. Curtis, S. P. Norton, R. A. Parker, and R. A. Wilson, *Atlas of Finite Groups*, Clarendon Press, 1985.

[3] W. M. Kantor, Primitive permutation groups of odd degree, and an application to finite projective planes, *J. Algebra* **106** (1987), 15–45.

[4] P. Kleidman, *The subgroup structure of some finite simple groups*, Ph.D. Thesis, Cambridge University, 1986.

[5] P. Kleidman, The maximal subgroup structure of the finite 8-dimensional orthogonal groups $P\Omega_8^+(q)$ and of their automorphism groups, *J. Algebra* **110** (1987), no. 1, 173–242.

[6] P. B. Kleidman and M. W. Liebeck, *The subgroup structure of finite classical groups*, Cambridge University Press, 1990.

[7] M. W. Liebeck and J. Saxl, The primitive permutation groups of odd degree, *J. London Math. Soc. (2)* **31** (1985), 250–264.

[8] N. V. Maslova, *Classification of maximal subgroups of odd index in finite simple classical groups*, Trudy IMM UrO RAN **14** (2008), 4, 100–118 (In Russian). English translation: Proceedings of the Steklov Institute of Mathematics, Suppl. **3** (2009), S164–S183.

[9] N. V. Maslova, Maximal subgroups of odd index in finite groups with simple linear, unitary or symplectic socle, *Algebra and Logic*, to appear.

[10] J. G. Thompson, Hall subgroups of the symmetric groups, *J. Combinatorial Theory* **1** (1966), 271–279.

SOME CLASSIC AND NEARLY CLASSIC PROBLEMS ON VARIETIES OF GROUPS

PETER M. NEUMANN

The Queen's College, Oxford, OX1 4AW, U.K.
Email: peter.neumann@queens.ox.ac.uk

1   Introduction

This lecture, delivered on B. H. Neumann day at the conference, was conceived as paying respects to both my parents. The main references are my mother's famous monograph *Varieties of groups* (1967) and my father's article that carries the same title and was published in the same year (see [21] and [20]).

We begin by establishing some notation and terminology and recalling some standard theory:

x_1, x_2, x_3, ... (often x, y, z, ...) are to be distinct letters;
$F := \langle x_1, x_2, x_3, \ldots \rangle$, the free group they generate;
$F_n := \langle x_1, x_2, \ldots, x_n \rangle$, the free group of rank n;
words $w(x_1, \ldots, x_n)$ are elements of F.

For a group G let

$$\mathrm{IdRel}(G) := \{ w \in F \mid \forall\, g_1, \ldots, g_n \in G : w(g_1, \ldots, g_n) = e \},$$

the set of all *identical relations* of G. Although the definition implies that identical relations are simply words, we often write them in the form $w(x_1, \ldots, x_n) = e$ to emphasise that they are relations. Then for $V \subseteq F$ set

$$\mathfrak{V} := \text{the class of all groups } X \text{ such that } V \subseteq \mathrm{IdRel}(X),$$

the *variety* defined by V; and for the group G define

$$\mathfrak{Var}(G) := \text{the class of all groups } X \text{ such that } \mathrm{IdRel}(G) \subseteq \mathrm{IdRel}(X),$$

the *variety generated by* G. The set of all identities in a variety $\mathfrak{V}$ or a group G is a *verbal subgroup* of F and this gives a one-one relationship between varieties and fully invariant subgroups of F. For a verbal subgroup V of F and a group G the verbal subgroup $V(G)$ is defined to be the subgroup generated by all the values $w(g_1, \ldots, g_n)$ for $w(x_1, \ldots, x_n) \in V$ and $g_1, \ldots, g_n \in G$. The (relatively) free group $F_n(\mathfrak{V})$ in the variety $\mathfrak{V}$ is the group $F_n/V(F_n)$, where V is the set of all identities that hold in $\mathfrak{V}$. Here are some important examples of varieties with typical defining identities.

$\mathfrak{O}$,   the variety of all groups: e.
$\mathfrak{E}$,   the variety containing only $\{e\}$: $x = e$.
$\mathfrak{B}_n$,   the *Burnside* variety (of exponent n): $x^n = e$.

$\mathfrak{A}$, the *abelian* variety: $[x, y] = e$ (where $[x, y] := x^{-1}y^{-1}xy$).

$\mathfrak{A}_n$, $\mathfrak{A} \cap \mathfrak{B}_n$, the abelian variety of exponent n: $[x, y]z^n = e$.

$\mathfrak{N}_c$, the *nilpotent* variety of class c: $[[\ldots[[x_1, x_2], x_3], \ldots], x_{c+1}] = e$.

$\mathfrak{S}_d$, the *soluble* variety of derived length d: $[\ldots[[x_1, x_2], [x_3, x_4]], \ldots] = e$.

$3\,\mathfrak{A}^2$, the *centre-by-metabelian* variety: $[[[x_1, x_2], [x_3, x_4]], x_5] = e$.

There is a product defined on the collection of varieties of groups: for varieties $\mathfrak{U}, \mathfrak{V}$,

$$\mathfrak{U}\mathfrak{V} := \{G \mid \exists K \trianglelefteq G : K \in \mathfrak{U}, G/K \in \mathfrak{V}\}.$$

For example $\mathfrak{S}_d = \mathfrak{A}^d$. It is well known (and easy to prove) that the product is associative. For the purposes of the lecture (and this note)

$\mathfrak{V}$ is called an $\mathfrak{A}\mathfrak{N}$-variety if $\exists c : \mathfrak{V} \subseteq \mathfrak{A}\,\mathfrak{N}_c$;

$\mathfrak{V}$ is called an $\mathfrak{N}\mathfrak{A}$-variety if $\exists c : \mathfrak{V} \subseteq \mathfrak{N}_c\mathfrak{A}$;

$\mathfrak{V}$ is called an $\mathfrak{N}\mathfrak{N}$-variety if $\exists c, d : \mathfrak{V} \subseteq \mathfrak{N}_c\mathfrak{N}_d$.

We use the terms $\mathfrak{A}\mathfrak{N}$-group, $\mathfrak{N}\mathfrak{A}$-group, $\mathfrak{N}\mathfrak{N}$-group similarly.

Acknowledgement. I am very grateful to a kind but anonymous referee for helpful suggestions, particularly in relation to the references.

2 Classic problems

For me the classic problems are the 25 problems formulated by my mother in her 1967 book [21] (see also Kovács & Newman [14]), together with the 18 problems formulated by my father in his paper [20].

The original classics were two questions that are implicit in my father's 1935 Cambridge PhD thesis and the version [19] published in 1937. A variety is said to be *finitely based* if it can be defined by just finitely many identical relations. It is evident that the examples listed above are finitely based. The two questions were: is it true that the variety generated by a finite group is finitely based, and is it perhaps true that every variety of groups is finitely based?

The affirmative answer to the first of these is supplied by the Oates–Powell Theorem of 1964 (see [23]). Negative answers to the second were supplied in 1969 by Adjan, Ol'shanskii, and Vaughan-Lee independently (see [1], [24], [32]). In fact there are $2^{\aleph_0}$ varieties of groups, but of course only $\aleph_0$ finite sets of words, so only $\aleph_0$ finitely based varieties.

My *almost classic* problems are other old and interesting questions. There was a lot of activity on varieties of groups from the late 1950s for about twenty years. In particular, it was an area to which I, and some of the Oxford doctoral students I supervised at that time, contributed. Our efforts produced quite as many problems as theorems, and not all of those problems were recorded in [21] or [20]. It is these that my lecture was designed to collect and advertise. A few of them have already appeared in print (see for example, [12] or [22]).

3 Finite basis problems and related questions

Recall that the variety $\mathfrak{V}$ is said to be *hereditarily finitely based* if all its subvarieties (including $\mathfrak{V}$ itself) are finitely based. By consideration of the corresponding word subgroups of F one easily sees that $\mathfrak{V}$ is hereditarily finitely based if and only if it itself is finitely based and also there are no infinite properly descending chains of subvarieties of $\mathfrak{V}$, that is, its subvarieties satisfy MIN. Observe that if the variety $\mathfrak{V}$ is hereditarily finitely based then it has only countably many subvarieties.

Problem 1. Does there exist a variety $\mathfrak{V}$ that has only countably many subvarieties, but which is not hereditarily finitely based?

With reference to Problem 1 we have

Lemma 3.1 (L. G. Kovács 1968) *Let $\mathfrak{V}$ be a variety in which there are $< 2^{\aleph_0}$ subvarieties. Suppose that for every finitely generated $G \in \mathfrak{V}$, the subvarieties of the variety $\mathfrak{Var}(G)$ satisfy* MIN. *Then also the subvarieties of the variety $\mathfrak{V}$ satisfy* MIN.

This is proved in [13]. It tells us, in particular, that if $\mathfrak{V}$ is locally finite or locally nilpotent and has fewer than $2^{\aleph_0}$ subvarieties then its subvarieties satisfy MIN. If the hypotheses of Kovács's lemma are satisfied, then every subvariety is finitely based *qua* subvariety of $\mathfrak{V}$ and therefore there can only be countably many subvarieties. This leads to the following question:

Problem 2. Does there exist a variety $\mathfrak{V}$ for which the number κ of subvarieties satisfies $\aleph_0 < \kappa < 2^{\aleph_0}$?

We return now to finite basis problems. The next has already been aired in [6] and [22].

Problem 3. Is it true that all varieties of exponent 4 are finitely based? That is, is $\mathfrak{B}_4$ hereditarily finitely based?

Note that $\mathfrak{B}_4$ is hereditarily finitely based if and only if $\mathfrak{B}_4$ has only $\aleph_0$ subvarieties. Here one implication is trivial; its converse comes from Lemma 3.1.

For context consider the following facts about the Burnside varieties $\mathfrak{B}_n$ for some small values of n:

$\mathfrak{B}_1$: it is trivial that this is hereditarily finitely based and there is just one subvariety;

$\mathfrak{B}_2$: it is almost trivial that this is hereditarily finitely based and there are just two subvarieties;

$\mathfrak{B}_3$: it is a consequence of the analysis by Levi and van der Waerden [16] that this is hereditarily finitely based and there are just four subvarieties, namely $\mathfrak{E}$, $\mathfrak{A}_3$, $\mathfrak{B}_3 \cap \mathfrak{N}_2$ and $\mathfrak{B}_3$ itself;

$\mathfrak{B}_4$: Problem 3;

$\mathfrak{B}_5$: nothing is known;

$\mathfrak{B}_6$: this variety, which is known to be locally finite by a theorem of Marshall Hall [8], was proved to be hereditarily finitely based by M. D. Atkinson in his 1970 Oxford DPhil thesis (see [2]).

Much is known about the variety $\mathfrak{B}_4$. For example, it has been known since 1940 that it is locally finite (Sanov [29]); it is not soluble and if $G := F_n(\mathfrak{B}_4)$ then $c(G) = 3n - 2$ where $c(G)$ denotes the nilpotency class of G (Razmyslov [28] building on, and substantially adding to, work that began with Sean Tobin's PhD thesis ([30], see also [31]), and continued through contributions by many authors, of which the most influential in their time were Wright [34], Gupta and Newman [7]). For further information and details see Vaughan-Lee [33, Chapter 6].

For comparison, note that the variety $\mathfrak{B}_8$ is not hereditarily finitely based. For example, responding to Problem 11 in [21] (p. 92), Roger Bryant has proved in [3] that the product variety $\mathfrak{B}_4\mathfrak{B}_2$ is not finitely based. Also, for every odd prime number p the variety $\mathfrak{B}_{p^2}$ is known not to be hereditarily finitely based. Indeed, C. K. Gupta and A. N. Krasil'nikov have proved in [6] that there exist $\mathfrak{N}\mathfrak{M}$-varieties of exponent p^2 that are not finitely based.

In his DPhil thesis (Oxford 1995) and in three articles [25, 26, 27], Martyn Quick has tackled the problem. I believe his methods and insights must lead to a solution. Nevertheless, although it has been around for a long time, Problem 3 remains open.

Problem 4. Are all $\mathfrak{A}\mathfrak{N}$-varieties finitely based?

This problem has been advertised before (Krasil'nikov, Problem 14.51 in the 1999 edition of [12], and before that in [22]). There is a wealth of context. According to a famous theorem of Philip Hall [9] all finitely generated $\mathfrak{A}\mathfrak{N}$-groups satisfy MAX-N, the maximal condition on normal subgroups. Therefore they are finitely presentable relative to any $\mathfrak{A}\mathfrak{N}$-variety in which they lie and there are only $\aleph_0$ of them (up to isomorphism). In 1990 A. N. Krasil'nikov published a wonderful paper [15] in which he developed methods of D. E. Cohen (1967) (see also Susan McKay [17] and Bryant & Newman [4]) to show that all $\mathfrak{N}\mathfrak{A}$-varieties are finitely based. From Philip Hall's theorems and proofs about infinite soluble groups (see, in particular, [9] and [10]) it is easy to acquire a feeling that $\mathfrak{A}\mathfrak{N}$-groups should be far more tractable than $\mathfrak{N}\mathfrak{A}$-groups: there are only $\aleph_0$ finitely generated $\mathfrak{A}\mathfrak{N}$-groups up to isomorphism whereas there are $2^{\aleph_0}$ non-isomorphic finitely generated $\mathfrak{N}\mathfrak{A}$-groups; finitely generated $\mathfrak{A}\mathfrak{N}$-groups satisfy MAX-N (maximal condition on normal subgroups), whereas plenty of finitely generated $\mathfrak{N}\mathfrak{A}$-groups do not; finitely generated $\mathfrak{A}\mathfrak{N}$-groups are residually finite, whereas plenty of finitely generated $\mathfrak{N}\mathfrak{A}$-groups are not; finitely generated $\mathfrak{A}\mathfrak{N}$-groups enjoy the Hopf property (for a reminder see §5), whereas plenty of finitely generated $\mathfrak{N}\mathfrak{A}$-groups do not. In light of Krasil'nikov's theorem it is therefore hard to bring oneself to believe that Problem 4 has a negative answer. Nevertheless, it has remained open for a very long time. Once it is solved it will complete a significant part of the picture because the examples created by Michael Vaughan-Lee in 1969 show that $\mathfrak{N}_2\mathfrak{N}_2$ has $2^{\aleph_0}$ subvarieties and therefore most of them are not finitely-based.

Problem 5. Let H be a subgroup of finite index in a group G. Is it true that if the laws of G are finitely based then the laws of H are finitely based?

Problem 6. Let H be a subgroup of finite index in a group G. Is it true that if the laws of H are finitely based then the laws of G are finitely based?

Merged into one 'if and only if' question, these too were advertised already in 1993 (see [22]) but had been around for a long time before then. The same goes for our next three problems, which are just special cases of Problem 6. They are, however, important enough to warrant individual mention and they may be considerably more tractable than the general problem.

Problem 7. Is it true that the laws of an abelian-by-finite group are finitely based?

Problem 8. Is it true that the laws of a nilpotent-by-finite group are finitely based?

Problem 9. Is it true that the laws of a polycyclic group are finitely based?

It is a consequence of Krasil'nikov's theorem that this last problem is a special case of Problem 6 because, as is well known, polycyclic groups are nilpotent-by-abelian-by-finite.

4   Small varieties

The variety $\mathfrak{V}$ will be said to be *small* if it contains only countably many finitely generated groups up to isomorphism. I owe to Professor Gabriel Sabbagh the observation that by an old theorem of Morley (see [18]) if $\mathfrak{V} \neq \mathfrak{E}$ then up to isomorphism the number of its finitely generated groups is one of $\aleph_0$, $\aleph_1$, $2^{\aleph_0}$. According to Sabbagh, Anatole Khelif has proved that the only possibilities are $\aleph_0$ and $2^{\aleph_0}$ [unpublished]. Thus $\mathfrak{V}$ is small if and only if the number of finitely generated groups (up to isomorphism of course) that it contains is strictly smaller than $2^{\aleph_0}$.

Clearly, every locally finite variety is small (these are the varieties in which every finitely generated group is finite). In particular, since there are $2^{\aleph_0}$ locally finite varieties, there are $2^{\aleph_0}$ small varieties. Further, if $\mathfrak{U}$ is small and $\mathfrak{V}$ is locally finite then the product variety $\mathfrak{U}\mathfrak{V}$ is small. By Philip Hall's 1954 theorem cited above every $\mathfrak{A}\mathfrak{N}$-variety is small. Note that if all finitely generated groups in $\mathfrak{V}$ satisfy MAX-N (the maximal condition on normal subgroups), which is what Hall proved for abelian-by-nilpotent groups, then the variety $\mathfrak{V}$ is small. It follows that every variety of locally abelian-by-nilpotent-by-finite groups is small.

An easy example of a variety that is not small is $\mathfrak{Z}\mathfrak{A}^2$, the centre-by-metabelian variety.

Problem 10. Find all small varieties. At least, find all small soluble varieties.

Problem 11. Classify the small varieties that are finitely based.

These are perhaps rather ambitious, but it seems just possible that a finitely based small variety has to be locally abelian-by-nilpotent-by finite.

Problem 12. Let $\mathfrak{V}$ be a small variety. Must all its finitely generated groups satisfy MAX-N?

This also looks quite ambitious, but there are special cases that look more tractable, in particular subvarieties of products of locally soluble and locally finite varieties. Since the subvariety generated by a group in a small variety is itself small, this special case may be neatly formulated (and slightly generalised) using the following terminology, which Olga Macedonska taught me. A group is said to be *locally graded* if every non-trivial finitely generated subgroup has a proper subgroup of finite index. Let's call a variety locally graded if all the groups in it are locally graded.

Problem 13. Let $\mathfrak{V}$ be a small locally graded variety. Must all its finitely generated groups satisfy MAX-N?

It is perhaps worth digressing away from groups briefly at this point. The problems in this section, except perhaps the last, have natural analogues for varieties of other algebraic systems. In particular, Gabriel Sabbagh asked (Logic Seminar, Paris, 16 March 2009) what could be said about Problem 12 if MAX-N was construed as being the maximal condition on congruences. Here is an example of a small variety of algebraic systems, rather different from groups, in which there are finitely generated algebras for which MAX-N fails.

Consider algebraic systems with four unary operators α, α^*, β, β^*. Let $\mathfrak{V}$ be the variety of such algebras defined by the identical relations

$$x\alpha\alpha^* = x, \quad x\alpha^*\alpha = x, \quad x\beta\beta^* = x, \quad x\beta^*\beta = x, \quad x\beta^*\alpha\beta = x\alpha\alpha.$$

These algebras may be construed as G-spaces where G is the group defined by

$$G := \langle a, b \mid b^{-1}ab = a^2 \rangle.$$

This group is well known to be metabelian. It is the split extension of the additive group A of dyadic rational numbers, which I'll write as $2^{-\infty}\mathbb{Z}$, by the infinite cyclic group B generated by b. It may be identified with the group of affine transformations of $2^{-\infty}\mathbb{Z}$ of the form $x \mapsto 2^\mu x + \nu$, where $\mu \in \mathbb{Z}$ and $\nu \in 2^{-\infty}\mathbb{Z}$. In this identification a becomes the map $x \mapsto x + 1$ and b the map $x \mapsto 2x$.

Let F_1 be the right-regular representation space of G. That is, the underlying set of F_1 is G itself, and it becomes an algebra in $\mathfrak{V}$ when one specifies that $x\alpha := xa$, $x\alpha^* := xa^{-1}$, $x\beta := xb$, and $x\beta^* := xb^{-1}$ for all $x \in F_1$. In fact F_1 is the free algebra of rank 1 in $\mathfrak{V}$. Congruences on F_1 are in one-to-one correspondence with subgroups of G, a congruence ρ corresponding with the subgroup H_ρ defined by

$$H_\rho := \{h \in G \mid 1 \equiv 1h \pmod{\rho}\},$$

and a subgroup H corresponding with the equivalence relation ρ_H given by

$$\rho_H := \{(x, hx) \in F_1 \times F_1 \mid h \in H\}.$$

Let us temporarily call a group 'small' if it has at most countably many subgroups. It is easy to see that an extension of a small group by a small group is small, that the infinite cyclic group Z is small, and that the Prüfer group C_{2^∞} is small. Since A is an extension of Z by C_{2^∞} it is small, and then, since G is an extension of A by Z, it too is small. Thus F_1 has only countably many congruences. It follows that, up to isomorphism there are only countably many 1-generator algebras in $\mathfrak{V}$. Any finitely generated algebra is the disjoint union of finitely many 1-generator algebras (orbits of the G-space) and therefore there are only countably many finitely generated algebras in $\mathfrak{V}$ up to isomorphism. Thus $\mathfrak{V}$ is a small variety.

On the other hand, there is a properly ascending chain

$$\langle a \rangle < \langle bab^{-1} \rangle < \langle b^2 ab^{-2} \rangle < \cdots < \langle b^k ab^{-k} \rangle < \cdots$$

of subgroups of G, and this translates into a properly ascending chain of congruences on F_1. Thus F_1 does not satisfy MAX-N.

5   Varieties and the Hopf property

This final section is intended as a reminder. Recall that the group G is said to have the Hopf property (or to be a Hopf group) if every surjection $G \to G$ is injective (that is, an automorphism).

Problem 14. Let $\mathfrak{V}$ be a small variety and let G be one of its finitely generated groups. Must G have the Hopf property? Is this true at least if $\mathfrak{V}$ is soluble?

Problem 15. Conversely, if all finitely generated $G \in \mathfrak{V}$ have the Hopf property, must it be true that the variety $\mathfrak{V}$ is small? Is this true at least if $\mathfrak{V}$ is soluble?

Again there is a context. On the one hand $\mathfrak{AN}$-varieties are small and contain no non-Hopf groups; on the other hand the centre-by-metabelian variety $\mathfrak{Z}\mathfrak{A}^2$ is not small, and it contains non-Hopf groups. Moreover, in a special case there is a direct connection: if the relatively free group $F_n(\mathfrak{V})$ is soluble and is not a Hopf group then it has $2^{\aleph_0}$ distinct normal subgroups, and therefore $\mathfrak{V}$ is not small.

Problem 16. Does there exist a soluble variety in which one of the finitely generated free groups does not have the Hopf property?

However,

$$\text{Problem 14 + Problem 15} \quad = \quad \text{Problem 16 of Hanna Neumann 1967}$$
$$= \quad \text{Conjecture on p. 610 of B. H. Neumann 1967}$$

and

$$\text{Problem 16} \quad = \quad \text{Problem 15 of Hanna Neumann 1967}$$
$$= \quad \text{Problems 12 \& 12}' \text{ of B. H. Neumann 1967}$$

(though with the added assumption of solubility). These problems therefore count as classics and do not really belong here.

Note added in proof: Olga Macedonska has kindly drawn my attention to [11], in which Ivanov and Storozhev solve the classic problems affirmatively. Their groups and varieties are, however, very far from being soluble. The whole point of Problem 16 is that solubility is of great interest in this context, and there is a great deal that can be said about a relatively free soluble group that is non-Hopf, which is not true in the non-soluble case.

References

[1] S. I. Adjan, 'Infinite irreducible systems of group identities' (Russian) *Izv. Akad. Nauk SSSR Ser. Mat.* **34** (1970), 715–734 and *Dokl. Akad. Nauk SSSR* **190** (1970), 499–501; English translation in *Soviet Math. Dokl.* **11** (1970), 113–115.

[2] M. D. Atkinson, 'Alternating trilinear forms and groups of exponent 6', *J. Austral. Math. Soc.* **16** (1973), 111–128.

[3] Roger M. Bryant, 'Some infinitely based varieties of groups' in *Collection of articles dedicated to the memory of Hanna Neumann. I. J. Austral. Math. Soc.* **16** (1973), 29–32.

[4] R. M. Bryant and M. F. Newman, 'Some finitely based varieties of groups', *Proc. London Math. Soc.* Ser. (3) **28** (1974), 237–252.

[5] D. E. Cohen, 'On the laws of a metabelian variety', *J. Algebra* **5** (1967), 267–273.

[6] C. K. Gupta and A. N. Krasilnikov, 'Metanilpotent varieties without torsion and varieties of groups of prime power exponent', *Internat. J. Algebra Comput.* **6** (1996), 325–338.

[7] N. D. Gupta and M. F. Newman, 'The nilpotency class of finitely generated groups of exponent four' *Proceedings of the Second International Conference on the Theory of Groups (Australian Nat. Univ., Canberra, 1973)*, Lecture Notes in Maths. **372**, Springer, Berlin 1974, 330–332.

[8] Marshall Hall, Jr., 'Solution of the Burnside problem for exponent six' *Illinois J. Math.* **2** (1958), 764–786.

[9] P. Hall, 'Finiteness conditions for soluble groups', *Proc. London Math. Soc.* (Ser. 3) **4** (1954), 419–436.

[10] P. Hall, 'On the finiteness of certain soluble groups', *Proc. London Math. Soc.* (Ser. 3) **9** (1959), 595–622.

[11] S. V. Ivanov & A. M. Storozhev, 'Non-Hopfian relatively free groups', *Geom. Dedicata* **114** (2005), 209–228

[12] E. I. Khukhro & V. D. Mazurov (editors) *Unsolved problems in group theory: The Kourovka Notebook.* 17th edition: Russian Academy of Sciences Siberian Division, Institute of Mathematics, Novosibirsk, 2009.

[13] L. G. Kovács, 'On the number of varieties of groups', *J. Austral. Math. Soc.* **8** (1968), 444–446.

[14] L. G. Kovács & M. F. Newman, 'Hanna Neumann's problems on varieties of groups', in *Procs. Conf. Theory of Groups, Canberra 1973*, Lecture Notes Math. **372** (Springer, Berlin, 1974), 417–431.

[15] A. N. Krasilnikov, 'On the finiteness of the basis of identities of groups with nilpotent commutators' [Russian], *Izv. Akad. Nauk SSSR Ser. Mat.* **54** (1990) 1181–1195; English translation in *Math. USSR-Izv.* **37** (1991), 539–553.

[16] F. W. Levi & B. L. van der Waerden, 'Über eine besondere Klasse von Gruppen', *Abhandl. Math. Sem. Univ. Hamburg* **9** (1932), 154–158.

[17] Susan McKay, 'On centre-by-metabelian varieties of groups', *Proc. London Math. Soc.* Ser. (3) **24** (1972), 243–256.

[18] Michael Morley, 'The number of countable models', *J. Symbolic Logic* **35** (1970),

14–18.

[19] B. H. Neumann, 'Identical relations in groups. I', *Math. Annalen* **114** (1937), 506–525.

[20] B. H. Neumann, 'Varieties of groups', *Bull. Amer. Math. Soc.* **73** (1967), 603–613.

[21] Hanna Neumann, *Varieties of groups*, Springer-Verlag 1967.

[22] Peter M. Neumann, Review of *Topics in Varieties of Group Representations* by Samuel M. Vovsi, *Bull. Amer. Math. Soc.* **29** (1993), 120–123.

[23] Sheila Oates and M. B. Powell, 'Identical relations in finite groups', *J. Algebra* **1** (1964), 11–39.

[24] A. Ju. Ol'šanskiĭ, 'The finite basis problem for identities in groups' (Russian) *Izv. Akad. Nauk SSSR Ser. Mat.* **34** (1970), 376–384.

[25] Martyn Quick, 'The module structure of some lower central factors of Burnside groups of exponent four', *Quart. J. Math. Oxford* **47** (1996), 469–492.

[26] Martyn Quick, 'A classification of some insoluble varieties of groups of exponent four', *J. Algebra* **197** (1997), 342–371.

[27] Martyn Quick, 'Varieties of groups of exponent 4', *J. London Math. Soc.* **60** (1999), 747–756.

[28] Ju. P. Razmyslov, 'The Hall–Higman problem' (Russian), *Izv. Akad. Nauk SSSR Ser. Mat.* **42** (1978), 833–847; English translation in *Math. USSR-Izv.* **13** (1979), 133–146.

[29] I. N. Sanov, 'Solution of Burnside's problem for exponent 4' (Russian), *Leningrad State Univ. Annals [Uchenye Zapiski] Math. Ser.* **10** (1940), 166–170.

[30] S. J. Tobin, *On groups with exponent* 4, Ph.D. Thesis, Victoria University of Manchester (1954).

[31] Seán Tobin, 'On a theorem of Baer and Higman', *Canad. J. Math.* **8** (1956), 263–270.

[32] M. R. Vaughan-Lee, 'Uncountably many varieties of groups', *Bull. London Math. Soc.* **2** (1970), 280–286.

[33] Michael Vaughan-Lee, *The Restricted Burnside Problem*, OUP Oxford 1990 (second edition 1993).

[34] C. R. B. Wright, 'On the nilpotency class of a group of exponent four', *Pacific J. Math.* **11** (1961), 387–394.

GENERALIZATIONS OF THE SYLOW THEOREM

DANILA O. REVIN and EVGENY P. VDOVIN

Sobolev Insitute of Mathematics, Siberian Branch of the Russian Academy of Sciences, Acad. Koptyug avenue, 4 and Novosibirsk State University, Pirogova street, 2; 630090 Novosibirsk, Russia
Email: revin@math.nsc.ru, vdovin@math.nsc.ru

Abstract

Let π be a set of primes. Generalizing the properties of Sylow p-subgroups, P. Hall introduced classes E_π, C_π, and D_π of finite groups possessing a π-Hall subgroup, possessing exactly one class of conjugate π-Hall subgroups, and possessing one class of conjugate maximal π-subgroups respectively. In this paper we discuss a description of these classes in terms of a composition and a chief series of a finite group G.

1 Introduction

In 1872, the Norwegian mathematician L. Sylow proved the following outstanding theorem.

Theorem 1.1 (L. Sylow [76]) *Let G be a finite group and p a prime. Assume $|G| = p^\alpha m$ and $(p, m) = 1$. Then the following statements hold:*
  (E) *G possesses a subgroup of order p^α (the, so-called, Sylow p-subgroup);*
  (C) *every two Sylow p-subgroups of G are conjugate;*
  (D) *every p-subgroup of G is included in a Sylow p-subgroup.*

A natural generalization of the concept of Sylow p-subgroups is the notion of π-Hall subgroups. We recall the definitions. Let G be a finite group and π be a set of primes. We denote by π' the set of all primes not in π, by $\pi(n)$ the set of all prime divisors of a positive integer n and for a finite group G we denote $\pi(|G|)$ by $\pi(G)$. A positive integer n with $\pi(n) \subseteq \pi$ is called a π-*number* and a group G with $\pi(G) \subseteq \pi$ is called a π-group. Given a positive integer n, denote by n_π the maximal divisor t of n with $\pi(t) \subseteq \pi$. A subgroup H of G is called a π-*Hall subgroup* if $\pi(H) \subseteq \pi$ and $\pi(|G : H|) \subseteq \pi'$. The set of all π-Hall subgroups of G is denoted by $\mathrm{Hall}_\pi(G)$. If $\pi \cap \pi(G) = \{p\}$, then a π-Hall subgroup H of G is a Sylow p-subgroup of G and we denote the set of all Sylow p-subgroups of G by $\mathrm{Syl}_p(G)$.

By analogy with the statements of Sylow's theorem, P. Hall introduced the following notation for finite groups.

The work is supported by RFBR, grant N 08-01-00322, RF President grant for scientific schools NSc-344.2008.1, and ADTP "Development of the Scientific Potential of Higher School" of the Russian Federal Agency for Education (Grant 2.1.1.419). The second author gratefully acknowledges the support from Deligne 2004 Balzan prize in mathematics.

Definition 1.2 (P. Hall [36]) We say that a finite group G *satisfies*
E_π if G possesses a π-Hall subgroup (i.e., $\mathrm{Hall}_\pi(G) \neq \varnothing$);
C_π if G satisfies E_π and every two π-Hall subgroups of G are conjugate;
D_π if G satisfies C_π and every π-subgroup of G is included in a π-Hall subgroup.

A group G satisfying E_π (resp. C_π, D_π) is said to be an E_π- (resp. a C_π-, a D_π-) *group*. Given a set of primes π, we denote also by E_π, C_π, and D_π the classes of all finite E_π-, C_π-, and D_π-groups respectively. Thus $G \in D_\pi$ if the complete analogue of Sylow's theorem for π-Hall subgroups of G holds, while $G \in C_\pi$ and $G \in E_\pi$ if weaker analogues of Sylow's theorem hold. In contrast with Sylow's theorem, there exists a set of primes π and a finite group G with $\mathrm{Hall}_\pi(G) = \varnothing$. There are also examples showing the inequalities $E_\pi \neq C_\pi$ and $C_\pi \neq D_\pi$ for suitable sets π of primes (see Examples 1.4–1.6).

In 1928 P. Hall proved:

Theorem 1.3 (P. Hall [37]) *If a finite group G is solvable then $G \in D_\pi$.*

Example 1.4 Alt_5 does not possess a subgroup of order 15, hence $\mathrm{Alt}_5 \notin E_{\{3,5\}}$.

Example 1.5 There are two classes of conjugate $\{2,3\}$-Hall subgroups of $\mathrm{GL}_3(2)$, the stabilizers of lines and planes respectively. So $\mathrm{GL}_3(2) \in E_{\{2,3\}} \setminus C_{\{2,3\}}$.

Example 1.6 Every subgroup of order 12 of Alt_5 is a point stabilizer, and all point stabilizers are conjugate, since Alt_5 is transitive. On the other hand, Alt_5 includes a $\{2,3\}$-subgroup $\langle(123),(12)(45)\rangle \simeq \mathrm{Sym}_3$, acting without stable points. Therefore $\mathrm{Alt}_5 \in C_{\{2,3\}} \setminus D_{\{2,3\}}$.

Later P. Hall [38] and, independently, S. A. Chunikhin [9], proved the reverse statement for Theorem 1.3. Their results can be summarized in:

Theorem 1.7 (P. Hall [38], S. A. Chunikhin [9]) *Let G be a finite group. The following statements are equivalent:*
(1) *G is solvable;*
(2) *$G \in D_\pi$ for every set π of primes;*
(3) *$G \in E_{p'}$ for every prime p.*

Thus $G \in E_\pi$ (resp. $G \in C_\pi$, $G \in D_\pi$) for every set of primes π if and only if G is solvable. If we fix π, then the classes E_π, C_π, D_π can be wider than the class of solvable finite groups. It is clear, for example, that each π- or π'-group satisfies D_π.

In the present article we consider the following:

Problem 1.8 Given a set π of primes and a finite group G, when does G satisfy E_π, C_π or D_π?

We show that the answer to this problem can be obtained in terms of composition and chief series of G.

2 Notation

Our notation for finite groups agrees with that of [7]. For groups A and B symbols $A \times B$ and $A \circ B$ denote direct and central products respectively. By $A : B$, $A \cdot B$, and $A . B$ we denote a split, a nonsplit, and an arbitrary extension of a group A by a group B. For a group G and a subgroup S of Sym_n by $G \wr S$ we always denote the permutation wreath product. We write $m \leqslant n$ if a real number m is not greater than n, while we use the notation $H \leq G$, $H \trianglelefteq G$, and $H \trianglelefteq\trianglelefteq G$ instead of "H is a subgroup of G", "H is a normal subgroup of G", and "H is a subnormal subgroup of G" respectively. For $M \subseteq G$ we set $M^G = \{M^g \mid g \in G\}$. For a group G we denote by $\mathrm{Aut}(G)$, $\mathrm{Inn}(G)$, and $\mathrm{Out}(G)$ the groups of all, of inner, and of outer automorphisms of G respectively.

Instead of $n \equiv m \pmod{k}$ we write $n \equiv_k m$ for brevity. Let r be an odd prime and q be an integer coprime to r. We denote by $e(q, r)$ the least number e such that $q^e \equiv_r 1$.

A group G acts by conjugation on the set $\mathrm{Hall}_\pi(G)$. We denote by $k_\pi(G)$ the number of orbits of this action. Given a π-Hall subgroup M of G denote by $k_M(G)$ the number of classes of conjugate subgroups of G isomorphic to M. Clearly, if either $|\pi \cap \pi(G)| \leqslant 1$, or $\pi(G) \subseteq \pi$, then $G \in D_\pi$. In this case a π-Hall subgroup H of G is called *standard*. Otherwise a π-Hall subgroup H of G is called *nonstandard*. If H is a standard π-Hall subgroup of G, then H is a trivial subgroup if $\pi \cap \pi(G) = \varnothing$, H is a Sylow p-subgroup if $\pi \cap \pi(G) = \{p\}$, and $H = G$ if $\pi(G) \subseteq \pi$. So, to solve Problem 1.8, we need to consider nonstandard π-Hall subgroups only.

For linear algebraic groups our notation agrees with that of [42]. For finite groups of Lie type we use notation from [5]. If $\overline{G}$ is a simple connected linear algebraic group over the algebraic closure $\overline{\mathbb{F}}_p$ of a finite field $\mathbb{F}_q$ of order q and characteristic p, then a surjective endomorphism $\sigma : \overline{G} \to \overline{G}$ is called a *Frobenius map* if the set of σ-stable points $\overline{G}_\sigma$ is finite. Every group G such that $O^{p'}(\overline{G}_\sigma) \leq G \leq \overline{G}_\sigma$ is called a *finite group of Lie type*. For all groups of Lie type $\mathbb{F}_q$ is called a *base field*. We also write $A_n^\eta(q)$, $D_n^\eta(q)$, and $E_n^\eta(q)$, where $\eta \in \{+, -\}$ and $A_n^+(q) = A_n(q)$, $A_n^-(q) = {}^2A_n(q)$, $D_n^+(q) = D_n(q)$, $D_n^-(q) = {}^2D_n(q)$, $E_n^+(q) = E_n(q)$, $E_n^-(q) = {}^2E_n(q)$.

Let $\overline{R}$ be a connected σ-stable subgroup of $\overline{G}$. Then we may consider $R = G \cap \overline{R}$ and $N(G, R) = G \cap N_{\overline{G}}(\overline{R})$. Note that $N(G, R) \neq N_G(R)$ in general, and $N(G, R)$ is called the *algebraic normalizer* of R. A group R is called a *maximal torus* if $\overline{R}$ is a maximal torus of $\overline{G}$.

Our notation for classical groups agrees with [44]. In order to unify statements and arguments we use the following notation $\mathrm{GL}_n^+(q) = \mathrm{GL}_n(q)$, $\mathrm{GL}_n^-(q) = \mathrm{GU}_n(q)$, $\mathrm{SL}_n^+(q) = \mathrm{SL}_n(q)$, $\mathrm{SL}_n^-(q) = \mathrm{SU}_n(q)$.

In this paper, if we consider groups of Lie type with the base field $\mathbb{F}_q$ of odd characteristic p and order $q = p^\alpha$, then we can choose $\varepsilon(q) \in \{+1, -1\}$ (usually we write just ε since q is defined by the group G) so that $q \equiv \varepsilon(q) \pmod{4}$, i.e., $\varepsilon(q) = (-1)^{(q-1)/2}$. Following [44], by $\mathrm{O}_n^\eta(q)$ we denote the general orthogonal group of degree n and of sign $\eta \in \{\circ, +, -\}$ over $\mathbb{F}_q$, while the symbol $\mathrm{GO}_n^\eta(q)$ denotes the group of similarities. Here $\circ$ is an empty symbol, and we use it if only if n is odd.

By η we always mean an element from the set $\{\circ, +, -\}$, and if $\eta \in \{+, -\}$, then we use η instead of $\eta 1$ as well. In classical groups the symbol P will also denote reduction modulo scalars. Thus for every subgroup H of $\mathrm{GL}_n(q)$ the image of H in $\mathrm{PGL}_n(q)$ is denoted by $\mathrm{P}H$.

3   Hall subgroups of finite simple groups

We start with a simple statement.

Proposition 3.1 ([36, Lemma 1]) *Let G be a finite group and A be a normal subgroup of G. If $G \in E_\pi$ and $H \in \mathrm{Hall}_\pi(G)$ then $A, G/A \in E_\pi$ and $H \cap A \in \mathrm{Hall}_\pi(A)$, $HA/A \in \mathrm{Hall}_\pi(G/A)$.*

Proposition 3.1 implies that if $G \in E_\pi$, then all chief and composition factors of G satisfy E_π. Furthermore, a π-Hall subgroup of G can be "constructed" from π-Hall subgroups in composition factors.[1] Thus the following problem is closely connected to Problem 1.8.

Problem 3.2 Find π-Hall subgroups in finite simple groups.

This problem has been studied by many mathematicians (see [46, 1, 2, 3, 6, 27, 28, 29, 30, 31, 32, 26, 74, 8, 36, 78, 48, 51, 52], for example). Already P. Hall realized the importance of Problem 3.2. In his paper [36] written in 1956 he found solvable π-Hall subgroups in symmetric groups. In 1966 J. Thompson found all nonsolvable π-Hall subgroups (see [78]). Problem 3.2 for groups of Lie type is mentioned in the survey paper [48]. In [46] L. S. Kazarin described r'-Hall subgroups in finite simple groups for all primes r. The classification of π-Hall subgroups in groups of Lie type with $p \in \pi$ was obtained by F. Gross (in case $2 \notin \pi$, [28]) and by D. O. Revin (in case $2 \in \pi$, [51]). The classification of π-Hall subgroups in sporadic groups was obtained by F. Gross in [29] ($2 \notin \pi$), and by D. O. Revin in [52] ($2 \in \pi$). The classification of π-Hall subgroups in groups of Lie type with $2, p \notin \pi$ was obtained by F. Gross for classical groups [29, 30, 31], and by D. O. Revin, E. P. Vdovin for exceptional groups [82]. The classification of π-Hall subgroups with $2 \in \pi$ and $3, p \notin \pi$ was obtained by F. Gross for linear and symplectic groups [32] and by D. O. Revin, E. P. Vdovin in the remaining cases [53].

In [53] the authors finished the classification of π-Hall subgroups in finite simple groups. Unfortunately Lemma 3.14 in [53] contains a wrong statement. Due to this gap we lost several series of $\{2, 3\}$-Hall subgroups in groups of Lie type. In [58] we correct this gap.[2]

---

[1]The problem, whether a finite group with all composition factors satisfying E_π possesses a π-Hall subgroup, we consider in the next section.

[2]Notice that if $G \in E_\pi$ and H is a π-Hall subgroup from a missed class, then it is shown that $G \notin D_\pi$, so all results concerning the D_π-property, obtained in [53, 82, 56, 55, 57] remain correct.

n	$\pi \cap \pi(\mathrm{Sym}_n)$	$H \in \mathrm{Hall}_\pi(Sym_n)$
prime	$\pi((n-1)!)$	Sym_{n-1}
7	$\{2,3\}$	$\mathrm{Sym}_3 \times \mathrm{Sym}_4$
8	$\{2,3\}$	$\mathrm{Sym}_4 \wr \mathrm{Sym}_2$

Table 1. π-Hall subgroups in symmetric groups

Symmetric, alternating, sporadic groups

Theorem 3.3 ([36, Theorem A4 and the note after it], [78], [53, Theorem 4.3 and Corollary 4.4]) *Let π be a set of primes.*

(1) *All possibilities for* Sym_n *to contain a nonstandard π-Hall subgroup are listed in Table 1.*

(2) *The following statements are equivalent:*
   (a) $\mathrm{Sym}_n \in C_\pi$;
   (b) $\mathrm{Sym}_n \in E_\pi$;
   (c) $\mathrm{Alt}_n \in E_\pi$;
   (d) $\mathrm{Alt}_n \in C_\pi$.

(3) $M \in \mathrm{Hall}_\pi(\mathrm{Alt}_n)$ *if and only if there exists* $M_0 \in \mathrm{Hall}_\pi(\mathrm{Sym}_n)$ *such that* $M = M_0 \cap \mathrm{Alt}_n$.

Each π-Hall subgroup of odd order of a sporadic group G has order rs, where $\pi \cap \pi(G) = \{r,s\}$, except the cases $G = J_4$ and $\pi = \{5,11\}$, and $G = O'N$ and $\pi = \{3,5\}$. In the first exceptional case $H \simeq 11_+^{1+2} : 5$, while in the latter $H \simeq 3^4 : 5$.

Theorem 3.4 ([29, Corollary 6.13], [52, Theorem 4.1]) *Let G be either one of the 26 sporadic simple groups or the Tits group. Then G possesses a nonstandard π-Hall subgroup H if and only if one of possibilities for G and π, listed in Tables 2 and 3, occurs.*

Groups of Lie type of characteristic $p \in \pi$

Consider a finite group of Lie type G over a field of characteristic p, and suppose that π is a set of primes with $p \in \pi$. If the index of a Borel subgroup in G is a π'-number, then $G \in E_\pi$, since a Borel subgroup is solvable. If G possesses a parabolic π-subgroup P such that $|G : P|$ is a π'-number, then P is a π-Hall subgroup of G, in particular, $G \in E_\pi$. The converse statement turns out to be true.

Theorem 3.5 ([51, Theorem 3.3]) *Let G be a finite group of Lie type over a field of characteristic $p \in \pi$. If H is a π-Hall subgroup of G, then either H is included in a Borel subgroup of G, or H is equal to a parabolic subgroup of G.*

In view of Theorem 3.5 we need to answer to the following two questions: whether $\pi(|G : B|) \subseteq \pi'$, and whether G possesses a parabolic π-subgroup P

G	$\pi \cap \pi(G)$	G	$\pi \cap \pi(G)$	G	$\pi \cap \pi(G)$
M_{11}	$\{5, 11\}$	M_{12}	$\{5, 11\}$	M_{22}	$\{5, 11\}$
Ru	$\{7, 29\}$	M_{24}	$\{5, 11\}$ $\{11, 23\}$	M_{23}	$\{5, 11\}$ $\{11, 23\}$
Fi_{23}	$\{11, 23\}$	Fi_{24}'	$\{11, 23\}$	Ly	$\{11, 67\}$
J_1	$\{3, 5\}$ $\{3, 7\}$ $\{3, 19\}$ $\{5, 11\}$	J_4	$\{5, 7\}$ $\{5, 11\}$ $\{5, 31\}$ $\{7, 29\}$ $\{7, 43\}$	$O'N$	$\{3, 5\}$ $\{5, 11\}$ $\{5, 31\}$
Co_1	$\{11, 23\}$	Co_2	$\{11, 23\}$	Co_3	$\{11, 23\}$
B	$\{11, 23\}$ $\{23, 47\}$	M	$\{23, 47\}$ $\{29, 59\}$		

Table 2. Sporadic E_π-groups, $2 \notin \pi$

G	π	Structure of H
M_{11}	$\{2, 3\}$ $\{2, 3, 5\}$	$3^2 : Q_8 . 2$ $\mathrm{Alt}_6 . 2$
M_{22}	$\{2, 3, 5\}$	$2^4 : \mathrm{Alt}_6$
M_{23}	$\{2, 3\}$ $\{2, 3, 5\}$ $\{2, 3, 5\}$ $\{2, 3, 5, 7\}$ $\{2, 3, 5, 7\}$ $\{2, 3, 5, 7, 11\}$	$2^4 : (3 \times A_4) : 2$ $2^4 : \mathrm{Alt}_6$ $2^4 : (3 \times \mathrm{Alt}_5) : 2$ $L_3(4) : 2_2$ $2^4 : \mathrm{Alt}_7$ M_{22}
M_{24}	$\{2, 3, 5\}$	$2^6 : 3 \cdot \mathrm{Sym}_6$
J_1	$\{2, 3\}$ $\{2, 7\}$ $\{2, 3, 5\}$ $\{2, 3, 7\}$	$2 \times \mathrm{Alt}_4$ $2^3 : 7$ $2 \times A_5$ $2^3 : 7 : 3$
J_4	$\{2, 3, 5\}$	$2^{11} : (2^6 : 3 \cdot \mathrm{Sym}_6)$

Table 3. π-Hall subgroups of the sporadic groups, $2 \in \pi$

G	$\pi(W)$
$A_{n-1}(q)$, $B_n(q)$, $C_n(q)$, $D_n(q)$	$\pi(n!)$
$^2A_{n-1}(q)$	$\pi([n/2]!)$
$^2D_n(q)$	$\pi((n-1)!)$
$E_7(q)$, $E_8(q)$	$\{2,3,5,7\}$
$E_6(q)$	$\{2,3,5\}$
$^2E_6(q)$, $F_4(q)$, $G_2(q)$, $^3D_4(q)$	$\{2,3\}$
$^2B_2(q)$, $^2G_2(q)$, $^2F_4(q)$	$\{2\}$

Table 4. Prime divisors of orders of Weyl groups for finite groups of Lie type

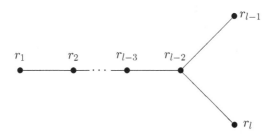

Figure 1. Dynkin diagram for the root system of type D_l.

with $\pi(|G : P|) \subseteq \pi'$. The answer to the first question is given in Theorem 3.6. The answer to the second question is given in Theorems 3.7, 3.8, and 3.9.

Theorem 3.6 ([29, Theorem 3.2], [32, Theorem 3.1], [51, Lemma 4.1, Theorem 4.2]) *Let π be a set of primes. Let G be a group of Lie type with the base field $\mathbb{F}_q$ of characteristic $p \in \pi$. Denote by B a Borel subgroup of G. Then $|G : B|$ is a π'-number if and only if $\pi \cap \pi(G) \subseteq \pi(q-1) \cup \{p\}$ and $\pi \cap \pi(W) \subseteq \{p\}$, where W is a Weyl group of G. Sets $\pi(W)$ for all finite groups of Lie type are given in Table 4.*

Theorem 3.7 ([51, Theorems 6.3 and 8.3]) *Let π be a set of primes. Let $G = D_n^\eta(q)$ be a group of Lie type with the base field $\mathbb{F}_q$ of characteristic $p \in \pi$. Then G possesses a proper parabolic π-subgroup P with $\pi(|G : P|) \subseteq \pi'$ if and only if one of the following statements holds:*

(a) *$G = D_l(q)$, $p = 2$, the Dynkin diagram for the root system of G is given in Figure 1, l is a Fermat prime, $(l, q-1) = 1$, P is conjugate to a parabolic subgroup G_J corresponding to the set $J = \{r_2, r_3, \ldots, r_l\}$ of fundamental roots and $\pi \cap \pi(G) = \pi(P) = \pi(G) \setminus \pi\left((q^l - 1)(q^{l-1} + 1)/(q-1)\right);$*

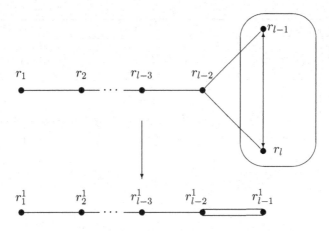

Figure 2. Dynkin diagram for the root system of type ^2D_l.

(b) $G = {}^2D_l(q)$, $p = 2$, the Dynkin diagram for the root system of G is given in Figure 2, $l - 1$ is a Mersenne prime, $(l - 1, q - 1) = 1$, P is conjugate to a parabolic subgroup G_J corresponding to the set $J = \{r_2^1, r_3^1, \ldots, r_{l-1}^1\}$ of fundamental roots and $\pi \cap \pi(G) = \pi(P) = \pi(G) \setminus \pi \left((q^{l-1} - 1)(q^l + 1)/(q - 1)\right)$.

Theorem 3.8 ([51, Theorems 10.2, 11.2, and 12.2]) *Let π be a set of primes, V an n-dimensional vector space over a field $\mathbb{F}_q$ of characteristic p, and let $G = \mathrm{SL}(V) \simeq \mathrm{SL}_n(q)$ be a special linear group. Each parabolic subgroup P of G can be obtained as a stabilizer of*

$$0 = V_0 < V_1 < \cdots < V_s = V$$

with $\dim V_i/V_{i-1} = n_i$, $i = 1, \ldots, s$. All cases, when P is a proper π-subgroup and $\pi(|G : P|) \subseteq \pi'$, are listed in Table 5.[3] Each parabolic π-Hall subgroup M of $\mathrm{PSL}_n(q) \simeq A_{n-1}(q)$ can be obtained as $P/Z(\mathrm{SL}_n(q))$, where P is a parabolic π-Hall subgroup of $\mathrm{SL}_n(q)$.

Theorem 3.9 ([51, Theorems 5.2 and 7.2]) *Let G be a finite group of Lie type not isomorphic to $A_n(q)$ or $D_n^\eta(q)$. Then G does not possess a proper parabolic subgroup P with $(|P|, |G : P|) = 1$.*

---

[3] Notice that in this table we give the set $\pi' \cap \pi(G)$, clearly $\pi \cap \pi(G) = \pi(G) \setminus (\pi' \cap \pi(G))$.

n	s	$\{n_1, \ldots, n_s\}$	$\pi' \cap \pi(G)$	Other conditions
odd prime	2	$\{1, n-1\}$	$\pi\left(\frac{q^n-1}{q-1}\right)$	$(n, q-1) = 1$
4	2	$\{2\}$	$\pi\left(\frac{(q^3-1)(q^4-1)}{(q-1)(q^2-1)}\right)$	$(6, q-1) = 1$
5	2	$\{2, 3\}$	$\pi\left(\frac{(q^4-1)(q^5-1)}{(q^2-1)(q-1)}\right)$	$(10, q-1) = 1$
5	3	$\{1, 2\}$	$\pi\left(\frac{(q^3-1)(q^4-1)(q^5-1)}{(q-1)(q^2-1)(q-1)}\right)$	$(30, q-1) = 1$
7	2	$\{3, 4\}$	$\pi\left(\frac{(q^3+1)(q^5-1)(q^7-1)}{(q+1)(q-1)(q-1)}\right)$	$(35, q-1) = 1$, $(3, q+1) = 1$
8	2	$\{4\}$	$\pi\left(\frac{(q^3+1)(q^4-1)(q^5-1)(q^7-1)}{(q+1)(q^2-1)(q-1)(q-1)}\right)$	$(70, q-1) = 1$, $(3, q+1) = 1$
11	2	$\{5, 6\}$	$\pi\left(\frac{(q^5+1)(q^7-1)(q^8-1)(q^9-1)(q^{11}-1)}{(q+1)(q-1)(q^4-1)(q^3-1)(q-1)}\right)$	$(462, q-1) = 1$, $(5, q+1) = 1$

Table 5. Parabolic π-Hall subgroups P in $\mathrm{SL}_n(q)$

Groups of Lie type of characteristic $p \notin \pi$

Theorem 3.10 ([30, 82]) *Let G be a group of Lie type with the base field $\mathbb{F}_q$ of characteristic p and π be a set of primes with $2, p \notin \pi$ and $|\pi \cap \pi(G)| \geqslant 2$. Set $r = \min \pi \cap \pi(G)$, $a = e(q, r)$, $\tau = \pi \cap \pi(G) \setminus \{r\}$, and $s \in \tau$. Then $G \in E_\pi$ if and only if one of the possibilities in Tables 6 and 7 occurs.*

Theorem 3.11 ([53, Lemma 5.1 and Theorem 5.2]) *Let G be a group of Lie type over a field $\mathbb{F}_q$ of characteristic p and π be a set of primes with $3, p \notin \pi$ and $2 \in \pi$. Set $\tau = \pi \cap \pi(G) \setminus \{2\}$, $\varepsilon = \varepsilon(q)$, $s \in \tau$. Then $G \in E_\pi$ if and only if one of the possibilities in Table 8 occurs. Moreover, with the exception of the case $G = {}^2G_2(q)$, $q = 3^{2n+1}$, $n \not\equiv_7 3$, and $\tau = \{7\}$, every π-Hall subgroup H of G has a normal τ-Hall subgroup, which is included in a maximal torus T such that $N(G, T)$ includes a Sylow 2-subgroup of G and H is a π-Hall subgroup of $N(G, T)$. All π-Hall subgroups of this type are conjugate in G. If $G = {}^2G_2(q)$, $q = 3^{2n+1}$, $n \not\equiv_7 3$, and $\tau = \{7\}$ there exist exactly two classes of conjugate π-Hall subgroups of G: a Frobenius groups of the order 56 and a π-Hall subgroup of $N(G, T)$, where T is a maximal torus of order $q + 1$.*

G	r	$a = e(q,r)$	$e(q,s)$	Other conditions
$A_{n-1}(q)$	—	—	a	$n < as$
	—	$r-1$	1	$(q^{r-1}-1)_r = r$, $[\frac{n}{r-1}] = [\frac{n}{r}]$, $n < s$
	—	$r-1$	r	$(q^{r-1}-1)_r = r$, $[\frac{n}{r-1}] = [\frac{n}{r}]$
	—	$r-1$	r	$(q^{r-1}-1)_r = r$, $[\frac{n}{r-1}] = [\frac{n}{r}]+1$, $n \equiv_r -1$
$^2A_{n-1}(q)$	—	$a \equiv_4 0$	a	$n < as$
	—	$a \equiv_4 2$	a	$2n < as$
	—	$a \equiv_2 1$	a	$n < 2as$
	$r \equiv_4 1$	$r-1$	2	$(q^{r-1}-1)_r = r$, $[\frac{n}{r-1}] = [\frac{n}{r}]$, $n < 2s$
	$r \equiv_4 3$	$\frac{r-1}{2}$	2	$(q^{r-1}-1)_r = r$, $[\frac{n}{r-1}] = [\frac{n}{r}]$, $n < 2s$
	$r \equiv_4 1$	$r-1$	$2r$	$(q^{r-1}-1)_r = r$, $[\frac{n}{r-1}] = [\frac{n}{r}]$
	$r \equiv_4 3$	$\frac{r-1}{2}$	$2r$	$(q^{r-1}-1)_r = r$, $[\frac{n}{r-1}] = [\frac{n}{r}]$
	$r \equiv_4 1$	$r-1$	$2r$	$(q^{r-1}-1)_r = r$, $[\frac{n}{r-1}] = [\frac{n}{r}]+1$, $n \equiv_r -1$
	$r \equiv_4 3$	$\frac{r-1}{2}$	$2r$	$(q^{r-1}-1)_r = r$, $[\frac{n}{r-1}] = [\frac{n}{r}]+1$, $n \equiv_r -1$
$B_n(q), C_n(q)$	—	$a \equiv_2 0$	a	$2n < as$
	—	$a \equiv_2 1$	a	$n < as$
$D_n(q)$	—	$a \equiv_2 1$	a	$n < as$
	—	$a \equiv_2 0$	a	$2n \leqslant as$
$^2D_n(q)$	—	$a \equiv_2 0$	a	$2n < as$
	—	$a \equiv_2 1$	a	$n \leqslant as$
	—	$a \equiv_2 1$	a or $2a$	$n = 2a$
	—	$a \equiv_4 2$	$\frac{a}{2}$ or a	$n = a$
$^3D_4(q)$	—	—	a	—
$E_6^\eta(q)$	—	—	a	$(q-\eta)_\pi \not\equiv_{15} 0$
$E_7(q)$, $E_8(q)$	—	—	a a	if $a = 1$ (resp. $a = 2$) then $(q-1)_\pi$ (resp. $(q+1)_\pi$) is not divisible by 15, 21, and 35
$F_4(q), G_2(q)$	—	—	a	—

Table 6. E_π-groups of Lie type, $2, p \notin \pi$

G	$\pi \cap \pi(G) \subseteq$
$^2B_2(2^{2m+1})$	$\pi(2^{2m+1} - 1)$
	$\pi(2^{2m+1} \pm 2^{m+1} + 1)$
$^2G_2(3^{2m+1})$	$\pi(3^{2m+1} - 1)$
	$\pi(3^{2m+1} \pm 3^{m+1} + 1)$
$^2F_4(2^{2m+1})$	$\pi(2^{2(2m+1)} \pm 1)$
	$\pi(2^{2m+1} \pm 2^{m+1} + 1)$
	$\pi(2^{2(2m+1)} \pm 2^{3m+2} \mp 2^{m+1} - 1),$
	$\pi(2^{2(2m+1)} \pm 2^{3m+2} + 2^{2m+1} \pm 2^{m+1} - 1)$

Table 7. Suzuki and Ree E_π-groups, $2, p \notin \pi$

G	$\tau \subseteq$	Other conditions
$A^\eta_{n-1}(q), B_n(q)$	$\pi(q - \varepsilon)$	$n < s$
$C_n(q)$, or $D^\eta_n(q)$		
$A^{-\varepsilon}_{n-1}(q)$	$\pi(q - \varepsilon)$	$(n+1)/2 < s$
$D^{-\varepsilon}_n(q)$	$\pi(q - \varepsilon)$	n is odd and $n - 1 < s$
$^3D_4(q), G_2(q),$	$\pi(q - \varepsilon)$	—
$F_4(q)$, or $E^{-\varepsilon}_6(q)$		
$E^\varepsilon_6(q)$	$\pi(q - \varepsilon)$	$5 \notin \tau$
$E_7(q), E_8(q)$	$\pi(q - \varepsilon)$	$5, 7 \notin \tau$
$^2G_2(q)$	$\pi(q - \varepsilon)$	—
	$\{7\}$	—

Table 8. E_π-groups of Lie type, $2 \in \pi$, $3, p \notin \pi$

Lemma 3.12 ([58]) *Assume that $G \simeq \mathrm{SL}_2(q) \simeq \mathrm{SL}^\eta_2(q) \simeq \mathrm{Sp}_2(q)$, where q is a power of an odd prime p, and $\varepsilon = \varepsilon(q)$. Then[4] $PG \in E_\pi$ if and only if one of the cases from Table 9 occurs. Moreover, if G satisfies E_π, then G contains one, two or three classes of conjugate π-Hall subgroups, i.e., $k_\pi(G) = k_\pi(PG) \in \{1, 2, 3\}$.*

In the last column of Table 9 we give the action of $\mathrm{PGL}_2(q)$ on the set of subgroups of $\mathrm{PSL}_2(q)$ isomorphic to M.

Corollary 3.13 ([58]) *Suppose $G = \mathrm{GL}^\eta_2(q)$, $PG = G/Z(G) = \mathrm{PGL}^\eta_2(q)$, where q is a power of a prime, and $\varepsilon = \varepsilon(q)$. Let π be a set of primes such that $2, 3 \in \pi$ and $p \notin \pi$. A subgroup M of G is a π-Hall subgroup if and only if $M \cap \mathrm{SL}^\eta_2(q)$ is a π-Hall subgroup of $\mathrm{SL}^\eta_2(q)$ and the set $M^{\mathrm{SL}^\eta_2(q)}$ is $\mathrm{GL}^\eta_2(q)$-invariant. More precisely, one of the following statements holds.*

(1) *$\pi \cap \pi(G) \subseteq \pi(q - \varepsilon)$, where $\varepsilon = \varepsilon(q)$, PM is a π-Hall subgroup in the dihedral group $D_{2(q-\varepsilon)}$ of order $2(q - \varepsilon)$ of PG. All π-Hall subgroups of this type are conjugate in G.*

---

[4]Recall that for a subgroup A of G we denote by PA reduction modulo scalars.

$\pi \cap \pi(G)$	M	$k_M(G)$	Conditions	Action
$\subseteq \pi(q - \varepsilon)$	$D_{q-\varepsilon}$	1	—	Sym_1
$\{2,3\}$	Alt_4	1	$(q^2 - 1)_{\{2,3\}} = 24$	Sym_1
$\{2,3\}$	Sym_4	2	$(q^2 - 1)_{\{2,3\}} = 48$	Sym_2
$\{2,3,5\}$	Alt_5	2	$(q^2 - 1)_{\{2,3,5\}} = 120$	Sym_2

Table 9. π-Hall subgroups M in $\mathrm{PSL}_2(q)$, $2,3 \in \pi$, $p \notin \pi$

(2) $\pi \cap \pi(G) = \{2,3\}$, $(q^2 - 1)_{\{2,3\}} = 24$, $PM \simeq \mathrm{Sym}_4$. All π-Hall subgroups of this type are conjugate in G.

Theorem 3.14 ([58]) *Assume $G = \mathrm{SL}_n^\eta(q)$ is a special linear or unitary group with the base field $\mathbb{F}_q$ of characteristic p and $n > 2$. Let π be a set of primes such that $2, 3 \in \pi$ and $p \notin \pi$. Then the following statements hold.*

(A) *Suppose $G \in E_\pi$, and M is a π-Hall subgroup of G. Then for G, M, and π one of the following statements holds.*

(1) *Either $q \equiv_{12} \eta$, or $n = 3$ and $q \equiv_4 \eta$, Sym_n satisfies E_π, $\pi \cap \pi(G) \subseteq \pi(q - \eta) \cup \pi(n!)$ and if $r \in (\pi \cap \pi(n!)) \setminus \pi(q - \eta)$, then $|G|_r = |\mathrm{Sym}_n|_r$. In this case M is included in*

$$H = L \cap G \simeq Z^{n-1} . \mathrm{Sym}_n,$$

where $L = Z \wr \mathrm{Sym}_n$ and $Z = \mathrm{GL}_1^\eta(q)$ is a cyclic group of order $q - \eta$. All π-Hall subgroups of this type are conjugate in G.

(2) *$n = 2m + k$, where $k \in \{0,1\}$, $m \geqslant 1$, $q \equiv -\eta \pmod 3$, $\pi \cap \pi(G) \subseteq \pi(q^2 - 1)$, groups Sym_m and $\mathrm{GL}_2^\eta(q)$ satisfy E_π.[5] In this case M is included in*

$$H = L \cap G \simeq ((\underbrace{\mathrm{GL}_2^\eta(q) \circ \cdots \circ \mathrm{GL}_2^\eta(q)}_{m \text{ times}}) . \mathrm{Sym}_m) \circ Z,$$

where $L = \mathrm{GL}_2^\eta(q) \wr \mathrm{Sym}_m \times Z$ and Z is a cyclic group of order $q - \eta$ for $k = 1$ and Z is trivial for $k = 0$. M acting by conjugation has at most two orbits on the set of factors in the product

$$\underbrace{\mathrm{GL}_2^\eta(q) \circ \cdots \circ \mathrm{GL}_2^\eta(q)}_{m \text{ times}},$$

while its intersection with each factor is a π-Hall subgroup in $\mathrm{GL}_2^\eta(q)$ and the intersection with factors from the same orbit satisfies to the same statement of Corollary 3.13. Two π-Hall subgroups of H are conjugate in G if and only if they are conjugate in H. Moreover H possesses one, two, or four classes of conjugate π-Hall subgroups, while all subgroups H are conjugate in G.

---

[5] Notice that in view of Lemma 3.12 conditions $\mathrm{GL}_2^\eta(q) \in E_\pi$ and $q \equiv -\eta \pmod 3$ imply that $q \equiv -\eta \pmod r$ for every odd $r \in \pi(q^2 - 1) \cap \pi$.

(3) $n = 4$, $\pi \cap \pi(G) = \{2, 3, 5\}$, $q \equiv 5\eta \pmod 8$, $(q+\eta)_3 = 3$, $(q^2+1)_5 = 5$, $M \simeq 4 \cdot 2^4 \cdot \mathrm{Alt}_6$. G possesses exactly two classes of conjugate π-Hall subgroups of this type and $\mathrm{GL}_4^\eta(q)$ interchanges these classes.

(4) $n = 11$, $(q^2 - 1)_{\{2,3\}} = 24$, $q \equiv -\eta \pmod 3$, $q \equiv \eta \pmod 4$, $\pi \cap \pi(G) = \{2, 3\}$, M is included in a subgroup $H = L \cap G$, where $L \simeq ((\mathrm{GL}_2^\eta(q) \wr \mathrm{Sym}_4) \times (\mathrm{GL}_1^\eta(q) \wr \mathrm{Sym}_3))$. In this case

$$M = (((X \circ \mathrm{GL}_2(3)) \wr \mathrm{Sym}_4) \times (X \wr \mathrm{Sym}_3)) \cap G,$$

where X is a Sylow 2-subgroup of a cyclic group $\mathrm{GL}_1^\eta(q)$. All π-Hall subgroups of this type are conjugate in G.

(B) Conversely, if conditions on π and q in one of statements (1)–(4) are satisfied, then $G \in E_\pi$.

(C) If $G \in E_\pi$, then $k_\pi(G) = k_\pi(PG) \in \{1, 2, 3, 4\}$.

(D) All π-Hall subgroups of PG have the form PM.

Theorem 3.15 ([58]) Let $G = \mathrm{Sp}_{2n}(q)$ be a symplectic group over a field $\mathbb{F}_q$ of characteristic p. Assume that π is a set of primes such that $2, 3 \in \pi$ and $p \notin \pi$. Then the following statements hold.

(A) Suppose $G \in E_\pi$ and $M \in \mathrm{Hall}_\pi(G)$. Then both Sym_n and $\mathrm{SL}_2(q)$ satisfy E_π and $\pi \cap \pi(G) \subseteq \pi(q^2 - 1)$. Moreover, M is a π-Hall subgroup of

$$H = \mathrm{Sp}_2(q) \wr \mathrm{Sym}_n \simeq \big(\underbrace{\mathrm{SL}_2(q) \times \cdots \times \mathrm{SL}_2(q)}_{n \text{ times}} \big) : \mathrm{Sym}_n \leq G.$$

(B) Conversely, if both Sym_n and $\mathrm{SL}_2(q)$ satisfy E_π and $\pi \cap \pi(G) \subseteq \pi(q^2 - 1)$, then $H \in E_\pi$ and every π-Hall subgroup M of H is a π-Hall subgroup of G, in particular $G \in E_\pi$.

(C) π-Hall subgroups of H are conjugate in G if and only if they are conjugate in H, while all such subgroups H are conjugate in G. In particular, if $G \in E_\pi$, then $k_\pi(G) = k_\pi(PG) \in \{1, 2, 3, 4, 9\}$.

(D) All π-Hall subgroups of PG have the form PM.

Theorem 3.16 ([58]) Assume that $G = \Omega_n^\eta(q)$, $\eta \in \{+, -, \circ\}$, q is a power of a prime p, $n \geqslant 7$, $\varepsilon = \varepsilon(q)$. Let π be a set of primes such that $2, 3 \in \pi$, $p \notin \pi$. Then the following statements hold.

(A) If G possesses a π-Hall subgroup M, then one of the following statements holds.

(1) $n = 2m + 1$, $\pi \cap \pi(G) \subseteq \pi(q - \varepsilon)$, $q \equiv \varepsilon \pmod{12}$, $\mathrm{Sym}_m \in E_\pi$. At that M is a π-Hall subgroup in $H = \big(\mathrm{O}_2^\varepsilon(q) \wr \mathrm{Sym}_m \times \mathrm{O}_1(q) \big) \cap G$. All π-Hall subgroups of this type are conjugate.

(2) $n = 2m$, $\eta = \varepsilon^m$, $\pi \cap \pi(G) \subseteq \pi(q - \varepsilon)$, $q \equiv \varepsilon \pmod{12}$, $\mathrm{Sym}_m \in E_\pi$. At that M is a π-Hall subgroup in $H = \big(\mathrm{O}_2^\varepsilon(q) \wr \mathrm{Sym}_m \big) \cap G$. All π-Hall subgroups of this type are conjugate.

(3) $n = 2m$, $\eta = -\varepsilon^m$, $\pi \cap \pi(G) \subseteq \pi(q-\varepsilon)$, $q \equiv \varepsilon$ (mod 12), $\mathrm{Sym}_{m-1} \in E_\pi$. At that M is a π-Hall subgroup of $H = \left(O_2^\varepsilon(q) \wr \mathrm{Sym}_{m-1} \times O_2^{-\varepsilon}(q) \right) \cap G$. All π-Hall subgroups of this type are conjugate.

(4) $n = 11$, $\pi \cap \pi(G) = \{2,3\}$, $q \equiv \varepsilon$ (mod 12), $(q^2 - 1)_\pi = 24$. At that M is a π-Hall subgroup of $H = \left(O_2^\varepsilon(q) \wr \mathrm{Sym}_4 \times O_1(q) \wr \mathrm{Sym}_3 \right) \cap G$. All π-Hall subgroups of this type are conjugate.

(5) $n = 12$, $\eta = -$, $\pi \cap \pi(G) = \{2,3\}$, $q \equiv \varepsilon$ (mod 12), $(q^2 - 1)_\pi = 24$. At that M is a π-Hall subgroup of $H = \left(O_2^\varepsilon(q) \wr \mathrm{Sym}_4 \times O_4^-(q) \right) \cap G$, and the projection of M on $O_4^-(q)$ is equal to a subgroup of type $O_1(q) \wr \mathrm{Sym}_3 \times O_1(q)$. There exist precisely two classes of conjugate subgroups of this type in G, and automorphism of order 2, induced by the group of similarities $\mathrm{GO}_{12}^-(q)$ of the natural module, interchanges these classes.

(6) $n = 7$, $\pi \cap \pi(G) = \{2,3,5,7\}$, $|G|_\pi = 2^9 \cdot 3^4 \cdot 5 \cdot 7$. At that $M \simeq \Omega_7(2)$. There exist precisely two classes of conjugate subgroups of this type in G, and $\mathrm{SO}_7(q)$ interchanges these classes.

(7) $n = 8$, $\eta = +$, $\pi \cap \pi(G) = \{2,3,5,7\}$, $|G|_\pi = 2^{13} \cdot 3^5 \cdot 5^2 \cdot 7$. At that $M \simeq 2.\Omega_8^+(2)$. There exist precisely four classes of conjugate subgroups of this type in G. The subgroup of $\mathrm{Out}(G)$ generated by diagonal and graph automorphisms is isomorphic to Sym_4 and acts on the set of these classes as Sym_4 in its natural permutation representation. Moreover every diagonal automorphism acts without stable points.

(8) $n = 9$, $\pi \cap \pi(G) = \{2,3,5,7\}$, $|G|_\pi = 2^{14} \cdot 3^5 \cdot 5^2 \cdot 7$. At that $M \simeq 2.\Omega_8^+(2):2$. There exist precisely two classes of conjugate subgroups of this type in G, and $\mathrm{SO}_9(q)$ interchanges these classes.

(B) *Conversely, if one of the statements (1)–(8) holds, then G possesses a π-Hall subgroup with given structure, i.e., $G \in E_\pi$.*

(C) *If $G \in E_\pi$, then $k_\pi(G) = k_\pi(\mathrm{P}G) \in \{1,2,3,4\}$.*

(D) *All π-Hall subgroups of $\mathrm{P}G$ have the form $\mathrm{P}M$.*

Theorem 3.17 ([58]) *Assume that $G \in \{E_6^\eta(q), E_7(q), E_8(q), F_4(q), G_2(q), {}^3D_4(q)\}$, q is a power of a prime p, $\varepsilon = \varepsilon(q)$. Let π be a set of primes such that $2,3 \in \pi$, $p \notin \pi$. Then $G \in E_\pi$ if and only if one of the following statements holds:*

(1) *G is a group from Table 10, $\pi(W) \subseteq \pi \cap \pi(G) \subseteq \pi(q-\varepsilon)$, M is a π-Hall subgroup of a group $T.W$, where T is a maximal torus of order $(q - \varepsilon)^l/|Z(G)|$. Values for l, $|Z(G)|$ and the structure of W are given in the Table 10. All π-Hall subgroups of this type are conjugate in G;*

(2) *$G = {}^3D_4(q)$, $\pi \cap \pi(G) \subseteq \pi(q - \varepsilon)$ and M is a π-Hall subgroup in $T.W(G_2)$, where T is a maximal torus of order $(q - \varepsilon)(q^3 - \varepsilon)$. All π-Hall subgroups of this type are conjugate in G;*

(3) *$G = E_6^{-\varepsilon}(q)$, $\pi \cap \pi(G) \subseteq \pi(q - \varepsilon)$ and M is a π-Hall subgroup in $T.W(F_4)$, where T is a maximal torus of order $(q^2 - 1)^2(q - \varepsilon)^2$. All π-Hall subgroups*

G	l	$|Z(G)|$	W	$|W|$
$E_6^{\eta}(q)$	6	$(3, q - \eta)$	$W(E_6) \simeq \mathrm{Sp}_4(3)$	$2^7.3^4.5$
$E_7(q)$	7	2	$W(E_7) \simeq 2 \times \mathrm{P\Omega}_7(2)$	$2^{10}.3^4.5.7$
$E_8(q)$	8	1	$W(E_8) \simeq 2\,.\,\mathrm{P\Omega}_8^+(2)\,.\,2$	$2^{14}.3^5.5^2.7$
$F_4(q)$	4	1	$W(F_4)$	$2^7.3^2$
$G_2(q)$	2	1	$W(G_2)$	$2^2.3$

Table 10. Weyl groups of exceptional root systems

of this type are conjugate in G;

(4) $G = G_2(q)$, $\pi \cap \pi(G) = \{2, 3, 7\}$, $(q^2 - 1)_{\{2,3,7\}} = 24$, $(q^4 + q^2 + 1)_7 = 7$, $M \simeq G_2(2)$, and all π-Hall subgroups of this type are conjugate in G.

The number of classes conjugate π-Hall subgroups in finite simple groups

The following theorem is a corollary to the classification of π-Hall subgroups in finite simple groups.

Theorem 3.18 ([58]) Let S be a finite simple group, π be a set of primes and $S \in E_\pi$. Then one of the following statements holds:
  (a) if $2 \notin \pi$ then $k_\pi(S) = 1$, i.e. $G \in C_\pi$;
  (b) if $3 \notin \pi$ then $k_\pi(S) \in \{1, 2\}$;
  (c) if $2, 3 \in \pi$ then $k_\pi(S) \in \{1, 2, 3, 4, 9\}$.
In particular, $k_\pi(S)$ is a π-number.

This theorem plays an important role in the solution of Problem 1.8.

4   A criterion of E_π

As we noted in Proposition 3.1, if a finite group satisfies E_π, then all its composition factors satisfy E_π. This property is a necessary, but not sufficient condition. Consider the following.

Example 4.1 Let $\pi = \{2, 3\}$, $G = \mathrm{GL}_3(2) = \mathrm{SL}_3(2)$ be a group of order $168 = 2^3 \cdot 3 \cdot 7$. From Theorem 3.8 it follows that G has exactly two classes $\mathscr{K}_1$ and $\mathscr{K}_2$ of π-Hall subgroups with representatives

$$\left(\begin{array}{c|c} \boxed{\mathrm{GL}_2(2)} & * \\ \hline 0 & \boxed{1} \end{array} \right) \quad \text{and} \quad \left(\begin{array}{c|c} \boxed{1} & * \\ \hline 0 & \boxed{\mathrm{GL}_2(2)} \end{array} \right)$$

respectively. The class $\mathscr{K}_1$ consists of line stabilizers in the natural representation of G, and the class $\mathscr{K}_2$ consists of plane stabilizers. The map $\iota : x \in G \mapsto (x^t)^{-1}$, where x^t means the transposed matrix to x, is an automorphism of order 2 of G. It interchanges classes of π-Hall subgroups. Consider the group $\widehat{G} = G : \langle \iota \rangle$. It interchanges classes $\mathcal{K}_1$ and $\mathcal{K}_2$. If $\widehat{G}$ would possess a π-Hall subgroup $\widehat{H}$, then Proposition 3.1 implies that $\widehat{H} \cap G$ lies in either $\mathcal{K}_1$ or $\mathcal{K}_2$, so $\widehat{H}$ stabilizes this class. On the other hand, $\widehat{G} = \widehat{H}G$, hence $\widehat{H}$ interchanges $\mathcal{K}_1$ and $\mathcal{K}_2$, a contradiction.

In [28], F. Gross obtained a sufficient condition for a finite group G to satisfy E_π, in terms of induced automorphisms of the composition factors of G.

Definition 4.2 Let A, B, H be subgroups of G such that $B \trianglelefteq A$ and $H \leq G$. Then $N_H(A/B) = N_H(A) \cap N_H(B)$ is the *normalizer* of A/B in H. If $x \in N_H(A/B)$, then x induces an automorphism of A/B by $Ba \mapsto Bx^{-1}ax$. Thus there exists a homomorphism $N_H(A/B) \to \operatorname{Aut}(A/B)$. The image of $N_H(A/B)$ under this homomorphism is denoted by $\operatorname{Aut}_H(A/B)$ and is called a *group of H-induced automorphisms* of A/B.

Theorem 4.3 ([28, Theorem 3.5]) *Let* $1 = G_0 < G_1 < \ldots < G_n = G$ *be a composition series of a finite group G which is a refinement of a chief series of G. Then the following are equivalent:*
(1) $H \in E_\pi$ *for all H such that $H \leq G$ and $H^{(\infty)} \trianglelefteq\trianglelefteq G$.*
(2) $\operatorname{Aut}_G(G_i/G_{i-1}) \in E_\pi$ *for all i, $1 \leqslant i \leqslant n$.*
(3) $\operatorname{Aut}_G(H/K) \in E_\pi$ *for all composition factors H/K of G.*

Here $H^{(\infty)}$ denotes the solvable coradical of H, i.e., the minimal normal subgroup N with H/N solvable. In particular, it follows that if groups of G-induced automorphisms of a composition series of G, which is a refinement of a chief series, satisfy E_π, then $G \in E_\pi$. It is natural to ask, whether this condition is necessary. The following theorem gives an affirmative answer to this question. Notice that this theorem uses the classification of finite simple groups and the classification of Hall subgroups in finite simple groups.

Theorem 4.4 ([59]) *Let* $1 = G_0 < G_1 < \ldots < G_n = G$ *be a composition series of a finite group G. If, for some i, $\operatorname{Aut}_G(G_i/G_{i-1}) \notin E_\pi$, then $G \notin E_\pi$.*

Corollary 4.5 ([59]) *Let* $1 = G_0 < G_1 < \ldots < G_n = G$ *be a composition series of a finite group G which is a refinement of a chief series of G. Then the following are equivalent:*
(1) $H \in E_\pi$ *for all H such that $H \leq G$ and $H^{(\infty)} \trianglelefteq\trianglelefteq G$;*
(2) $\operatorname{Aut}_G(G_i/G_{i-1}) \in E_\pi$ *for all i, $1 \leqslant i \leqslant n$;*
(3) $\operatorname{Aut}_G(H/K) \in E_\pi$ *for all composition factors H/K of G;*
(4) $G \in E_\pi$.

Corollary 4.5 shows a way to give the complete description of finite E_π-groups. It remains to solve the following:

Problem 4.6 Find all almost simple E_π-groups and π-Hall subgroups in these groups.

Recall that a finite group G is called *almost simple* if G possesses a unique minimal normal subgroup S and S is nonabelian simple. We can say, equivalently, that an almost simple group is isomorphic to an automorphism group G of a finite nonabelian simple group S such that, $\mathrm{Inn}(S) \leq G \leq \mathrm{Aut}(S)$. At the present time the authors are preparing a paper that solves Problem 4.6. Since π-Hall subgroups for every finite simple group S are known, Proposition 3.1 implies that we need to understand which π-Hall subgroups can be lifted to π-Hall subgroups of the corresponding almost simple group G. Moreover, since G/S is solvable, Hall Theorem implies that we may assume G/S to be a π-group. The following proposition gives a criterion whether such lifting can be done (cf. Example 4.1).

Proposition 4.7 *Let S be a normal E_π-subgroup of G such that G/S is a π-group and let M be a π-Hall subgroup of S. Then a π-Hall subgroup H of G with $H \cap S = M$ exists if and only if the class M^G is G-invariant, i.e., $M^S = M^G$.*

Thus in order to solve Problem 4.6 it is important to study the action of an almost simple goup G on the set of classes of conjugate π-Hall subgroups of its socle S, and the value $k_\pi(S)$ is also important.

We are able to give a criterion of E_π-property in a particular case, when 2 or 3 is not in π.

Theorem 4.8 ([31, 53]) *Suppose the set of primes π is chosen so that $2 \notin \pi$ or $3 \notin \pi$. Then a finite group G satisfies E_π if and only if each composition factor S of G satisfies E_π.*

Results from previous section imply that, if 2 or 3 is not in π, then for every finite group G with known composition factors one can check whether $G \in E_\pi$.

By Corollary 4.5 one more interesting statement follows. Proposition 3.1 states that if A is a normal subgroup of an E_π-group G, then there exist maps $\mathrm{Hall}_\pi(G) \to \mathrm{Hall}_\pi(G/A)$ and $\mathrm{Hall}_\pi(G) \to \mathrm{Hall}_\pi(A)$, given by $H \mapsto HA/A$ and $H \mapsto H \cap A$ respectively. The first map appears to be surjective.

Corollary 4.9 ([59]) *Every π-Hall in a homomorphic image of an E_π-group G is the image of a π-Hall subgroup of G.*

The former map $H \mapsto H \cap A$, in general, is not surjective, as we will see in Example 5.3.

5   A criterion of C_π

In the previous sections we noted that the class of E_π-groups is closed under normal subgroups and homomorphic images (see Proposition 3.1), but is not closed under extensions (see Example 4.1). In contrast, one of the most important properties of C_π is that this class is closed under extensions (this result was proved by S. A. Chunikhin, see also [36, Theorems C1 and C2]):

Proposition 5.1 *Let A be a normal subgroup of G. If both A and G/A satisfy C_π, then $G \in C_\pi$.*

Corollary 4.9 implies

Proposition 5.2 ([60]) *If $G \in C_\pi$ and $A \trianglelefteq G$, then $G/A \in C_\pi$.*

The following example shows that C_π is not closed under normal subgroups.

Example 5.3 Suppose $\pi = \{2,3\}$. Let $G = \mathrm{GL}_5(2) = \mathrm{SL}_5(2)$. Consider the map $\iota : x \in G \mapsto (x^t)^{-1}$ and the natural semidirect product $\widehat{G} = G : \langle \iota \rangle$. Theorem 3.8 implies that G possesses π-Hall subgroups, and each subgroup is the stabilizer in G of a series $\{0\} = V_0 < V_1 < V_2 < V_3 = V$, where V is a natural module for G, and $\dim V_k/V_{k-1} \in \{1,2\}$ for each $k = 1,2,3$. Therefore there exists exactly three classes of conjugate π-Hall subgroups of G with representatives

$$H_1 = \begin{pmatrix} \boxed{\mathrm{GL}_2(2)} & & * \\ & \boxed{1} & \\ 0 & & \boxed{\mathrm{GL}_2(2)} \end{pmatrix},$$

$$H_2 = \begin{pmatrix} \boxed{1} & & * \\ & \boxed{\mathrm{GL}_2(2)} & \\ 0 & & \boxed{\mathrm{GL}_2(2)} \end{pmatrix} \quad \text{and} \quad H_3 = \begin{pmatrix} \boxed{\mathrm{GL}_2(2)} & & * \\ & \boxed{\mathrm{GL}_2(2)} & \\ 0 & & \boxed{1} \end{pmatrix}.$$

Notice that $N_G(H_k) = H_k$ for $k = 1,2,3$, since each H_k is parabolic. In view of Proposition 3.1, for each π-Hall subgroup H of $\widehat{G}$, the intersection $H \cap G$ is conjugate to one of H_1, H_2, H_3. The class H_1^G is ι-invariant, so by Proposition 4.7 there exists a π-Hall subgroup H of $\widehat{G}$ with $H \cap \widehat{G} = H_1$ and, moreover $H = N_{\widehat{G}}(H_1)$. Now ι interchanges H_2^G and H_3^G. Thus Propositions 3.1 and 4.7 imply that H_2 and H_3 are not included in π-Hall subgroups of $\widehat{G}$. Therefore $\widehat{G}$ possesses exactly one class of conjugate π-Hall subgroups, so $\widehat{G} \in C_\pi$, while its normal subgroup G does not satisfy C_π.

This example also shows that the map $H \mapsto H \cap A$ for a normal subgroup A of an E_π-group G is not surjective in general (cf. Corollary 4.9).

The condition $2 \in \pi$ is essential in Example 5.3, since F. Gross in [29, 31] proves (mod CFSG) the following:

Theorem 5.4 ([30, Theorem A]) *If π is a set of primes and $2 \notin \pi$ then $E_\pi \Leftrightarrow C_\pi$.*

By Proposition 3.1 this theorem implies that for each set π of odd primes the class $E_\pi = C_\pi$ is closed under normal subgroups, homomorphic images and extensions.

By using Theorem 3.18 on the number of classes of conjugate π-Hall subgroups in finite simple groups it is possible to show that also in the case $2 \in \pi$ some normal subgroups of a C_π-groups satisfy C_π. Moreover, a criterion for C_π is obtained.

Theorem 5.5 ([60]) *Let π be a set of primes, A a normal and H a π-Hall subgroup of a C_π-group G. Then $HA \in C_\pi$.*

Corollary 5.6 (C_π-criterion, [60]) *Let π be a set of primes and A be a normal subgroup of G. Then $G \in C_\pi$ if and only if $G/A \in C_\pi$ and, for $K/A \in \mathrm{Hall}_\pi(G/A)$, its complete preimage K satisfies C_π.*

Corollary 5.7 ([60]) *Let π be a set of primes, A a normal subgroup of G. If $|G : A|$ is a π'-number, then $G \in C_\pi$ if and only if $A \in C_\pi$.*

If the classification of finite almost simple C_π-groups were known, then by using Corollary 5.6 it is not difficult to answer whether a given group satisfies C_π. Indeed the following Lemma holds.

Lemma 5.8 ([60]) *Suppose $G = HA$, where H is a π-Hall and A a normal subgroups of G, and $A = S_1 \times \cdots \times S_k$ is a direct product of finite simple groups. Then $G \in C_\pi$ if and only if $\mathrm{Aut}_G(S_i) \in C_\pi$ for each $i = 1, \ldots, k$.*

Now assume that

$$G = G_0 > G_1 > \ldots > G_n = 1 \qquad (1)$$

is a chief series of G. Set $H_1 = G = G_0$ and suppose that, for some $i = 1, \ldots, n$, the subgroup H_i satisfies $G_{i-1} \le H_i$ and $H_i/G_{i-1} \in \mathrm{Hall}_\pi(G/G_{i-1})$. Since (1) is a chief series, the decomposition $G_{i-1}/G_i = S_1^i \times \ldots \times S_{k_i}^i$, where $S_1^i, \ldots, S_{k_i}^i$ are simple groups, holds. We check whether

$$\mathrm{Aut}_{H_i}(S_1^i) \in C_\pi, \ldots, \mathrm{Aut}_{H_i}(S_{k_i}^i) \in C_\pi.$$

If so, then Lemma 5.8 implies $H_i/G_i \in C_\pi$, and we can choose a complete preimage of a π-Hall subgroup of H_i/G_i as H_{i+1}. Otherwise Corollary 5.6 implies that $G \notin C_\pi$ and we stop the process. Corollary 5.6 implies that $G \in C_\pi$ if and only if H_{n+1} can be constructed. Note that H_{n+1} is automatically a π-Hall subgroup of G.

Problem 5.9 Describe all finite almost simple groups satisfying C_π.

As in the case of the E_π-property, Problem 5.9 is reduced to the investigation of the action of $G \le \mathrm{Aut}(S)$ on the set of classes of conjugate π-Hall subgroups of S, where S is a nonabelian finite simple group, and $S \simeq \mathrm{Inn}(S) \le G \le \mathrm{Aut}(S)$. In view of Corollary 5.6 we may also assume that G/S is a π-group. Now $G \in C_\pi$ if and only if S possesses exactly one G-invariant class of conjugate π-Hall subgroups. Since for each finite simple group S classes of conjugate π-Hall subgroups are known, the solution to Problem 5.9 will be obtained soon.

6 A criterion of D_π

The theory of D_π-groups seems to be the most interesting. First, in D_π-groups the structure of all π-subgroups is defined by the structure of a π-Hall subgroup. Many properties of a π-Hall subgroup, such as solvability, nilpotency, commutativity, etc., hold for every π-subgroup of a D_π-group. Second, D_π means that the complete analogue of the Sylow theorem for π-subgroups holds. So many specialists have investigated this property. Third, the theory of D_π-groups is the most complete at the moment, and the main results can be formulated in a quite simple way.

We start with a historical survey. By Schur–Zassenhaus theorem (and Feit–Thompson Odd Order Theorem [25]) if a finite group is equal to an extension of a π-group by a π'-group, then it satisfies both D_π and $D_{\pi'}$ [94, Chapter IV, Satz 27] (see also [36, Theorems D6, D7]).

S. A. Chunikhin generalized this theorem and include Hall theorem in this result. S. A. Chunikhin introduced the notions of a π-solvable, a π-separable, and a π-selected group. Later P. Hall generalized some of Chunikhin's results. Recall that G is called

- π-separable if each its chief factor is either a π- or a π'-group;
- π-selected if the order of each its chief factor is divisible by at most one prime from π;
- π-solvable if it is both π-separable and π-selected.

In view of the Feit–Thompson theorem [25], each π-separable group is either π- or π'-solvable.

Theorem 6.1 *Let G be a finite group and π be a set of primes. Then the following statements hold:*

(1) (Chunikhin, [13]) *if G is π-solvable then $G \in D_\pi \cap D_{\pi'}$;*

(2) (Hall, [36, Corollary D5.2]) *if G is π-separable, then $G \in D_\tau$ for each $\tau \subseteq \pi$.*

In this direction S. A. Chunikhin [9]–[23] offered the following method of finding D_π-groups. If $\mathfrak{D}_\pi$ is a class of known D_π-groups, then new D_π-groups are constructed as groups with factors of a subnormal series in $\mathfrak{D}_\pi$.

An important step was a paper by H. Wielandt [83].

Theorem 6.2 (H. Wielandt [83]) *Let π be a set of primes and G be a finite group. If G possesses a nilpotent π-Hall subgroup, then $G \in D_\pi$.*

Combining Wielandt's result and Chunikhin's method, P. Hall proved in [36] the following:

Theorem 6.3 (P. Hall [36, Theorem D5]) *An extension of a D_π-group A, possessing a nilpotent π-Hall subgroup, by a D_π-group B, possessing a solvable π-Hall subgroup, satisfies D_π.*

The following problem arises in a natural way.

Problem 6.4 Does an extension of a D_π-group A by a D_π-group B satisfy D_π?

This problem was formulated for the first time by H. Wielandt in a one-hour survey talk given at the XIII International mathematical congress in Edinburgh in 1958 [86]. During the following fifty years this problem has been investigated by many mathematicians. It was mentioned in surveys [22, 70, 87], monographs by L. A. Shemetkov [71, Problem 22] and M. Suzuki [75], and was included by L. A. Shemetkov in the "Kourovka notebook" [95, Problem 3.62].

L. A. Shemetkov himself made significant progress on this problem. He solved in [69] Problem 6.4 in the case when the Sylow p-subgroups of A are cyclic for all $p \in \pi$, and also obtained in [64]–[68] a series of other important results, mentioned in his monograph [71]. A recent survey of result concerning this problem can be found in [73]. Without doubt, L. A. Shemetkov solved Problem 6.4 in all possible cases which can be handled without using the classification of finite simple groups (see [71, Lemma 18.3, Theorem 18.14]).

In 1971 B. Hartley [40] showed that the solvability of a π-Hall subgroup of B in Hall's Theorem 6.3 is not necessary. This condition can be substituted by a Schreier hypothesis (on the solvability of $\text{Out}(S)$ for S simple) for composition factors of A. Earlier this result was announced by H. Wielandt [91].

L. S. Kazarin in his paper [45] written in 1981 for the first time used the classification of finite simple groups. He strengthened previous results by L. A. Shemetkov on, the so-called, Wielandt π-classes.

V. D. Mazurov and D. O. Revin in [50] shown that Problem 6.4 has an affirmative solution, if Sylow 2-subgroups of all composition factors of A are abelian. There also exist quite recent results by V. N. Tyutyanov in [79, 80].

In 1997 paper [50] by V. D. Mazurov and D. O. Revin appeared. This paper gives an idea, how this problem can be solved by using the classification of finite simple groups. Namely, in [50] Problem 6.4 was reduced to the case, when A is a simple group and B is a subgroup of $\text{Out}(A)$. The authors in the series of papers [52, 53, 82] have given an affirmative answer to this problem.

Problem 6.4 is closely connected with the following:

Problem 6.5 Does a normal subgroup of a D_π-group satisfy D_π?

V. D. Mazurov included this problem in the "Kourovka notebook" [95, Problem 13.33]. Earlier this problem was mentioned in the paper [28] written by F. Gross. Elementary arguments show that a factor group of a D_π-group satisfies D_π. Therefore Problems 6.4 and 6.5 complement each other in some sense. These problems are useful in obtaining necessary and sufficient conditions for D_π. An affirmative answer to both problems is obtained in [52, 53] and uses the classification of finite simple groups. Notice also that the solution to Problem 6.5 can be derived from Theorem 3.18 on the number of classes of conjugate π-Hall subgroups.

The solution to Problems 6.4 and 6.5 can be summarized in the following:

Theorem 6.6 ([53, Theorem 7.7]) Let G be a finite group, $A \trianglelefteq G$ and π be a set of primes. Then $G \in D_\pi$ if and only if $A \in D_\pi$ and $G/A \in D_\pi$.

An equivalent form of this theorem is:

Corollary 6.7 *Let G be a finite group and π a set of primes. Then $G \in D_\pi$ if and only if each composition factor of G satisfy D_π.*

Thus the problem of describing finite groups satisfying D_π is reduced to the following:

Problem 6.8 *Given a set π of primes find all finite simple groups satisfying D_π.*

According to the classification theorem, each finite simple group is either sporadic or belongs to an infinite series of groups characterized by one or two arithmetic parameters. For example, the alternating group Alt_n is completely determined by its degree n, while a group of Lie type $A_n(q)$ is determined by its Lie rank n and the order of the base field q, etc. The description of D_π-groups in Problem 6.8 should be obtained in terms of these arithmetic parameters.

The description of π-Hall subgroups in symmetric groups obtained by P. Hall in [36, Theorem 4] and J. Thompson in [78] (see Theorem 3.3) implies that an alternating group Alt_n satisfies D_π if and only if either $|\pi \cap \pi(\mathrm{Alt}_n)| \leqslant 1$, or $\pi(\mathrm{Alt}_n) \subseteq \pi$. F. Gross [29, Corollary 6.13 and Theorem 6.14] found all sporadic groups satisfying D_π for every set π of odd primes. He stated in [29] a conjecture concerning an arithmetic criterion on D_π for finite groups of Lie type with the base field of characteristic $p \in \pi$, and proved this criterion in some particular cases. In [52] the first author completed the description of sporadic groups satisfying D_π and confirmed Gross's conjecture. The remaining cases are considered in [54]–[57].

Theorem 6.9 ([55, 57]) *Let π be a set of primes and let S be a known finite simple group. Then $S \in D_\pi$ if and only if the pair (S, π) satisfies one of conditions I–VII formulated below.*

Condition I. We say that (S, π) *satisfies Condition I if $\pi(S) \subseteq \pi$ or $|\pi \cap \pi(S)| \leqslant 1$.*

Condition II. We say that (S, π) *satisfies Condition II if one of the following cases holds:*

    (1) $S \simeq M_{11}$ and $\pi \cap \pi(S) = \{5, 11\}$;

    (2) $S \simeq M_{12}$ and $\pi \cap \pi(S) = \{5, 11\}$;

    (3) $S \simeq M_{22}$ and $\pi \cap \pi(S) = \{5, 11\}$;

    (4) $S \simeq M_{23}$ and $\pi \cap \pi(S)$ coincides with one of the sets $\{5, 11\}$, $\{11, 23\}$;

    (5) $S \simeq M_{24}$ and $\pi \cap \pi(S)$ coincides with one of the sets $\{5, 11\}$, $\{11, 23\}$;

    (6) $S \simeq J_1$ and $\pi \cap \pi(S)$ coincides with one of the sets $\{3, 5\}$, $\{3, 7\}$, $\{3, 19\}$, $\{5, 11\}$;

    (7) $S \simeq J_4$ and $\pi \cap \pi(S)$ coincides with one of the sets $\{5, 7\}$, $\{5, 11\}$, $\{5, 31\}$, $\{7, 29\}$, $\{7, 43\}$;

    (8) $S \simeq O'N$ and $\pi \cap \pi(S)$ coincides with one of the sets $\{5, 11\}$, $\{5, 31\}$;

    (9) $S \simeq Ly$ and $\pi \cap \pi(S) = \{11, 67\}$;

    (10) $S \simeq Ru$ and $\pi \cap \pi(S) = \{7, 29\}$;

    (11) $S \simeq Co_1$ and $\pi \cap \pi(S) = \{11, 23\}$;

(12) $S \simeq Co_2$ and $\pi \cap \pi(S) = \{11, 23\}$;

(13) $S \simeq Co_3$ and $\pi \cap \pi(S) = \{11, 23\}$;

(14) $S \simeq M(23)$ and $\pi \cap \pi(S) = \{11, 23\}$;

(15) $S \simeq M(24)'$ and $\pi \cap \pi(S) = \{11, 23\}$;

(16) $S \simeq B$ and $\pi \cap \pi(S)$ coincides with one of the sets $\{11, 23\}$, $\{23, 47\}$;

(17) $S \simeq M$ and $\pi \cap \pi(S)$ coincides with one of the sets $\{23, 47\}$, $\{29, 59\}$.

Condition III. Let S be a group of Lie type with the field $\mathbb{F}_q$ of characteristic $p \in \pi$. We set $\tau = \pi \cap \pi(S) \setminus \{p\}$. We say that (S, π) *satisfies Condition III* if $\tau \subseteq \pi(q - 1)$ and every number from π does not divide $|W|$ (the prime divisors of the corresponding Weyl group's order are given in Table 4).

Condition IV. Let S be a group of Lie type, with the base field $\mathbb{F}_q$ of characteristic p, not isomorphic to ${}^2B_2(q)$, ${}^2F_4(q)$ and ${}^2G_2(q)$. Let $2, p \notin \pi$. Denote by r the number $\min(\pi \cap \pi(S))$. We set $\tau = \pi \cap \pi(S) \setminus \{r\}$ and $a = e(q, r)$. We say that (S, π) *satisfies Condition IV* if there exists $t \in \tau$ with $b = e(q, t) \neq a$ and one of the following conditions holds:

(1) $S \simeq A_{n-1}(q)$, $a = r - 1$, $b = r$, $(q^{r-1} - 1)_r = r$, $\left[\dfrac{n}{r-1}\right] = \left[\dfrac{n}{r}\right]$ and for every $s \in \tau$ hold $e(q, s) = b$ and $n < bs$;

(2) $S \simeq A_{n-1}(q)$, $a = r - 1$, $b = r$, $(q^{r-1} - 1)_r = r$, $\left[\dfrac{n}{r-1}\right] = \left[\dfrac{n}{r}\right] + 1$, $n \equiv -1 \pmod r$ and, for every $s \in \tau$, $e(q, s) = b$ and $n < bs$;

(3) $S \simeq {}^2A_{n-1}(q)$, $r \equiv 1 \pmod 4$, $a = r - 1$, $b = 2r$, $(q^{r-1} - 1)_r = r$, $\left[\dfrac{n}{r-1}\right] = \left[\dfrac{n}{r}\right]$ and $e(q, s) = b$ for every $s \in \tau$;

(4) $S \simeq {}^2A_{n-1}(q)$, $r \equiv 3 \pmod 4$, $a = \dfrac{r-1}{2}$, $b = 2r$, $(q^{r-1} - 1)_r = r$, $\left[\dfrac{n}{r-1}\right] = \left[\dfrac{n}{r}\right]$ and $e(q, s) = b$ for every $s \in \tau$;

(5) $S \simeq {}^2A_{n-1}(q)$, $r \equiv 1 \pmod 4$, $a = r - 1$, $b = 2r$, $(q^{r-1} - 1)_r = r$, $\left[\dfrac{n}{r-1}\right] = \left[\dfrac{n}{r}\right] + 1$, $n \equiv -1 \pmod r$ and $e(q, s) = b$ for every $s \in \tau$;

(6) $S \simeq {}^2A_{n-1}(q)$, $r \equiv 3 \pmod 4$, $a = \dfrac{r-1}{2}$, $b = 2r$, $(q^{r-1} - 1)_r = r$, $\left[\dfrac{n}{r-1}\right] = \left[\dfrac{n}{r}\right] + 1$, $n \equiv -1 \pmod r$ and $e(q, s) = b$ for every $s \in \tau$;

(7) $S \simeq {}^2D_n(q)$, $a \equiv 1 \pmod 2$, $n = b = 2a$ and for every $s \in \tau$ either $e(q, s) = a$ or $e(q, s) = b$;

(8) $S \simeq {}^2D_n(q)$, $b \equiv 1 \pmod 2$, $n = a = 2b$ and for every $s \in \tau$ either $e(q, s) = a$ or $e(q, s) = b$.

Condition V. Let S a group of Lie type, with the base field $\mathbb{F}_q$ of characteristic p, not isomorphic to ${}^2B_2(q)$, ${}^2F_4(q)$ and ${}^2G_2(q)$. Let $2, p \notin \pi$. Denote by r the number $\min(\pi \cap \pi(S))$. We set $\tau = \pi \cap \pi(S) \setminus \{r\}$ and $c = e(q, r)$. We say that (S, π) *satisfies Condition V* if $e(q, t) = c$ for every $t \in \tau$ and one of the following conditions holds:

(1) $S \simeq A_{n-1}(q)$ and $n < cs$ for every $s \in \tau$;

(2) $S \simeq {}^2A_{n-1}(q)$, $c \equiv 0 \pmod 4$ and $n < cs$ for every $s \in \tau$;

(3) $S \simeq {}^2A_{n-1}(q)$, $c \equiv 2 \pmod 4$ and $2n < cs$ for every $s \in \tau$;

(4) $S \simeq {}^2A_{n-1}(q)$, $c \equiv 1 \pmod 2$ and $n < 2cs$ for every $s \in \tau$;

(5) S is isomorphic to one of the groups $B_n(q)$, $C_n(q)$, or ${}^2D_n(q)$, c is odd and $2n < cs$ for every $s \in \tau$;

(6) S is isomorphic to one of the groups $B_n(q)$, $C_n(q)$, or $D_n(q)$, c is even and $n < cs$ for every $s \in \tau$;

(7) $S \simeq D_n(q)$, c is even and $2n \leqslant cs$ for every $s \in \tau$;

(8) $S \simeq {}^2D_n(q)$, c is odd and $n \leqslant cs$ for every $s \in \tau$;

(9) $S \simeq {}^3D_4(q)$;

(10) $S \simeq E_6(q)$ and if $r = 3$ and $c = 1$ then $5, 13 \notin \tau$;

(11) $S \simeq {}^2E_6(q)$ and if $r = 3$ and $c = 2$ then $5, 13 \notin \tau$;

(12) $S \simeq E_7(q)$, if $r = 3$ and $c \in \{1, 2\}$ then $5, 7, 13 \notin \tau$, and if $r = 5$ and $c \in \{1, 2\}$ then $7 \notin \tau$;

(13) $S \simeq E_8(q)$, if $r = 3$ and $c \in \{1, 2\}$ then $5, 7, 13 \notin \tau$, and if $r = 5$ and $c \in \{1, 2\}$ then $7, 31 \notin \tau$;

(14) $S \simeq G_2(q)$;

(15) $S \simeq F_4(q)$ and if $r = 3$ and $c = 1$ then $13 \notin \tau$.

Condition VI. We say that (S, π) *satisfies Condition VI* if one of the following conditions holds:

(1) S is isomorphic to ${}^2B_2(2^{2m+1})$ and $\pi \cap \pi(S)$ is contained in one of the sets $\pi(2^{2m+1} - 1)$ or $\pi(2^{2m+1} \pm 2^{m+1} + 1)$;

(2) S is isomorphic to ${}^2G_2(3^{2m+1})$ and $\pi \cap \pi(S)$ is contained in one of the sets $\pi(3^{2m+1} - 1) \setminus \{2\}$ or $\pi(3^{2m+1} \pm 3^{m+1} + 1) \setminus \{2\}$;

(3) S is isomorphic to ${}^2F_4(2^{2m+1})$ and $\pi \cap \pi(S)$ is contained in one of the sets $\pi(2^{2(2m+1)} \pm 1)$, $\pi(2^{2m+1} \pm 2^{m+1} + 1)$, $\pi(2^{2(2m+1)} \pm 2^{3m+2} \mp 2^{m+1} - 1)$, or $\pi(2^{2(2m+1)} \pm 2^{3m+2} + 2^{2m+1} \pm 2^{m+1} - 1)$.

Condition VII. Let S be a group of Lie type with the base field $\mathbb{F}_q$ of characteristic p. Let $2 \in \pi$ and $3, p \notin \pi$. We set $\tau = \pi \cap \pi(S) \setminus \{2\}$ and $\varphi = \{t \in \tau \mid t \text{ is Fermat number}\}$. We say that (S, π) *satisfies Condition VII* if $\tau \subseteq \pi(q - \varepsilon)$, where $\varepsilon = \varepsilon(q)$, and one of the following conditions holds:

(1) $S = A_{n-1}^\eta(q)$, $s > n$ for every $s \in \tau$ and $t > n + 1$ for every $t \in \varphi$;

(2) $S = B_n(q)$ and $s > 2n + 1$ for every $s \in \tau$;

(3) $S = C_n(q)$, $s > n$ for every $s \in \tau$ and $t > 2n + 1$ for every $t \in \varphi$;

(4) $S = D_n^\eta(q)$ and $s > 2n$ for every $s \in \tau$.

	E_π	C_π	D_π
normal subgroups	Yes	No	Yes (mod CFSG)
homomorphisms	Yes	Yes (mod CFSG)	Yes
extensions	No	Yes	Yes (mod CFSG)

Table 11. Hall properties under normal subgroups, extensions and homomorphism

(5) S is either $G_2(q)$ or $^2G_2(q)$ and $7 \notin \tau$;

(6) $S = F_4(q)$ and $5, 7 \notin \tau$;

(7) $S = E_6^\eta(q)$ and $5, 7 \notin \tau$;

(8) $S = E_7(q)$ and $5, 7, 11 \notin \tau$;

(9) $S = E_8(q)$ and $5, 7, 11, 13 \notin \tau$;

(10) $S = {}^3D_4(q)$ and $7 \notin \tau$.

Corollary 6.10 *Let π be a set of primes. Then a finite group satisfies D_π if and only if the pair (S, π) satisfies one of the conditions* I–VII *for each of its composition factors S.*

This corollary solves Problem 1.8 for D_π.

7 Open problems

As we already mentioned, Problems 4.6 and 5.9 remain open, although them both are closed to being solved. But a complete solution of these problems would not imply that the investigation of classes of E_π, C_π, and D_π is finished. Of course, given a finite group G and a set π of primes we can answer whether $G \in \Psi$, where $\Psi \in \{E_\pi, C_\pi, D_\pi\}$, but we still do not know many properties of these classes. In this section we give a survey of interesting problems.

In Sections 4–6 we discussed whether a class $\Psi \in \{E_\pi, C_\pi, D_\pi\}$ is closed under extensions, normal subgroups and homomorphic images. These results are assembled in Table 11.

The natural question arising here is: Find subgroups H of $G \in \Psi$ (except normal subgroups) such that $H \in \Psi$. For example, Sylow's theorem implies that $G \in D_p$ for every finite group G and prime p. Moreover, every subgroup H of G satisfies D_p. There exist examples of D_π-groups possessing a subgroup not satisfying D_π.

Example 7.1 Let $\pi = \{3, 5\}$, $G = \mathrm{SL}_2(2^4) \simeq A_1(2^4)$ and $H = \mathrm{SL}_2(2^2) \simeq \mathrm{Alt}_5$. Clearly, G possesses a subgroup isomorphic to H, so we may assume that $H \leq G$. Now (G, π) satisfies Condition V(1) of Theorem 6.9, so $G \in D_\pi$. On the other hand $H \simeq \mathrm{Alt}_5 \notin E_\pi$ (see Example 1.4), hence $H \notin D_\pi$.

We say that a finite group G is in W_π, if for every $H \leq G$ we have $H \in D_\pi$. Example 7.1 shows that $W_\pi \neq D_\pi$ in general. Theorem 6.6 implies that $G \in W_\pi$ if and only if each composition factor of G satisfies W_π. Thus the following problem arises in a natural way.

Problem 7.2 Let π be a set of primes. Find all finite simple groups in which every subgroup is a D_π-group.

Notice that Problem 7.2 is formulated by H. Wielandt at the conference in Santa Cruz in 1979 [92, Section 6, Frage (h)].

In Example 7.1 H does not satisfy E_π. Consider the class V_π of finite D_π-groups in which every E_π-subgroup satisfies D_π. It is clear that $W_\pi \subseteq V_\pi \subseteq D_\pi$. Proposition 3.1 and Theorem 6.6 imply that $G \in V_\pi$ if and only if $S \in V_\pi$ for every composition factor S of G. So an analogue of Problem 7.2 for V_π is also interesting.

Problem 7.3 Let π be a set of primes. Find all finite simple D_π-groups in which every E_π-subgroup is a D_π-group.

Given an E_π-group G, every overgroup of a π-Hall subgroup of G clearly satisfies E_π. The same statements for C_π and D_π are not evident and it is possible to formulate the analogous problems for C_π- and D_π- groups.

Problem 7.4 In a C_π-group, is an overgroup of a π-Hall subgroup always a C_π-group?

Problem 7.5 In a D_π-group, is an overgroup of a π-Hall subgroup always a D_π-group?

Theorem 5.6 solves Problem 7.4 in a particular case.

Problem 7.4 has an equivalent form: *Are the π-Hall subgroups of C_π-groups pronormal?* Recall that a subgroup H of G is called *pronormal* if, for every $g \in G$, subgroups H and H^g are conjugate in $\langle H, H^g \rangle$. The following example shows that π-Hall subgroups are not pronormal in general.

Example 7.6 Suppose $\pi = \{2, 3\}$, $X = \mathrm{GL}_3(2)$. As we already noted, X possesses exactly two classes of conjugate π-Hall subgroups (cf. Example 4.1). Let Y be a cyclic subgroup of order 7 in Sym_7, $G = X \wr Y$ and

$$H \simeq \underbrace{X \times \ldots \times X}_{7 \text{ times}}$$

be the base of this wreath product. Clearly H possesses 2^7 classes of conjugate π-Hall subgroups. Since H is normal in G and $|G : H| = 7$ is a π'-number, $\mathrm{Hall}_\pi(G) = \mathrm{Hall}_\pi(H)$. Now Y acts on the set of classes of conjugate π-Hall subgroups of H. By using this fact, it is not difficult to check that $k_\pi(G) = 20 < 128 = k_\pi(H)$. So there exist π-Hall subgroups of G which are conjugate in G but are not conjugate even in their common normal closure H, whence these subgroups are not pronormal.

This example also shows that in arbitrary (nonsimple) finite group G the number $k_\pi(G)$, in general, is not a π-number (cf. Theorem 3.18).

Problem 7.7 In finite simple groups, are Hall subgroups always pronormal?

We say that a subgroup H of G is *strongly pronormal* if K^g is conjugate to a subgroup of H in $\langle H, K^g \rangle$ for every $K \leq H$ and $g \in G$. Thus the equivalent variant of Problem 7.5 can be formulated in the following way. *Are π-Hall subgroups in D_π-groups strongly pronormal?* By analogue with Problem 7.7 we may state the following:

Problem 7.8 In finite simple groups, are Hall subgroups always strongly pronormal?

It is clear that an affirmative answer to Problem 7.8 implies an affirmative answer to Problem 7.7.

Finally we state several "arithmetical" problems. Denote by $\mathfrak{G}$ the class of all finite groups, and by B_π the class of groups such that each composition factor is either a π-group, or is divisible by at most one prime from π (i.e., each composition factor possesses a standard π-Hall subgroup). The general problem can be formulated in the following form.

Problem 7.9 Find sets π of primes such that one or more inclusions in

$$B_\pi \subseteq D_\pi \subseteq C_\pi \subseteq E_\pi \subseteq \mathfrak{G}$$

are identities?

Trivial examples when

$$B_\pi = D_\pi = C_\pi = E_\pi = \mathfrak{G}$$

arise from the cases when π is the empty set, the set of all primes, and a one-element subset.

Corollary 6.10 implies that if $2, 3 \in \pi$, then $B_\pi = D_\pi$.

In [1] for $\pi = 2'$ the identities $D_\pi = C_\pi = E_\pi$ are obtained by using the classification of finite simple groups. This result (and Problem 7.9) appeared because of the influence of Hall's hypothesis, whether E_π implies D_π for every set π of odd primes.

Condition $2 \notin \pi$ in Hall's hypothesis is essential.

Example 7.10 Suppose $2, 3 \in \pi$ and π is not equal to the set of all primes. Set $p = \min \pi'$. Consider $G = \mathrm{Sym}_p$. Then G possesses exactly one class of conjugate π-Hall subgroups isomorphic to Sym_{p-1}. So $G \in C_\pi$. Each π-Hall subgroup of G is a point stabilizer in the natural permutation representation. Consider a π-subgroup $K = \langle (1, 2, \ldots, p-2)(p-1, p) \rangle$ of G. K acts without stable points, hence is not included in a π-Hall subgroup of G.

Thus we derive the following:

Proposition 7.11 *Let π be a set of primes containing 2 and 3, and $\pi' \neq \varnothing$. Then $D_\pi \neq C_\pi$.*

F. Gross [27] obtained a negative answer to the original Hall's hypothesis and proved the following theorem.

Proposition 7.12 ([27]) *For every finite set π of odd primes with $|\pi| \geqslant 2$ there exists a finite E_π-, but not D_π-group.*

A weaker statement that $E_\pi = C_\pi$ for every set of odd primes π was already mentioned in Theorem 5.4. Propositions 5.4 and 7.12 imply the following:

Proposition 7.13 *There exists a continuum number of sets π of odd primes such that $D_\pi \neq C_\pi$.*

Indeed, for every pair (G, π), where π is a finite set of odd primes and $G \in (C_\pi \setminus D_\pi)$ (the existence of G follows from Propositions 7.12 and 5.4), consider all sets τ of odd primes such that $\pi \subseteq \tau$ and $\pi(G) \cap (\tau \setminus \pi) = \varnothing$. Clearly there exists a continuum number of such sets τ and $G \in C_\tau \setminus D_\tau$ for each τ.

We formulate a problem which is a weaker variant of the above mentioned Hall's hypothesis.

Problem 7.14 Does there exist a continuum number of sets π such that $E_\pi = D_\pi$?

Clearly, the number of such sets π is infinite since the set of one-element subsets of primes is infinite. If Problem 7.14 has an affirmative answer, then there exists a continuum number of *infinite* sets π of primes such that $E_\pi = C_\pi = D_\pi$. But, up to now, only a few examples of infinite sets π with $E_\pi = D_\pi$ are known. We construct countably many infinite sets of primes π with $E_\pi = D_\pi$.

Theorem 7.15 *Given a real number x, denote by π_x the set of primes $\{p \mid p > x\}$. If x is sufficiently large, then $E_{\pi_x} = D_{\pi_x}$.*

References

[1] Z. Arad and M. B. Ward, New criteria for the solvability of finite groups, *J. Algebra* **77** (1982), N1, 234–246.

[2] Z. Arad and D. Chilag, A criterion for the existence of normal π-complements in finite groups, *J. Algebra* **87** (1984), N2, 472–482.

[3] Z. Arad and E. Fisman, On finite factorizable groups, *J. Algebra* **86** (1984), N2, 522–548.

[4] R. Baer, Verstreute Untergruppen endlicher Gruppen, *Arch. Math.* **9** (1958), N1–2, 7–17.

[5] R. W. Carter, *Simple groups of Lie type*, John Wiley and Sons, London, 1972.

[6] P. Cobb, Existence of Hall subgroups and embedding of π-subgroups into Hall subgroups, *J. Algebra* **127** (1989), N1, 229–243.

[7] J. H. Conway, R. T. Curtis, S. P. Norton, R. A. Parker and R. A. Wilson, *Atlas of Finite Groups* (OUP, Oxford 1985)

[8] M. P. Coone, The nonsolvable Hall subgroups of the general linear groups, *Math. Z.* **114** (1970), N4, 245–270.

[9] S. A. Chunikhin, On solvable groups, *Izv. NIIMM Tom. Univ.* **2** (1938), 220–223 (in Russian).

[10] S. A. Chunikhin, On Sylow-regular groups, *Doklady Akad. Nauk SSSR* **60** (1948), N5, 773–774 (in Russian).

[11] S. A. Chunikhin, On Π-properties of finite groups, *Amer. Math. Soc. Translation* (1952). no. 72. Translated from *Mat. Sbornik* **25** (1949), N3, 321–346.

[12] S. A. Chunikhin, On the conditions of theorems of Sylow's type, *Doklady Akad. Nauk SSSR* **69** (1949), N6, 735–737 (in Russian).

[13] S. A. Chunikhin, On Sylow properties of finite groups, *Doklady Akad. Nauk SSSR* **73** (1950), N1, 29–32 (in Russian).

[14] S. A. Chunikhin, On weakening the conditions in theorems of Sylow type, *Doklady Akad. Nauk SSSR* **83** (1952), N5, 663–665 (in Russian).

[15] S. A. Chunikhin, On subgroups of a finite group, *Doklady Akad. Nauk SSSR* **86** (1952), N1, 27–30 (in Russian).

[16] S. A. Chunikhin, On existence and conjugateness of subgroups of a finite group, *Mat. Sbornik* **33** (1953), N1, 111–132 (in Russian).

[17] S. A. Chunikhin, On Π-solvable subgroups of finite groups, *Doklady Akad. Nauk SSSR* **103** (1955), N3, 377–378 (in Russian).

[18] S. A. Chunikhin, Some trends in the development of the theory of finite groups in recent years, *Uspehi Mat. Nauk* **16** (1961), N4 (100), 31–50 (in Russian).

[19] S. A. Chunikhin, On indexials of finite groups, *Soviet Math. Dokl.* **121** (1961), N4 (100), 70–71. Translated from the Russian *Doklady Akad. Nauk SSSR* **136**, 299–300.

[20] S. A. Chunikhin, A π-Sylow theorem following from the hypothesis of solvability of groups of odd order, *Dokl. Akad. Nauk BSSR* **6** (1962), N6, 345–346 (in Russian).

[21] S. A. Chunikhin, *Subgroups of finite groups*. Wolters-Noordhoff Publishing, Groningen 1969.

[22] S. A. Chunikhin and L. A. Shemetkov, Finite groups, *J. Soviet Math.* **1** (1973), no. 3., Plenum Press, New York, 1973. 291–390, Translated from the Russian *Algebra. Topology. Geometry, 1969*, 7–70, Akad. Nauk SSSR Vsesojuz. Inst. Nauchn. i Tehn. Informacii, Moscow, 1971.

[23] S. A. Chunikhin, On the definition of π-solvable and π-separable groups, *Dokl. Akad. Nauk SSSR* **212** (1973), 1078–1081 (in Russian).

[24] K. Doerk and T. Hawks, *Finite soluble groups*, Berlin, New York, Walter de Gruyter, 1992.

[25] W. Feit and J. G. Thompson, Solvability of groups of odd order, *Pacif. J. Math.* **13** (1963) N3, 775–1029.

[26] P. A. Ferguson and P. Kelley, Hall π-subgroups which are direct product of nonabelian simple groups, *J. Algebra* **120** N1 (1989), 40–46.

[27] F. Gross, Odd order Hall subgrous of $GL(n, q)$ and $Sp(2n, q)$, *Math. Z.* **187** (1984), N2, 185–194.

[28] F. Gross, On the existence of Hall subgroups, *J. Algebra* **98** (1986), N1, 1–13.

[29] F. Gross, On a conjecture of Philip Hall, *Proc. London Math. Soc. Ser. III* **52** (1986), N3, 464–494.

[30] F. Gross, Odd order Hall subgroups of the classical linear groups, *Math. Z.* **220** (1995), N3, 317–336.

[31] F. Gross, Conjugacy of odd order Hall subgroups, *Bull. London Math. Soc.* **19** (1987), N4, 311–319.

[32] F. Gross, Hall subgroups of order not divisible by 3, *Rocky Mt. J. Math.* **23** (1993), N2, 569–591.

[33] W. Guo and B. Li, On the Shemetkov problem for Fitting classes, *Beiträge Algebra Geom.* **48** (2007), N1, 281–289.

[34] W. Guo, Formations determined by Hall subgroups, *J. Appl. Algebra Discrete Struct.* **4** (2006), N3, 139–147.

[35] W. Guo, Some problems and results for the research on Sylow objects of finite groups. *J. Xuzhou Norm. Univ. Nat. Sci. Ed.* **23** (2005), N3, 1–6, 40 (in Chinese).

[36] P. Hall, Theorems like Sylow's, *Proc. London Math. Soc. (3)* **6** (1956), 286–304.

[37] P. Hall, A note on soluble groups *J. London Math. Soc.* **3** (1928), 98–105.

[38] P. Hall, A characteristic property of soluble groups, *J. London Math. Soc.* **12** (1937), 198–200.

[39] P. Hall, On the Sylow system of a soluble group, *Proc. London Math. Soc. Ser. II* **43** (1937), 316–320.

[40] B. Hartley, A theorem of Sylow type for a finite groups, *Math. Z.* **122** (1971), N4, 223–226.

[41] B. Hartley, Helmut Wielandt on the π-structure of finite groups, *Mathematische Werke = Mathematical Works / Helmut Wielandt*, ed. by B. Huppert and H. Sneider, vol. 1, Walter de Gruyter, Berlin, 1994, 511–516.

[42] J. E. Humphreys, *Linear algebraic groups*, Springer-Verlag, New York, 1972.

[43] N. Îto On π-structures of finite groups, *Tôhoku Math. Journ.* **4** (1952) N1, 172–177.

[44] P. B. Kleidman and M. Liebeck, *The subgroups structure of finite classical groups*, Cambridge Univ. Press, 1990.

[45] L. S. Kazarin, Sylow type theorems for finite groups, *Sructure properties of algebraic systems*, Nalchik, Kabardino-Balkarskii Univ., 1981, 42–52 (in Russian).

[46] L. S. Kazarin, On the product of finite groups, *Doklady Akad. Nauk SSSR* **269** (1983), N3, 528–531 (in Russian).

[47] M. I. Kargapolov, Factorization of Π-separable groups, *Doklady Akad. Nauk SSSR* **114** (1957), N6, 1155–1157 (in Russian).

[48] A. S. Kondrat'ev, Subgroups of finite Chevalley groups, *Russian Math. Surveys* **41**, N1 (1986), 65–118.

[49] V. D. Mazurov, On a question of L. A. Shemetkov, *Algebra and Logic* **31** (1992), N6, 360–366.

[50] V. D. Mazurov and D. O. Revin, On the Hall D_π-property for finite groups, *Siberian Math. J.* **38** (1997), N1, 106–113.

[51] D. O. Revin, Hall π-subgroups of finite Chevalley groups whose characteristic belongs to π, *Siberian Adv. Math.* **9** (1999), N2, 25–71.

[52] D. O. Revin, The D_π-property in a class of finite groups, *Algebra and Logic* **41** (2002) N3, 187–206.

[53] D. O. Revin and E. P. Vdovin, Hall subgroups of finite groups, *Contemporary Mathematics* **402** (2006), 229–265.

[54] D. O. Revin, The D_π property of finite groups in the case $2 \notin \pi$, *Proc. Steklov Inst. Math.* **257** (2006) Suppl. 1, S164–S180.

[55] D. O. Revin, A characterization of finite D_π-groups, *Doklady Math.* **76** (2007), N3, 925–928.

[56] D. O. Revin, The property D_π in linear and unitary groups, *Sib. Math. J.* **49** (2008), N2, 353–361.

[57] D. O. Revin, The D_π-property in finite simple groups, *Algebra and Logic* **47** (2008), N3, 210–227.

[58] D. O. Revin and E. P. Vdovin, On the number of classes of conjugate Hall subgroups in finite simple groups, in preparation.

[59] D. O. Revin and E. P. Vdovin, Existence criterion for Hall subgroups of finite groups, in preparation.

[60] D. O. Revin and E. P. Vdovin, Conjugacy criterion for Hall subgroups of finite groups, *Siberian Math. J.*, to appear.

[61] S. A. Rusakov, Analogues of Sylow's theorem on inclusion of subgroups, *Doklady Akad. Nauk SSSR* **5** (1961), 139–141 (in Russian).

[62] S. A. Rusakov, Analogues of Sylow's theorem on the existence and imbedding of subgroups, *Sibirsk. Mat. Zh.* **4** (1963), N5, 325–342 (in Russian).

[63] S. A. Rusakov, *C*-theorems for *n*-groups, *Vesci Akad. Navuk BSSR Ser. Fiz.-Mat. Navuk*, 1972, N3, 5–9 (in Russian).

[64] L. A. Shemetkov, On a theorem of Hall, *Dokl. Akad. Nauk SSSR* **147** (1962), 321–322 (in Russian).

[65] L. A. Shemetkov, *D*-structure of finite groups, *Mat. Sb.* **67** (1965), N3, 384–497 (in Russian).

[66] L. A. Shemetkov, A new *D*-theorem in the theory of finite groups, *Dokl. Akad. Nauk SSSR* **160** (1965), N2, 290–293 (in Russian).

[67] L. A. Shemetkov, Sylow properties of finite groups, *Mat. Sb.* **76** (1968), N2, 271–287 (in Russian).

[68] L. A. Shemetkov, Conjugacy and imbedding of subgroups, *Finite Groups*, Nauka i Tehnika, Minsk, 1966, 881–883 (in Russian).

[69] L. A. Shemetkov, Sylow properties of finite groups, *Dokl. Akad. Nauk BSSR* **16** (1972), N10, 881–883.

[70] L. A. Shemetkov, Two trends in the development of the theory of nonsimple finite groups, *Uspehi Mat. Nauk* **30** (1975), N2(182), 179-198 (in Russian).

[71] L. A. Shemetkov, *Formatsii konechnykh grupp (Formations of finite groups)*, Monographs in Modern Algebra, ("Nauka", Moscow, 1978) (in Russian).

[72] L. A. Shemetkov and A. F. Vasil'ev, Nonlocal formations of finite groups *Dokl. Akad. Nauk Belarusi* **39** (1995), 5–8 (in Russian).

[73] L. A. Shemetkov, A generalization of Sylow's theorem, *Siberian Math. J.* **44** (2003), N6, 1127–1132.

[74] E. L. Spitznagel, Hall subgroups of certain families of finite groups, *Math. Z.* **97** (1967), N4, 259–290.

[75] M. Suzuki, *Group theory II*, NY, Berlin, Heidelberg, Tokyo, Springer-Verl. 1986.

[76] M. L. Sylow, Théorèmes sur les groupes de substitutions, *Math. Ann.* **5** (1872), N4, 584–594.

[77] M. C. Tibiletti, Sui prodotti ordinati di gruppi finiti, *Boll. Un. Mat. Ital.* (3) **13** (1958), 46-57.

[78] J. G. Thompson, Hall subgroups of the symmetric groups, *J. Comb. Th.* **1** (1966) N2, 271–279.

[79] V. N. Tyutyanov, The D_π-theorem for finite groups having composition factors such that the 2-length of any solvable subgroup does not exceed one, *Vestsi Nats. Akad. Navuk Belarusi Ser. Fiz.-Mat. Navuk*, 2000 N1, 12–14 (in Russian).

[80] V. N. Tyutyanov, Sylow-type theorems for finite groups, *Ukrainian Math. J.* **52** (2000), N10, 1628–1633.

[81] V. N. Tyutyanov, On the Hall conjecture, *Ukrainian Math. J.* **54** (2002), N7, 1181–1191.

[82] E. P. Vdovin and D. O. Revin, Hall subgroups of odd order in finite groups, *Algebra and Logic* **41** (2002) N1, 8–29.

[83] H. Wielandt, Zum Satz von Sylow, *Math. Z.* **60** (1954), N4. 407–408.

[84] H. Wielandt, Sylowgruppen und Kompositoin-Struktur, *Abh. Math. Sem. Univ. Hamburg* **22** (1958), 215–228.

[85] H. Wielandt, Zum Satz von Sylow. II, *Math. Z.* **71** (1959), N4. 461–462.

[86] H. Wielandt, Entwicklungslinien in der Strukturtheorie der endlichen Gruppen, *Proc. Intern. Congress Math., Edinburg, 1958*. London: Cambridge Univ. Press, 1960, 268–278.

[87] H. Wielandt, Arithmetische Struktur und Normalstuktur endlicher Gruppen, *Conv. Internaz. di Teoria dei Gruppi Finitine Applicazioni, Firenze, 1960*. Roma: Edizioni

Cremonese, 1960, 56–65.

[88] H. Wielandt, Der Normalisator einer subnormalen Untergruppe, *Acta Sci. Math. Szeged* **21** (1960) 324–336.

[89] H. Wielandt, Sylowtürme in subnormalen Untergruppen, *Math. Z.* **73** (1960), N4. 386–392.

[90] H. Wielandt and B. Huppert, Arithmetical and normal sructureof finite groups, *Proc. Symp. Pure Math.* **6** (1962), Providence RI: Amer. Math. Soc., 17–38.

[91] H. Wielandt, Sur la Stucture des groupes composés, *Séminare Dubriel-Pisot(Algèbre et Théorie des Nombres),* 17e anée, 10 pp. 1963/64. N17.

[92] H. Wielandt, Zusammenghesetzte Gruppen: Hölder Programm heute, *The Santa Cruz conf. on finite groups, Santa Cruz, 1979.* Proc. Sympos. Pure Math. **37**, Providence RI: Amer. Math. Soc., 1980, 161–173.

[93] G. Zappa, Sopra unestensione di Wielandt del teorema di Sylow, *Boll. Un. Mat. Ital.* (3) **9**, (1954), N4, 349–353.

[94] H. Zassenhaus, *Lehrbuch der Gruppentheorie,* Leipzig, Berlin. 1937.

[95] *The Kourovka notebook. Unsolved problems in group theory.* Edited by V. D. Mazurov and E. I. Khukhro. 16-th. ed., Russian Academy of Sciences Siberian Division, Institute of Mathematics, Novosibirsk, 2006.

ENGEL GROUPS

GUNNAR TRAUSTASON

Department of Mathematical Sciences, University of Bath, Bath BA2 7AY, UK
Email: gt223@bath.ac.uk

Abstract

We give a survey on Engel groups with particular emphasis on the development during the last two decades.

Introduction

We define the n-Engel word $e_n(x, y)$ as follows: $e_0(x, y) = x$ and $e_{n+1}(x, y) = [e_n(x, y), y]$. We say that a group G is an Engel group if for each pair of elements $a, b \in G$ we have $e_n(a, b) = 1$ for some positive integer $n = n(a, b)$. If n can be chosen independently of a, b then G is an n-Engel group.

One can also talk about Engel elements. An element $a \in G$ is said to be a left Engel element if for all $g \in G$ there exists a positive integer $n = n(g)$ such that $e_n(g, a) = 1$. If instead one can for all $g \in G$ choose $n = n(g)$ such that $e_n(a, g) = 1$ then a is said to be a right Engel element. If in either case we can choose n independently of g then we talk about left n-Engel or right n-Engel element respectively.

So to say that a is left 1-Engel or right 1-Engel element is the same as saying that a is in the center and a group G is 1-Engel if and only if G is abelian. Every group that is locally nilpotent is an Engel group. Furthermore for any group G we have that all the elements of the locally nilpotent radical are left Engel elements and all the elements in the hyper-center are right Engel elements. For certain classes of groups the converse is true and the locally nilpotent radical and the hyper-center can be characterised as being the set of the left Engel and right Engel elements respectively. There was in particular much activity in this area in the 50's and the 60's.

In this paper we will for the most part omit discussion on Engel elements and focus instead on Engel groups. In particular we are interested in the varieties of n-Engel groups. Our survey is by no means meant to be complete. The main aim is to discuss some central results obtained in the last two decades on n-Engel groups as well as giving some general background to these. The material is organised as follows. As the Engel groups and the Burnside problems are closely related and both originate from the 1901 paper of W. Burnside [7], our survey begins with the work of Burnside on 2-Engel groups. Apart from this the first main result on Engel groups is Zorn's Theorem that tells us that any finite Engel group is nilpotent. In Section 2 and 3 we look at a number of generalisations of this. In Section 4 we look into the structure of n-Engel groups. The main open question here is whether n-Engel groups are locally nilpotent and we first demonstrate that if this is not the case then there must exist a finitely generated non-abelian simple n-Engel group. For the remainder of Section 4 we describe some results on the structure of locally nilpotent n-Engel groups. Sections 5 and 6 are then devoted to 3-Engel and 4-Engel groups. We end the survey by looking at some recent generalisations of Engel groups. Although this is a survey, some of the material is new. In particular much of our treatment of 3-Engel groups differs from the original one. The proof of the local nilpotence of 3-Engel groups is new and we have made use of Lie ring methods to give a shorter proof of another of Heineken's central results (Theorem 5.6). Our hope is that this survey will be a useful starting point for a graduate student entering this area and whenever possible we have tried to include proofs when they are short.

1   Origin and early results

> ... Although some useful facts have been brought to light about Engel groups by K. W. Gruenberg and also by the attempts on the Burnside problem, the word problem for E_n remains unsolved for $m > 2$. Problems such as these still seem to present a formidable challenge to the ingenuity of algebraists. In spite of, or perhaps because of, their relatively concrete and particular character, they appear, to me at least, to offer an amiable alternative to the ever popular pursuit of abstractions.
>
> (P. Hall [26], 1958)

As indicated by P. Hall's paragraph above, the theory of Engel groups and the Burnside problems are closely related. The common origin is the famous 1901 paper [7] of W. Burnside where he formulated the Burnside problems. In particular he asked whether a finitely generated group of bounded exponent must be finite. Burnside proves that this is the case for groups of exponent 3 and observes that these groups have the property that any two conjugates a, a^b commute. Let us see briefly why this is the case. Let $a, b \in G$. Then $1 = (ba)^3 = b^3 a^{b^2} a^b a$ which implies that

$$a^{b^2} a^b a = 1. \tag{1}$$

Replacing b by b^2 in (1) gives

$$a^b a^{b^2} a = 1 \tag{2}$$

From (1) and (2) it is clear that a and a^b commute or equivalently that a and $[b, a] = a^{-b} a$ commute. Thus every group of exponent 3 satisfies the 2-Engel identity

$$[[y, x], x] = 1.$$

It is clear that every finitely generated abelian periodic group is finite and it seems natural to think that the fact that groups of exponent 3 satisfy this weak commutativity property is the reason that they are locally finite. It comes thus hardly as a surprise that Burnside wrote a sequel to this paper [8] a year later where he singles out this weak commutativity property. This paper seems to have received surprisingly little attention, being the first paper written on Engel groups. In this paper Burnside proves that any 2-Engel group satisfies the laws

$$[x, y, z] = [y, z, x] \tag{3}$$

$$[x, y, z]^3 = 1. \tag{4}$$

In particular every 2-Engel group without elements of order 3 is nilpotent of class at most 2. Burnside failed however to observe that these groups are in general nilpotent of class at most 3, although he proved (in modern terminology) that any periodic 2-Engel group is locally nilpotent. It was C. Hopkins [29] that seems to have been the first to show that the class is at most 3. So any 2-Engel group also satisfies

$$[x, y, z, t] = 1. \tag{5}$$

Hopkins also observes that (3)–(5) characterize 2-Engel groups. This transparent description of the variety of 2-Engel groups is usually attributed to Levi [33], although his paper appears much later.

Of course this settles the study of 2-Engel groups no more than knowing that the variety of abelian groups is characterised by the law $[x, y] = 1$ settles the study of abelian groups. For example, the following well known problems raised by Caranti [37] still remain unsolved.

Problem 1   (a) Let G be a group of which every element commutes with all its endomorphic images. Is G nilpotent of class at most 2?

(b) Does there exist a finite 2-Engel 3-group of class three such that $\text{Aut}\, G = \text{Aut}_c G \cdot \text{Inn}\, G$ where $\text{Aut}_c G$ is the group of central automorphisms of G?

2   Zorn's Theorem and some generalisations

Moving on from the work of Burnside and Hopkins on 2-Engel groups the first main general result on Engel groups is Zorn's Theorem [63].

Theorem 2.1 (Zorn) *Every finite Engel group G is nilpotent.*

To prove this one argues by contradiction and taking G to be a minimal non-nilpotent Engel group one first makes use of a well known result of Schmidt that G must be solvable and then the shortest way of finishing the proof is to apply a result that appeared later, namely Gruenberg's Theorem (see Theorem 2.2).

Thus the Engel condition is a generalised nilpotence property. It is not difficult to find examples that show that in general it is weaker than nilpotence. For example for any given prime p the standard wreath product

$$G(p) = C_p \,\text{wr}\, C_p^{\infty}$$

is a $(p + 1)$-Engel p-group that is non-nilpotent. This example also shows that for any given n and prime $p < n$ there exists a non-nilpotent metabelian n-Engel p-group. As the group

$$H(p) = C_p \,\text{wr}\, C_p^m$$

is nilpotent of class n it is also clear that for a given n the nilpotence class of a finite n-Engel p-groups is not bounded when $p < n$ and this is even true under the further assumption that the group is metabelian.

The question that now arises is to what extent can Zorn's Theorem be generalised. The most natural question to ask here is whether one can replace "finite" by "finitely generated". One neccessary condition for nilpotence of a finitely generated group is residual finiteness. However we have the famous examples of Golod [16] that give for each prime p an infinite 3-generator residually finite p-group with all 2-generator subgroups finite. In particular these groups are non-nilpotent Engel groups. On the positive side we have that some other necessary conditions, namely solvability [18] and the max condition [3], are sufficient.

Theorem 2.2 (Gruenberg) *Every finitely generated solvable Engel group is nilpotent.*

Theorem 2.3 (Baer) *Every Engel group satisfying the max condition is nilpotent.*

So according to Theorem 2.3 we have that if every subgroup of the Engel group G is finitely generated then it must be nilpotent. Let us see briefly why Theorem 2.2 holds. For the proof we will apply the notion of a group being restrained. This is a very useful property introduced by Kim and Rhemtulla [32].

Definition A group G is said to be *restrained* if

$$\langle a \rangle^{\langle b \rangle}$$

is finitely generated for all $a, b \in H$.

Notice that in every n-Engel group $\langle a \rangle^{\langle b \rangle}$ is generated by $a, a^b, a^{b^2}, \ldots, a^{b^{n-1}}$ so every n-Engel group is restrained. We next prove an elementary but very useful lemma [32].

Lemma 2.4 *Let G be a finitely generated restrained group. If H is a normal subgroup of G such that G/H is cyclic, then H is finitely generated.*

Proof As G is finitely generated we have $G = \langle h_1, \ldots, h_r, g \rangle$ with $h_1, \ldots, h_r \in H$. Then

$$H = \langle h_1, \ldots, h_r \rangle^G \cdot \langle g^m \rangle$$

where m is the order of gH in G/H. So H is generated by g^m and

$$\langle h_1 \rangle^{\langle g \rangle} \cup \cdots \cup \langle h_r \rangle^{\langle g \rangle}.$$

As G is restrained each subset in this union is finitely generated. Hence H is finitely generated. □

From this we get the following easy corollary.

Lemma 2.5 *Let G be a finitely generated restrained group. Then G' is finitely generated.*

From this one can easily prove Gruenberg's Theorem. By a well known result of P. Hall we have that if N is normal subgroup of G such that $G/[N, N]$ and N are nilpotent, it follows that G is nilpotent. We use this to prove Gruenberg's Theorem by induction on the derived length of G. Of course this is obvious when G is abelian. For the induction step suppose that G has derived length $r > 1$. By the lemma above we have that $N = [G, G]$ is a finitely generated subgroup of derived length $r - 1$ and thus nilpotent by the induction hypothesis. By P. Hall's result we now only need to show that $G/[N, N]$ is nilpotent but it is an easy exercise to show that a finitely generated metabelian Engel group is nilpotent.

Remark There is a generalisation of Baer's Theorem due to Peng [42] that says that it suffices to have the max condition on the abelian subgroups. Further generalisations of Baer's and Gruenberg's theorem can be found in [10,11].

We end this section by mentioning another generalisation of Zorn's Theorem that is analogous to a theorem of Burnside on periodic linear groups. Burnside proved that any linear group of bounded exponent is finite. The corresponding result for Engel groups is a theorem of Suprenenko and Garščuk [15].

Theorem 2.6 *Any linear Engel group is nilpotent.*

By the examples of Golod there are finitely generated Engel groups that are not nilpotent. The following problem is the most important open question on Engel groups. We will come back to it later.

Problem 2 Is every finitely generated n-Engel group nilpotent?

3 More recent generalisations of Zorn's Theorem

Many of the results proved in the last two decades rely on Zel'manov's solution to the restricted Burnside problem that is based on some deep results on Engel Lie rings.

3.1 Engel Lie rings

These are the Lie algebra analogs of Engel groups.

We say that a Lie algebra L is an Engel Lie algebra if for each $u, v \in L$ there exists an integer $n = n(u, v)$ such

$$u \underbrace{v \cdots v}_{n} = 0.$$

(We are using here the left bracketing convention.) If n can be chosen independently of u, v then we say that L is an n-Engel Lie algebra. The connection between

Engel groups and Engel Lie algebras is established through the associated Lie ring. Consider the lower central series

$$G = G_1 \geq G_2 \geq \ldots$$

where $G_{i+1} = [G_i, G]$. It is well known that $[G_i, G_j] \leq G_{i+j}$. Let $L_i = G_i/G_{i+1}$ and consider the abelian group

$$L(G) = L_1 \oplus L_2 \oplus \cdots$$

that we turn into a Lie ring by first letting $aG_{i+1} \cdot bG_{j+1} = [a, b] \cdot G_{i+j+1}$ for $aG_{i+1} \in L_i$ and $bG_{j+1} \in L_j$ and then extending linearly to the whole of L. The standard commutator identities including the Hall–Witt identity imply that $L(G)$ is a Lie ring.

Suppose now that G is an n-Engel group and let $y, x_1, \ldots, x_n$ be variables. Expanding

$$1 = [y, \underbrace{x_1 \cdots x_n, \ldots, x_1 \cdots x_n}_{n}]$$

gives

$$1 = \left(\prod_{\sigma \in S_n} [y, x_{\sigma(1)}, \ldots, x_{\sigma(n)}] \right) \cdot z$$

where z is a product of commutators involving all of $y, x_1, \ldots, x_n$ and at least one of them twice. Take any elements $v = bG_{j+1} \in L_j$, $u_1 = a_1 G_{i_1+1} \in L_{i_1}, \ldots,$ $u_n = a_n G_{i_n+1} \in L_{i_n}$. Then for any $\sigma \in S_n$ we have that the product $vu_{\sigma(1)} \cdots u_{\sigma(n)}$ is in $L_{j+i_1+\cdots+i_n}$. Furthermore

$$\sum_{\sigma \in S_n} vu_{\sigma(1)} \cdots u_{\sigma(n)} = \prod_{\sigma \in S_n} [b, a_{\sigma(1)}, \ldots, a_{\sigma(n)}] G_{j+i_1+\cdots+i_n+1}$$

$$= 1 \cdot G_{j+i_1+\cdots+i_n+1}$$

$$= 0.$$

As $L(G)$ is generated by $L_1 \cup L_2 \cup \cdots$ as an abelian group, multilinearity gives us that $L(G)$ satisfies the "linearised n-Engel identity"

$$\sum_{\sigma \in S_n} yx_{\sigma(1)} \cdots x_{\sigma(n)} = 0. \tag{6}$$

Notice that when the characteristic of L is not divisible by any of the primes $p \leq n$ then the linearized Engel identity is equivalent to the n-Engel identity $yx^n = 0$. In the general situation if we take any $v = bG_{j+1} \in L_j$ and $u = aG_{i+1} \in L_i$ then $vu^n \in L_{j+ni}$ and as

$$vu^n = [b, \underbrace{u, \ldots, u}]G_{j+ni+1} = 1 \cdot G_{j+ni+1} = 0,$$

it follows that $L(G)$ satisfies

$$vu^n = 0 \text{ if } v \in L \text{ and } u \in L_i \text{ for some integer } i \geq 1. \tag{7}$$

The relevance of all this comes from the following two celebrated theorems of Zel'manov [58,60,61] that have had profound impact on the theory of Engel groups as we will see later.

Theorem Z1 *Let* $L = \langle a_1, \ldots, a_r \rangle$ *be a finitely generated Lie ring and suppose that there exist positive integers* s, t *such that*

$$\sum_{\sigma \in \mathrm{Sym}(s)} x x_{\sigma(1)} x_{\sigma(2)} \cdots x_{\sigma(s)} = 0 \tag{i}$$

$$xy^t = 0 \tag{ii}$$

for all $x, x_1, \ldots, x_s \in L$ *and all Lie products* y *of the generators* $a_1, a_2, \ldots, a_r$. *Then* L *is nilpotent.*

Theorem Z2 *Any torsion free n-Engel Lie ring is nilpotent.*

The first of these theorems was the main ingredient in Zel'manov's solution to the restricted Burnside problem for groups of prime power exponent and before we leave this section we discuss some analogs to the Burnside problems in the theory of Engel groups. Let us start by stating the main questions that originated from Burnside's paper.

b1) The general Burnside problem. Is every finitely generated periodic group finite?

b2) The Burnside problem. Let n be a given positive integer. Is every finitely generated group of exponent n finite?

b3) The restricted Burnside problem. Let r and n be given positive integers. Is there a largest finite r-generator group of exponent n?

These have the following analogs for Engel groups:

e1) The general local nilpotence problem. Is every finitely generated Engel group nilpotent?

e2) The local nilpotence problem. Let n be a given positive integer. Is every finitely generated n-Engel group nilpotent?

e3) The restricted local nilpotence problem. Let r and n be given positive integers. Is there a largest nilpotent r-generator n-Engel group?

As we have seen the answer to e1 is negative like the answer to b1 and in both cases Golod's examples provide the counterexamples. Theorem Z1 provides a positive answer to both b3 and e3 and we will deal with e3 in the next section. There is however a difference between b2 and e2 in that whereas there are well known counter examples to b2 no such are known for problem e2. We will have a closer look at this later.

3.2 Some consequences for Engel groups

From theorems Z1 and Z2 we are now going to derive strong consequences for Engel groups. We start with a consequence of Theorem Z2 on torsion free n-Engel groups that is due to Zel'manov [59].

As a preparation we first observe that the nilpotence class of a torsion free n-Engel Lie ring is n-bounded and in fact more is true. Let F be the relatively free Lie ring satisfying the linearised n-Engel identity on countably many free generators $x_1, x_2, \ldots$ and let I be the torsion ideal of F. By Theorem Z2 the Lie ring F/I is nilpotent of class say $m = m(n)$. Now if L is any torsion free Lie ring satisfying the linearised n-Engel identity and $u_1, \ldots, u_{m+1}$ any $m + 1$ elements in L, there is a homomorphism $\phi : F \to L$ that maps x_i to u_i for $i = 1, \ldots, m + 1$ and the remaining generators to 0. Now $x_1 \cdots x_{m+1} \in I$, say $kx_1 \cdots x_{m+1} = 0$. It follows that $ku_1 \cdots u_{m+1} = 0$. This shows that any Lie ring satisfying the linearised n-Engel identity satisfies the law $kx_1 \cdots x_{m+1} = 0$ and in particular if L is a torsion-free n-Engel Lie ring, we must have that L is nilpotent of class at most m. Let $\pi(n)$ be the set of all prime divisors of k.

Theorem 3.1 *There exists a finite set of primes $\pi = \pi(n)$ so that any locally nilpotent n-Engel group G without π-elements is nilpotent of n-bounded class. In particular any torsion-free n-Engel group that is locally nilpotent is nilpotent of n-bounded class.*

Proof Let $m = m(n), k = k(n)$ and $\pi = \pi(n)$ be as above and let $a_1, \ldots, a_{m+1}$ be any $m + 1$ elements of G. Let $H = \langle a_1, \ldots, a_{m+1} \rangle$ and consider the descending central series

$$H = \gamma_1(H) \geq \gamma_2(H) \geq \cdots .$$

Let $L(H)$ be the associated Lie ring. Then as we have seen above $L(H)$ satisfies the linearised n-Engel identity. By what we have observed above $L = L(H)$ satisfies the law $kx_1 \cdots x_{m+1} = 0$. I claim that the class of H is at most m. We argue by contradiction and suppose that the class is $c > m$. As $kL^c = 0$ it follows that $\gamma_c(H)^k \leq \gamma_{c+1}(H) = 1$. But by our assumption G has no elements of order dividing k. Hence $\gamma_c(H) = 1$ which contradicts the assumption that the class is c. This shows that $\gamma_{m+1}(H) = 1$ and as $a_1, \ldots, a_{m+1}$ where arbitrary it follows that G has class at most m. $\qquad \square$

We now move on to a second application that makes use of both Theorem Z1 and Z2 and gives a positive solution to problem e3. This is a theorem of Wilson [56]. Our proof is based on arguments from [5].

Theorem 3.2 *Any finitely generated residually nilpotent n-Engel group is nilpotent.*

Proof It suffices to show that there exists a positive integer $l(r, n)$ such that any nilpotent r-generator n-Engel group is nilpotent of class at most $l(r, n)$. Let $G = \langle a_1, \ldots, a_r \rangle$ be any nilpotent r-generator n-Engel group and let $L = L(G)$ be the associated Lie ring. Now $L = L(G)$ is generated by $u_1 = a_1 G_2, \ldots, u_r = a_r G_2$ and satisfies conditions (i) and (ii) of Theorem Z1. It follows that L is nilpotent of class at most l and thus that G is nilpotent of class at most l. $\square$

We mention without proof three other strong results that rely on Theorems Z1 and Z2. The first two [57,38] are strong generalisations of Zorn's Theorem for Engel groups without the assumption that the Engel degree is bounded. The second theorem is a generalisation of the first.

Theorem 3.3 (Wilson, Zel'manov) *Every profinite Engel group is locally nilpotent.*

Theorem 3.4 (Medvedev) *Every compact Engel group is locally nilpotent.*

The last result is on ordered groups. We say that a pair $(G, \leq)$ is an ordered group if $\leq$ is a total order on G and

$$a \leq b \Rightarrow xay \leq xby$$

for all $a, b, x, y \in G$. If we only assume the weaker condition

$$a \leq b \Rightarrow ay \leq by$$

for all $a, b, y \in G$ the group is said to be a right ordered group. Clearly every ordered and right ordered group is torsion-free and so it follows from Theorem 3.1 that a right ordered n-Engel group is locally nilpotent if and only if it is nilpotent. It is known that any torsion-free nilpotent group is orderable that this is also a sufficient condition follows from the following result of Kim and Rhemtulla [31].

Theorem 3.5 *Every orderable n-Engel group is nilpotent.*

The following is still an open question.

Problem 3 Is every right orderable n-Engel group nilpotent?

This is known to be case for 4-Engel groups [34,35]. Before leaving this section we mention another recent result of H. Smith [46]. This has to do with another generalised nilpotence property, namely subnormality. When G is n-Engel this turns out to be a sufficient condition for nilpotence.

Theorem 3.6 (H. Smith) *An n-Engel group with all subgroups subnormal is nilpotent.*

4 The structure of n-Engel groups

In this section we discuss the structure of n-Engel groups. The main question is whether n-Engel groups need to be locally nilpotent. We begin by proving a result that shows that if for a given n there exists a finitely generated n-Engel group that is not nilpotent, there must exist a finitely generated non-abelian simple n-Engel group. This result is well known among specialists in this area but as there doesn't seem to be a proof in the literature we include one here.

Theorem 4.1 *Let G be a finitely generated n-Engel group that is non-nilpotent. There exists a finitely generated section S of G that is simple non-abelian.*

Proof Let $R = \bigcap_{i=1}^{n} \gamma_i(G)$ be the nilpotent residual of G. By Gruenberg's Theorem, R is equal to the solvable residual. In particular R has no proper solvable quotient. By Wilson's Theorem, we know that G/R is nilpotent and as R was the solvable residual we know that $R = G^{(r)}$ for some positive integer r. By Lemma 2.5 we know then that R is finitely generated. As G is non-solvable, R is non-nilpotent. We claim that R has a maximal normal subgroup S with respect to R/S being non-nilpotent. Let $S_1 \subseteq \ldots \subseteq S_i \subseteq \ldots$ be an ascending chain of normal subgroups of R for which R/S_i is non-nilpotent. Let $S_0 = \bigcup_i S_i$. We claim that R/S_0 is non-nilpotent. We argue by contradiction and suppose that R/S_0 is nilpotent of class m. Equivalently all left-normed commutators in the generators of R of weight $m+1$ are in S_0. But there are only finitely many such commutators and so they would all be contained in some S_i. This however gives the contradiction that R/S_i is nilpotent. Hence R/S_0 is non-nilpotent. By Zorn's Lemma, R has a maximal normal subgroup S with respect to R/S being non-nilpotent. We finish the proof by showing that R/S is simple. Otherwise there would be a normal subgroup T of R lying strictly between S and R. By maximality of S we must then have that R/T is nilpotent. But we had already seen that R has no proper nilpotent quotient. Hence no such T exists and R/S is simple. □

Remark. Notice that the situation here is however different from the Burnside problem. By Baer's Theorem every n-Engel group that satisfies the max condition is nilpotent. Thus there are no n-Engel Tarsky monster and the structure of a simple n-Engel group, if it exists, is likely to be complicated. This is perhaps the underlying reason why it seems so difficult to solve the local nilpotence question for n-Engel groups. This question remains open except for $n \leq 4$. We will discuss the local nilpotence of 3-Engel and 4-Engel groups later.

As every finitely generated n-Engel group that is residually nilpotent is nilpotent it follows that the locally nilpotent n-Engel groups form a subvariety. We will study this variety for the remainder of this section. We start by proving a result that strengthens Theorem 3.1.

Proposition 4.2 *There exist numbers $l = l(n)$ and $m = m(n)$ such that the law*

$$[x_1, x_2, \ldots, x_{m+1}]^l = 1$$

holds in all locally nilpotent n-Engel groups.

Proof By Theorem 3.1, every torsion free locally nilpotent n-Engel group is nilpotent of bounded class, say $m = m(n)$. Let F be the free n-Engel group on $m + 1$ generators, say $x_1, \ldots, x_{m+1}$. Let $R = \cap_{i=1}^{\infty} \gamma_i(F)$. It is clear that F/R is residually nilpotent and as every finitely generated nilpotent group is residually finite it follows that F/R is residually finite. By Theorem 3.2 and Theorem 2.1, F/R is then nilpotent. Let T/R be the torsion group of F/R. Now F/T is a torsion free nilpotent n-Engel group and by the remark at the beginning of the proof, F/T is thus nilpotent of class at most m. So $[x_1, \ldots, x_{m+1}]^l \in R$ for some positive integer $l = l(n)$. Now let G be any locally nilpotent n-Engel group and let $g_1, \ldots, g_{m+1} \in G$. There is a homomorphism $\phi : F \to G$, $\phi(x_i) = g_i$, $i = 1, 2, \ldots, m + 1$. As $\langle g_1, \ldots, g_{m+1} \rangle = \phi(F)$ is nilpotent, we have that $R \leq \mathrm{Ker}\,(\phi)$. Hence $1 = \phi([x_1, \ldots, x_{m+1}]^l) = [g_1, \ldots, g_{m+1}]^l$. □

We will use this proposition to prove two strong structure results for locally nilpotent n-Engel groups. The first one is due to Burns and Medvedev [5] and can be derived from the proposition and Theorem 3.2.

Theorem 4.3 (Burns, Medvedev) *There exist positive integers m and r such that for any locally nilpotent n-Engel group G we have*

$$\gamma_{m+1}(G)^r = \{1\}.$$

Proof Let m and l be as in the proposition. Let F be the relatively free nilpotent n-Engel group on $m + 2$ generators $x_1, \ldots, x_{m+2}$. Let $H = \langle [x_1, \ldots, x_{m+1}], x_{m+2} \rangle$. Now $H'/[H', H']$ is abelian of exponent dividing l by Proposition 4.2. If H' has class c it follows that H' has exponent diving l^c. Take large enough integer e such that $f = \binom{l^e}{k}$ is divisible by l^c for $k = 1, \ldots, c$.

Now let $g = a_1 \cdots a_t$ be any product of left normed commutators of weight $m+1$. We prove by induction on t that $g^f = 1$. For $t = 1$ this follows from the proposition. Now suppose that $t \geq 2$ and that the result holds when t has a smaller value. Let $h = a_2 \cdots a_t$. We apply the Hall–Petrescu identity to see that

$$a_1^f h^f = (a_1 h)^f w_2^{\binom{f}{2}} w_3^{\binom{f}{3}} \cdots w_c^{\binom{f}{c}}$$

By the induction hypothesis the left hand side is trivial and by the choice of f we have that all $w_i^{\binom{f}{i}}$ are trivial as well. Hence $g^f = 1$. This finishes the inductive proof and hence $\gamma_{m+1}(G)^f = 1$. □

The second result is recent and due to Crosby and Traustason [9]. It generalises another result of Burns and Medvedev. It says that modulo the hyper-centre, every locally nilpotent n-Engel group is of bounded exponent.

Theorem 4.4 (Crosby, T) *There exists positive integers e and m so that any locally nilpotent n-Engel group satisfies the law*

$$[x^e, x_1, \ldots, x_m] = 1.$$

Proof Let l and m be as in Proposition 4.2. Now let G be the relatively free nilpotent n-Engel group on $m + 1$ generators. Suppose that the nilpotence class of G is $r = r(n)$. Notice that $r \geq m$. We finish the proof by showing by reverse induction on c that G satisfies the law

$$[x^{l^{r-m-c}}, x_1, \ldots, x_m, y_1, \ldots, y_c] = 1$$

for $c = 0, \ldots, r - m$. Letting $c = 0$ then gives us the theorem with $e = l^{r-m}$.

We turn to the inductive proof. The statement is clearly true for $c = r - m$. Now suppose that $0 \leq c \leq r - m - 1$ and that the result holds for larger values of c in $\{0, 1, \ldots, r - m\}$. By the induction hypothesis we have that $x^{l^{r-m-c-1}}$ is in the $(c + m + 1)$st centre. Hence

$$[x^{l^{r-m-c}}, x_1, \ldots, x_m, y_1, \ldots, y_c] = [(x^{l^{r-m-c-1}})^l, x_1, \ldots, x_m, y_1, \cdots, y_c]$$
$$= [x^{l^{r-m-c-1}}, x_1, \ldots, x_m, y_1, \ldots, y_c]^l$$
$$= 1.$$

This finishes the proof.   □

Remark. Burns and Medvedev had proved that there exist integers $e = e(n)$ and $m = m(n)$ such that for any locally nilpotent n-Engel group G we have that G^e is nilpotent of class at most m. The result above shows that we can choose e and m such that G^e is always in the mth centre.

We next turn to the structure of locally finite n-Engel p-groups. The following results are due to Abdollahi and Traustason [1]. We have seen that in general there is no upper bound for the nilpotency class of n-Engel p-groups for a given n and p. The situation is different when one restricts oneself to powerful p-groups. We remind the reader of some definitions. Let G be a finite p-group. If p is odd then G is said to be powerful if $[G, G] \leq G^p$ and if $p = 2$ then G is powerful if $[G, G] \leq G^4$. We also need the notion of powerful embedding. Let H be a subgroup of G. If p is odd then H is said to be powerfully embedded in G if $[H, G] \leq H^p$ and if $p = 2$ then we require instead that $[H, G] \leq H^4$.

Now we list some of the properties that we will be using. Let G be a powerful p-group. If a subgroup H is powerfully embedded in G then H^p is also powerfully embedded. We also have that $(G^{p^i})^{p^j} = G^{p^{i+j}}$. Furthermore, if G is generated by $x_1, \ldots, x_d$ then G^p is and generated by $x_1^p, \ldots, x_d^p$. It follows that if G is generated by elements of order dividing p^m then G has exponent dividing p^m. We also have that the terms of the lower central series are powerfully embedded in G.

Theorem 4.5 (Abdollahi, T) *There exists a positive integer $s = s(n)$ such that any powerful n-Engel p-group is nilpotent of class at most s.*

Proof By Proposition 4.2, $[g_1, \ldots, g_{m+1}]^l = 1$ for all $g_1, \ldots, g_{m+1} \in G$, where m and l are the integers given in Proposition 4.2. Suppose that $v = v(n)$ is the largest exponent of the primes that appear in the decomposition of l. Then

$[g_1, \ldots, g_{m+1}]^{p^v} = 1$. So $\gamma_{m+1}(G)$ is generated by elements of order dividing p^v. But $\gamma_{m+1}(G)$ is powerfully embedded in G and therefore it follows that $\gamma_{m+1}(G)$ is powerful and has exponent dividing p^v.

On the other hand, since $\gamma_{m+1}(G)$ is powerfully embedded in G, we have that $[\gamma_{m+1}(G), G] \leq \gamma_{m+1}(G)^p$ if p odd, and $[\gamma_{m+1}(G), G] \leq \gamma_{m+1}(G)^4$, if $p = 2$. Using some basic properties of powerful groups we see inductively that

$$[\gamma_{m+1}(G), \,_v G] \leq \gamma_{m+1}(G)^{p^v} = 1.$$

Hence, G is nilpotent of class at most $s(n) = m + v$. □

Building on this result one obtains [1] the following result. We omit the proof here.

Theorem 4.6 (Abdollahi, T) *Let p a prime and let $r = r(p, n)$ be the integer satisfying $p^{r-1} < n \leq p^r$. Let G be a locally finite n-Engel p-group.*
(a) If p is odd, then G^{p^r} is nilpotent of n-bounded class.
(b) If $p = 2$ then $(G^{2^r})^2$ is nilpotent of n-bounded class.

Remark. The r given in Theorem 4.6 is close to be the best lower bound. Suppose that n, p are such that $r = r(p, n) \geq 2$ and

$$G = \mathbb{Z}_p \text{wr} \left(\bigoplus_{i=1}^{\infty} \mathbb{Z}_{p^{r-1}} \right).$$

Then G is a metabelian n-Engel p-group of exponent p^r such that $G^{p^{r-2}}$ is not nilpotent. Therefore, if t is the least non-negative integer such that G^{p^t} is nilpotent, for all locally finite n-Engel p-groups G, then we have $t \in \{r - 1, r\}$ if p is odd and $t \in \{r - 1, r, r + 1\}$ if $p = 2$.

We now turn to some other related structural properties of n-Engel groups. By Gruenberg's theorem all solvable n-Engel groups are locally nilpotent and we thus have the following implications for n-Engel groups:

$$G \text{ nilpotent} \implies G \text{ solvable} \implies G \text{ locally nilpotent}.$$

By Theorem 3.1, we know that there are only finitely many primes that we need to exclude in order to have global nilpotence for a locally nilpotent n-Engel group. The answer to the following questions is therefore essential for clarifying the structure of locally nilpotent n-Engel groups:

Question 1. Which primes need to be excluded to have nilpotence?

Question 2. Which primes need to be excluded to have solvability?

Remark. As we have seen above, for each prime $p < n$ there exists a metabelian n-Engel p-group that is non-nilpotent. Under the assumption that these primes are excluded we however have that nilpotence and solvability are equivalent [19] (see also [23] for a close analysis of the metabelian case).

Theorem 4.7 (Gruenberg) *Let G be a solvable n-Engel group with derived length d. If G has no elements of prime order $p < n$ then G is nilpotent of class at most $(n + 1)^{d-1}$.*

Let us next consider another natural question. Recall that G is said to be a Fitting group if $\langle x \rangle^G$ is nilpotent for all $x \in G$. (That the class is m means that every commutator of the form $[\ldots, \underbrace{x, \ldots, x, \ldots, x}_{m+1}, \ldots]$ is trivial). If there is a bound for the nilpotence class of $\langle x \rangle^G$ and the lowest bound is n then we say that G has Fitting degree n.

Question 3. Are all locally nilpotent n-Engel groups Fitting groups?

It is clear that a group G is 2-Engel if and only if $\langle x \rangle^G$ is abelian for all $x \in G$ and by a well known result of L.C. Kappe and W. Kappe [30] that we will come back to later we have that G is 3-Engel if and only if $\langle x \rangle^G$ is nilpotent of class at most 2 for all $x \in G$. Every Fitting group with Fitting degree n is an n-Engel group and as we have just seen the converse is true for $n = 2$ and $n = 3$. However this fails for larger values of n. In [22] Gupta and Levin gave an example of a 4-Engel group that has Fitting degree 4 they also gave the following example that in particular shows that there are 5-Engel groups that are not Fitting groups.

Example. Let p be any prime greater than or equal to 3. Let G be the free nilpotent group of class 2 and exponent p that is of countable rank and let $\mathbb{Z}_p G$ be the group ring over the integers modulo p. Let M_p be the multiplicative group of 2×2 matrices over $\mathbb{Z}_p G$ of the form

$$\begin{pmatrix} g & 0 \\ r & 1 \end{pmatrix}, \quad g \in G, \ r \in \mathbb{Z}_p G.$$

One can check that M_p is a $(p + 2)$-Engel group that is not a Fitting group.

More recently [50] it has been established that 4-Engel groups are Fitting groups of Fitting degree at most 4 we will come back to this result later when we discuss 4-Engel groups. In view of the examples of Gupta and Levin the following question arises.

Problem 4 Let G be an n-Engel p-group where $p > n$. Is G a Fitting group?

5   3-Engel groups

5.1   The local nilpotence

It was Heineken [28] who proved in 1961 that 3-Engel are locally nilpotent. We give here a short new proof based on some arguments that were used to deal with 4-Engel groups [52]. The idea is to minimize the use of commutator calculus and try to apply arguments based on the symmetry of the problem and to make use of the Hirsch–Plotkin radical. Another short proof can also be found in [39].

It is not difficult to show that it suffices to show that any 3-generator 3-Engel group is nilpotent. In fact we will prove later a stronger result, namely the analogous result for 4-Engel groups (see Proposition 6.1). So we are left with the three-generator groups. We first prove two useful lemmas. The first one shows that within the class of n-Engel groups the locally nilpotent radical has the radical property.

Lemma 5.1 *Let G be an n-Engel group and let R be the Hirsch–Plotkin radical of G. Then the Hirsch–Plotkin radical of G/R is trivial.*

Proof Let S/R be the Hirsch-Plotkin radical of G/R. It suffices to show that S is locally nilpotent. Let H be a finitely generated subgroup of S. Then $H/(H \cap R) \cong HR/R$ is nilpotent and thus solvable of derived length, say m. As H is restrained we have by Lemma 2.5 that that $H^{(m)}$ is finitely generated subgroup of $H \cap R$ and thus nilpotent. Therefore H is solvable and thus nilpotent by Gruenberg's Theorem. $\square$

Lemma 5.2 *Let G be a 3-Engel group and $d, c, x \in G$. If d commutes with c and c^x then d commutes with any element in $\langle c \rangle^{\langle x \rangle}$.*

Proof It follows from $1 = [c, x, x, x]$ that

$$\langle c \rangle^{\langle x \rangle} = \langle c, c^x, c^{x^2} \rangle.$$

So it suffices to show that d commutes with c^{x^2}. But this follows from

$$\begin{aligned}
1 &= [c, dx, dx, dx] \\
&= [c, x, dx, dx] \\
&= [c, x, x, dx] \\
&= [c, x, x, d]^x.
\end{aligned}$$

So d commutes with $[c, x, x] = c^{-x} c c^{-x} c^{x^2}$ and thus c^{x^2}. $\square$

Lemma 5.3 *Every two-generator 3-Engel group is nilpotent.*

Proof Let $G = \langle a, b \rangle$ be a two-generator 3-Engel group and let R be the Hirsch–Plotkin radical. We know from Lemma 5.1 that G/R has a trivial Hirsch–Plotkin radical. Replacing G by G/R we can thus assume that G has a trivial Hirsch–Plotkin radical and the aim is then to show that G is trivial. As G is three-Engel we have that $[a, a^b]$ commutes with both a and a^b and $\langle a, a^b \rangle$ is nilpotent of class at most 2. By last lemma $Z(\langle a, a^b \rangle)$ is contained in the centre of $\langle a \rangle^G$ which is a abelian normal subgroup of G and thus trivial. Hence $\langle a, a^b \rangle$ is a nilpotent subgroup with a trivial centre and therefore trivial itself. It follows that $a = 1$ and G is cyclic and again as the Hirsch–Plotkin radical is trivial it follows that $G = \{1\}$. $\square$

Theorem 5.4 (Heineken) *Every 3-Engel group is locally nilpotent.*

Proof As we said previously it suffices to show that any 3-generator 3-Engel group is nilpotent. Thus let $G = \langle x, y, z \rangle$ be a 3-Engel group. As before we let $R(G)$ the Hirsch-Plotkin radical of G. We argue as in the proof of Lemma 5.3 and replacing G by $G/R(G)$ we can assume that $R(G) = 1$. We want to show that $G = 1$. Consider the following setting:

$$H = \langle xy^{-1}, yz^{-1} \rangle \qquad\qquad Z(H) \ni c$$
$$\langle c, c^x \rangle \qquad\qquad Z(\langle c, c^x \rangle) \ni d.$$

Note that

$$c^x = c^y = c^z.$$

By Lemma 5.2 we have that

$$C_G(\langle c, c^x \rangle) = C_G(\langle c \rangle^{\langle x \rangle})$$

which shows that $C_G(\langle c, c^x \rangle) = C_G(\langle c, c^y \rangle) = C_G(\langle c, c^z \rangle)$ is normalised by G. Thus in particular d^g commutes with c for all $g \in G$ or equivalently d commutes with c^g for all $g \in G$. Hence $d \in Z(\langle c \rangle^G)$ and thus trivial. Thus

$$Z(\langle c, c^x \rangle) = 1 \Rightarrow \langle c, c^x \rangle = 1 \Rightarrow Z(H) = 1 \Rightarrow H = 1 \Rightarrow G = \langle x \rangle \Rightarrow G = 1.$$

$\square$

5.2 Other structure results

We start by proving global nilpotence for 3-Engel groups that are $\{2, 5\}$-free. The short proof given here differs from Heineken's original argument [28] and is based on Lie methods. We start with an easy preliminary lemma.

Lemma 5.5 *Let L be a 2-Engel Lie ring. Then $3L^3 = 0$ and $L^4 = 0$.*

Proof Let $x, x_1, x_2, x_3 \in L$. The 2-Engel identity implies the linearised 2-Engel identity

$$xx_1x_2 + xx_2x_1 = 0,$$

or equivalently

$$xx_2x_1 = -xx_1x_2.$$

We use this to derive the result. Firstly

$$xx_1x_2 = -x_1xx_2 = x_1x_2x = -x(x_1x_2) = -2xx_1x_2$$

that gives $3L^3 = 0$. Secondly

$$xx_1x_2x_3 = -x_3(xx_1x_2) = x_3x_2(xx_1) = -xx_1(x_3x_2) = 2xx_1x_2x_3$$

that gives $xx_1x_2x_3 = 0$ and thus $L^4 = 0$. $\square$

Theorem 5.6 (Heineken) *Let G be a 3-Engel group that is $\{2, 5\}$-free. Then G is nilpotent of class at most 4.*

Proof Let G be any finitely generated $\{2,5\}$-free 3-Engel group. We know already that G is nilpotent and thus residually a finite p-group, $p \notin \{2,5\}$. We can thus assume that G is a finite 3-Engel p-group where $p \neq 2,5$. Let L be the associated Lie ring then, as we have seen previously, L satisfies the linearized 3-Engel identity. For $x,y \in L$ we let $X = \operatorname{ad} x$ and $Y = \operatorname{ad} y$. Since $p \neq 2$ we have that

$$XY^2 + YXY + Y^2X = 0. \tag{8}$$

For any $a \in L$ we also have

$$\begin{aligned} 0 &= -xay^2 - xyay - xy^2a \\ &= axy^2 + a(xy)y + a(xy^2) \\ &= 3axy^2 - 3ayxy + ay^2x, \end{aligned}$$

and therefore

$$3XY^2 - 3YXY + Y^2X. \tag{9}$$

We first deal with the case when $p = 3$. Replace L by $\tilde{L} = L/3L$. Now as the characteristic of $\tilde{L}$ is 3 we have $Y^2X = 0$. Consider the subspace

$$I = \operatorname{Sp}\langle uv^2 \mid u,v \in \tilde{L}, \rangle.$$

It is not difficult to see that I is an ideal. The quotient algebra $\tilde{L}/I$ is an 2-Engel algebra and thus nilpotent of class at most 3 by Lemma 5.5. As $\tilde{L}$ is centre-by-2-Engel it follows that $\tilde{L}$ is nilpotent of class at most 4. Now L is graded and this therefore implies that $L^5 \leq 3L^5 \leq 3^2L^5 \leq \cdots = 0$ and L is nilpotent of class at most 4. We can thus assume that char $L \neq 2,3,5$. From (8) and (9) we have

$$XY^2 = 2YXY \quad \text{and} \quad Y^2X = -3YXY. \tag{10}$$

It also follows that $3XY^2 = -2Y^2X$. If we interchange X and Y in (10) we get

$$YX^2 = 2XYX \quad \text{and} \quad X^2Y = -3XYX. \tag{11}$$

Now multiply (10) by X on the left and (11) on the right by Y. We then get

$$X^2Y^2 = 2XYXY \quad \text{and} \quad X^2Y^2 = -3XYXY.$$

It follows that $5X^2Y^2 = 0$ so $X^2Y^2 = 0$ since $p \neq 5$. As 2-Engel Lie algebras of characteristic $p \neq 3$ are nilpotent of class at most 2 it follows that that $X_1X_2Y^2 = 0$. Using $3XY^2 = -2Y^2X$ we get from this

$$4Y^2X_1X_2 = -9X_1X_2Y^2 = 0$$

and it follows that $Y_1Y_2X_1X_2 = 0$ and $L^5 = 0$. $\qquad\square$

We next turn to the Fitting property of 3-Engel groups [30]. The proof that we sketch comes from [24].

Theorem 5.7 (Kappe, Kappe) *Let G be a 3-Engel group then $\langle x \rangle^G$ is nilpotent of class at most 2 for all $x \in G$.*

Proof In order to prove the theorem it suffices to show that $[a^b, a, a^c] = 1$ for all $a, b, c \in G$. It therefore suffices to work with 3-generator groups $G = \langle a, b, c \rangle$. By Heineken's result every 3-Engel group is nilpotent and detailed analysis shows that it is nilpotent of class at most 5 (see for example [24] for details). It can now be checked using this detailed analysis that $[a^b, a, a^c] = [a[a, b], a, a[a, c]]$ is trivial. $\square$

The primes 2 and 5 that are not covered by Theorem 5.6 turn out to be exceptional. For the prime 2 this is simply because 2 is smaller than the Engel degree 3 and we had remarked earlier that for a prime $p < n$ there exists a metabelian n-Engel p-group that is non-nilpotent. Only the prime 5 turns however out to be exceptional with respect to solvability. In order to see this we need to show that any 3-Engel 2-group is solvable [21]. By Theorem 4.5 it suffices to show that every 3-Engel group of exponent 4 is solvable [25].

Theorem 5.8 (Gupta, Weston) *Every 3-Engel group of exponent 4 is solvable.*

Proof Let G be a 3-Engel group of exponent 4 and let $x, y \in G$. By Theorem 5.7 we have that
$$[x^2, x^{2y}] = [x, x^y]^4 = 1.$$
Hence $\langle x^2 \rangle^G$ is abelian for all $x \in G$. This implies that G^2 is generated by elements $a_1, a_2, \ldots$, where, for each i, $\langle a_i \rangle^G$ is abelian. As G/G^2 is abelian it suffices to show that G^2 is solvable. Let x_1, x_2, x_3 be any three of the generators. Using the 3-Engel identity we have for all $x \in G^2$
$$1 = [x, x_1 x_2 x_3, x_1 x_2 x_3, x_1 x_2 x_3]$$
$$= [[x, x_1], [x_2, x_3]] \cdot [[x, x_2], [x_3, x_1]] \cdot [[x, x_3], [x_1, x_2]].$$

Now let y be any element in G^2. Replacing x by xy gives
$$1 = [[x, x_1, y], [x_2, x_3]] \cdot [[x, x_2, y], [x_3, x_1]] \cdot [[x, x_3, y], [x_1, x_2]]. \tag{12}$$

Equipped with this identity and the Jacobi identity we will now see that G^2 is centre-by-metabelian. In order to see this, let x_1, x_2, x_3, x_4, x_5 be any five of the generators of G^2. We have
$$1 = [x_5, x_1 x_2 x_3 x_4, x_1 x_2 x_3 x_4, x_1 x_2 x_3 x_4, x_1 x_2 x_3 x_4]$$
$$= [[x_5, x_1, x_2], [x_3, x_4]] \cdot [[x_5, x_2, x_1], [x_3, x_4]]$$
$$\quad [[x_5, x_1, x_3], [x_2, x_4]] \cdot [[x_5, x_3, x_1], [x_2, x_4]]$$
$$\quad [[x_5, x_1, x_4], [x_2, x_3]] \cdot [[x_5, x_4, x_1], [x_2, x_3]]$$
$$\quad [[x_5, x_2, x_3], [x_1, x_4]] \cdot [[x_5, x_3, x_2], [x_1, x_4]]$$
$$\quad [[x_5, x_2, x_4], [x_1, x_3]] \cdot [[x_5, x_4, x_2], [x_1, x_3]]$$
$$\quad [[x_5, x_3, x_4], [x_1, x_2]] \cdot [[x_5, x_4, x_3], [x_1, x_2]] \quad \text{(by the Jacobi identity)}$$

$$= [[x_1, x_2, x_5], [x_3, x_4]] \cdot [[x_1, x_3, x_5], [x_2, x_4]] \cdot [[x_1, x_4, x_5], [x_2, x_3]]$$
$$[[x_2, x_3, x_5], [x_1, x_4]] \cdot [[x_2, x_4, x_5], [x_1, x_3]] \cdot [[x_2, x_1, x_5], [x_3, x_4]]$$
$$[[x_1, x_2, x_5], [x_3, x_4]] \cdot [[x_3, x_4, x_5], [x_1, x_2]]$$

(by (12) and the Jacobi identity)

$$= [[x_1, x_2], [x_3, x_4], x_5].$$

Hence the result. □

In fact one can show more than this. Gupta and Newman [24] have shown that any 5-torsion free 3-Engel group satisfies the identity $[[x_1, x_2, x_3], [x_4, x_5], x_6] = 1$ and so in particular they are solvable of derived length at most 3.

To show that the prime 5 is exceptional with respect to solvability is much harder to prove. This had been conjectured by Macdonald and Neumann [36] and was confirmed by Bachmuth and Mochizuki [2] using ring theoretic methods.

We leave the 3-Engel groups by mentioning again the article [24] by Gupta and Newman. This paper almost completes the description of the variety of 3-Engel groups. The authors in particular get a normal from theorem for the relatively free 3-Engel group of infinite countable rank without elements of order 2. They prove also that every n-generator 3-Engel group is nilpotent of class at most $2n - 1$ which is the best upper bound, that the fifth term of the lower central series of a 3-Engel group has exponent dividing 20 (again the best possible result) and they show that the subgroup generated by fifth powers satisfies the law $[[x_1, x_2, x_3], [x_4, x_5], x_6]$.

Problem 5 To obtain a normal form theorem for the relatively free 3-Engel group of infinite countable rank.

6   4-Engel groups

6.1   The local nilpotence theorem

In this section we give a broad outline of the proof of the local nilpotence theorem. The proof is quite intricate and spread over a few papers. This makes the complete proof very long and technical. The aim here is only to present the broad outline of the proof in linear order skipping over most of the technical details.

Let G be a finitely generated 4-Engel group. We want to show that G is nilpotent. In fact it is quite easy to see that it suffices to show that all three generator 4-Engel groups are nilpotent. Let us see why this is the case. The short proof presented here comes from [52].

Proposition 6.1 *A 4-Engel group is locally nilpotent if and only if all its 3-generator subgroups are nilpotent.*

Proof One inclusion is obvious. Suppose now that G is a 4-Engel group with all its 3-generator subgroups nilpotent. Let $a, b, c \in G$. By our assumption $H = \langle a, b, c \rangle$ is nilpotent and one can read from a polycyclic presentation of the free nilpotent 3-generator 4-Engel group [41] that $[[a, b, b, b], [a, b, b, b]^c] = 1$. It follows that $\langle [a,b,b,b] \rangle^G$ is abelian for all $a, b \in G$ and thus contained in the Hirch–Plotkin radical R of G. It follows that G/R is a 3-Engel group and thus locally nilpotent. But by Lemma 5.1, the Hirsch–Plotkin radical of G/R is trivial. Hence $G/R = \{1\}$ and $G = R$ and thus locally nilpotent. □

Even proving that the 2-generator 4-Engel groups are nilpotent turned out to be a major obstacle. Some good initial progress was however made because of the following much weaker result. This was first proved only for torsion groups [48] and then later for torsion-free groups [34] by Maj and Longobardi. Two proofs of the general result appeared about the same time. Vaughan-Lee gave a computer proof using the Knuth–Bendix algorithm (unpublished) and a machine free proof appeared in 1997 [49].

Proposition 6.2 *Let G be a 4-Engel group and let $a, b \in G$. Then $\langle a, a^b \rangle$ is metabelian and nilpotent of class at most 4.*

We skip over the proof which although not very long is quite intricate. Let us see how this proposition can be applied to make some progress on the local nilpotence problem. The following two lemmas come from [48].

Lemma 6.3 *Let G be a 4-Engel group. The 2-elements form a subgroup.*

Proof Suppose that $a^{2^i} = 1$. We then have

$$(ab)^{2^i} = a^{-2^i}(ab)^{2^i} = b^{a^{2^i-1}} b^{a^{2^i-2}} \cdots b^a b.$$

Since $\langle u, u^x \rangle$ is nilpotent for all $u, x \in G$, we have that $u^x u$ is a 2-element whenever u is a 2-element. Therefore $b^a b$ is a 2-element and then also $b^{a^3} b^{a^2} b^a b = (b^a b)^{a^2} (b^a b)$. By induction we get that

$$b^{a^{2^i-1}} b^{a^{2^i-2}} \cdots b^a b = (ab)^{2^i}$$

is a 2-element and hence ab is a 2-element. □

Lemma 6.4 *Let G be a 4-Engel group. Suppose that $a^{p^i} = 1$ where p is a prime and either $i \geq 2$ and p is odd or $i \geq 3$ and $p = 2$. Then*

$$[a^{p^{i-1}}, a^{p^{i-1}b}] = 1$$

for all $b \in G$.

Proof From the 4-Engel identity we have that $1 = [a, [a, [a, a^b]]]$. Now $\langle a, a^b \rangle$ is metabelian and it follows from this that

$$[a, a^b]^{a^2} = [a, a^b]^{2a-1}.$$

To deduce the result from this are straightforward calculations, we refer to [48] for the details. Notice also that when $p \neq 2, 3$ this is a consequence of $\langle a, a^b \rangle$ being a regular p-group as p is greater than the class. □

From this we can deduce the following main ingredient towards the proof of the local nilpotence result.

Proposition 6.5 *Let G be a 4-Engel p-group. If $p = 2$ or $p = 3$ then G is locally finite. For $p \geq 5$ we have that G/R is of exponent dividing p where R is the locally nilpotent radical.*

Proof Consider $H = G/R$. First assume that p is an odd prime number. Let $a \in G$. We claim that the order of $\bar{a} \in H$ divides p. We argue by contradiction and suppose that $\bar{a}$ has order p^i for some $i \geq 2$. By last lemma $\langle \bar{a}^{p^{i-1}} \rangle^H$ is abelian and thus contained in the Hirsch–Plotkin radical of H. But by Lemma 5.1, this radical is trivial. Hence we in particular get the contradiction that $\bar{a}^{p^{i-1}} = 1$. In the case when $p = 3$ we know that groups of exponent 3 are locally finite so we conclude that all 4-Engel 3-groups are locally finite. This leaves us with the case $p = 2$. Arguing in the same manner as for the odd case we see from the last lemma that G/R has exponent dividing 4. As groups of exponent 4 are known to be locally finite [44] it follows from this and Lemma 5.1 that G is locally finite. $\square$

In fact with a bit more work one can deduce from Proposition 6.2 that in any 4-Engel group the torsion elements form a subgroup T and that $T/Z(T)$ is a direct product of p-groups [48]. However next result proved about decade later [52] gives us a stronger result.

Proposition 6.6 *All 2-generator 4-Engel groups are nilpotent.*

Dealing with this major obstacle turned out to give one of the main tools for obtaining the local nilpotence result. The proof is similar in spirit to the proof that we gave of the local nilpotence of 3-Engel groups although it is much more complicated. The proof although relatively short, is quite intricate and tricky and we only give the general idea. Let $G = \langle x, y \rangle$ be a 2-generator 4-Engel group. Replacing G with G/R, where R is the Hirsch–Plotkin radical, one can assume that G has a trivial Hirsch–Plotkin radical and the aim is to show that if follows that G itself is trivial. As xy^{-1} and $y^{-1}x$ are conjugates it follows from Proposition 6.2 that $H = \langle xy^{-1}, y^{-1}x \rangle$ is nilpotent. Let $c \in Z(H)$, then $c^x = c^y$ and $c^{x^{-1}} = c^{y^{-1}}$. This symmetrical property of c turns out to be very useful and one is able after several reduction steps to show that c must be in the Hirsch–Plotkin radical of G. Hence $c = 1$. But then H is a nilpotent group with a trivial centre and thus trivial itself. This implies that $x = y$ and G is cyclic. As the Hirsch–Plotkin radical of G is trivial it follows that G is trivial.

From this we get the following easy corollaries.

Proposition 6.7 *Let G be a 4-Engel group. The torsion elements form a subgroup that is a direct product of p-groups. If R is the Hirsch-Plotkin radical of G then the torsion subgroup of G/R is a direct product of groups of prime exponent $p \neq 2, 3$.*

Lemma 6.8 *Let G be a 4-Engel group and $a, b \in G$. Then $\langle a, a^b \rangle$ is nilpotent of class at most 3.*

The proposition reduces the local nilpotence question for 4-Engel groups to groups that are either of prime exponent $p \neq 2, 3$ or torsion free. The lemma can be read off from the polycyclic structure of the free two-generator 4-Engel group [41] that we now know is nilpotent. Reducing the bound 4 to the correct bound 3 turned out to be significant for finishing the proof of the local nilpotence theorem.

We saw earlier on that in order to prove that 4-Engel groups are locally nilpotent it suffices to show that all three-generator 4-Engel groups are nilpotent. In fact this is not proved directly. Instead one obtains the weaker result that any subgroup generated by three conjugates is nilpotent. Before we discuss the proof of this result we introduce a notation. We say that a three-generator group $G = \langle a, b, c \rangle$ is of type (r, s, t) if

$$\langle a, b \rangle \text{ is nilpotent of class at most } r$$
$$\langle a, c \rangle \text{ is nilpotent of class at most } s$$
$$\langle b, c \rangle \text{ is nilpotent of class at most } t.$$

By Lemma 6.8, every 4-Engel group generated by three conjugates is of type $(3, 3, 3)$. The idea is roughly speaking to use induction on the complexity of the type. The next lemma provides the induction basis but is also another important tool to obtain other results.

Lemma 6.9 *Let* $G = \langle a, b, c \rangle$ *be a 4-Engel group of type* $(1, 2, 3)$. *Then* G *is nilpotent.*

This was first proved for 4-Engel groups of exponent 5 using coset enumeration [54]. The general result was first proved with an aid of a computer [27] using the Knuth Bendix procedure. A short machine proof was later given in [53]. With the aid of the machinery obtained so far one can prove the following [54,27].

Proposition 6.10 *Let* G *be a 4-Engel group that is either of exponent 5 or without* $\{2, 3, 5\}$-*elements and let* $a, x, y \in G$. *Then* $\langle a, a^x, a^y \rangle$ *is nilpotent.*

The proof is very technical and tricky and we skip over it here. We describe only the outline. The proof consists of few steps.

Step 1. $\langle a, a^{a^{a^x}}, a^{a^y} \rangle$ and $\langle a, a^{a^{a^x}}, (a^{a^{a^x}})^{a^y} \rangle$ are nilpotent.

Step 2. $\langle a, a^{a^{a^x}}, a^y \rangle$ is nilpotent.

Step 3. $\langle a, a^{a^x}, a^{a^y} \rangle$ and $\langle a, a^{a^x}, (a^{a^x})^{a^y} \rangle$ are nilpotent.

Step 4. $\langle a, a^{a^x}, a^y \rangle$ is nilpotent.

Step 5. $\langle a, a^x, a^y \rangle$ is nilpotent.

Notice that the groups in Step 1 are of type $(1, 2, 3)$ and therefore nilpotent by last lemma. The group in Step 2 is of type $(1, 3, 3)$, those in Step 3 of type $(2, 2, 3)$, the one in Step 4 of type $(2, 3, 3)$ and finally the one in Step 5 of type $(3, 3, 3)$. The proof of each step consists of clever commutator calculus building on the previous steps to obtain nilpotence. When the nilpotence has been established one can get a precise information about the structure using either machine or hand calculations. With the aid of last proposition one can then prove the following [54,27].

Proposition 6.11 *Let G be a 4-Engel group with a trivial Hirsch–Plotkin radical. Suppose that either G is of exponent 5 or without $\{2, 3, 5\}$-elements. Let $a, x, c \in G$. Then*

$$[c, [x, a, a, a], [x, a, a, a], [x, a, a, a]] = 1.$$

We again skip over the proof which makes use of several commutator identities that hold in the free 4-Engel group of rank 2. The final key step is the following result [27].

Proposition 6.12 *Let G be a 4-Engel group with trivial Hirsch–Plotkin radical, and let $a \in G$. Suppose that if a_1, a_2 are any conjugates of a in G, then $\langle a_1, a_2 \rangle$ has class at most 2. Then the normal closure of a in G is locally nilpotent.*

Let us see how we can now deduce the local nilpotence of 4-Engel groups from these results. First let G be a 4-Engel group of exponent 5 and let R be the Hirsch–Plotkin radical of G. Replacing G by G/R we can suppose that G has a trivial Hirsch–Plotkin radical. Let $x, a \in G$. By Proposition 6.11 we have that any two conjugates of $[x, a, a, a]$ in G generate a subgroup that is nilpotent of class at most 2. By Proposition 6.12 it follows that the normal closure of $[x, a, a, a]$ is locally nilpotent and therefore trivial as the Hirsch–Plotkin radical was trivial. It follows that G is a 3-Engel group and therefore locally nilpotent by Heineken's result. Hence G is trivial. Now we move on to the general case. Let G be a 4-Engel group. As 4-Engel groups of exponent 5 are locally finite we can deduce from Proposition 6.7 that G/R has no elements of order $2, 3$ or 5. Now the same argument as before using Proposition 6.11 and 6.12 shows that G/R is trivial.

6.2   Other structure results

In this section we will discuss various structure results for locally nilpotent 4-Engel groups. Our knowledge today is pretty good and almost on a level with our knowledge on locally nilpotent 3-Engel groups. We start with the question of global nilpotence [48].

Theorem 6.13 *Let G be a 4-Engel group without $\{2, 3, 5\}$-elements. Then G is nilpotent of class at most 7.*

Like the analogous result for 3-Engel groups, this can be proved using Lie ring methods. It suffices to show that any finite 4-Engel p-group G, where $p \neq 2, 3, 5$ is nilpotent of class at most 7. Let $L(G)$ be the associated Lie ring of G. As $p > 4$ we have that $L(G)/pL(G)$ is a 4-Engel Lie algebra over the field of p elements. But any 4-Engel Lie algebra over a field of characteristic $p \notin \{2,3,5\}$ is known to be nilpotent of class at most 7 [17,47]. □

Remark. The primes $2,3,5$ are genuinely exceptional. The primes 2 and 3 are exceptional because they are less than 4 and 5 is exceptional as it was already exceptional for 3-Engel groups. By considering a power-commutator presentation of the relatively free 4-Engel group on three generators [41], one can see that 7 is the best upper bound for the class.

We next turn to solvability. As 5 was exceptional for 3-Engel groups it remains so for 4-Engel groups. The structure of 4-Engel groups of exponent 5 is studied in quite some detail in [40]. In particular the authors obtain a normal form theorem for the relatively free group in this variety.

Groups of exponent 4 are known to be center-by-4-Engel and by a well known result of Razmyslov [44] there are non-solvable groups of exponent 4. Thus we have that 2 is an exceptional prime as well with respect to solvability. This leaves out the prime 3. That 4-Engel 3-groups are solvable was proved by Abdollahi and T [1].

Theorem 6.14 *Every 4-Engel 3-group G is solvable.*

We don't include the proof here as it is quite technical. The general strategy is the same as for the analogous result for 3-Engel groups although the details are much harder. Notice that as $3 < 4 \le 3^2$ it follows from Theorem 4.5 that G^9 is nilpotent. One can therefore first reduce the problem to groups of exponent 9.

As we mentioned in section 4 there are 5-Engel groups that are not Fitting groups. There remains the question whether 4-Engel groups are Fitting groups. The best possible result would be if the Fitting degree was always at most 3. By the example of Gupta and Levin mentioned earlier, we know however that this is not the case. However 4-Engel groups are always Fitting groups of degree at most 4 [50].

Theorem 6.15 *Let G be a 4-Engel group. Then G is a Fitting group of Fitting degree at most 4. Furthermore if G has no $\{2,5\}$ elements then G has Fitting degree at most 3.*

Although there are 4-Engel groups with Fitting degree greater than 3 there is another way of characterizing 4-Engel groups in terms of the normal closures of elements. Notice that a group G is a 3-Engel group if and only if the normal closure of every element is 2-Engel. Vaughan-Lee [55] has shown recently that the analog holds for 4-Engel groups. It should be noted that this property fails to hold for 5-Engel groups.

Theorem 6.16 (Vaughan-Lee) *G is a 4-Engel group if and only if $\langle x \rangle^G$ is 3-Engel for all $x \in G$.*

We leave this section with two challenging problems. M. Newell has obtained a remarkable generalisation of Heineken's local nilpotence theorem for 3-Engel groups by showing that in any group G, the right 3-Engel elements belong to the locally nilpotent radical. Whether the analog holds for 4-Engel groups is an open problem.

Problem 6 Let G be a group. Do the right 4-Engel elements belong to the locally nilpotent radical of G? Do the left 3-Engel elements belong to the locally nilpotent radical of G?

What about the structure of 5-Engel groups? At present we hardly know anything about them.

Problem 7 Describe the structure of 5-Engel groups?

7 Generalisations

As we mentioned in the the beginning, the theory of Engel groups and the Burnside problems are closely related and in this last section we look at some recent generalised settings. First we will discuss what we call generalised Burnside varieties. These are natural generalisations of the Burnside varieties and the Engel varieties and share many properties with these. There are two types of these that we will refer to as the strong generalised Burnside varieties and the weak generalised Burnside varieties where the latter include the former. Then we will discuss a further generalisation where instead of varieties we work with certain classes of groups satisfying weaker conditions.

7.1 Generalised Burnside varieties

Let $\mathcal{V}$ be a variety of groups. It is well known that the following are equivalent:

A1) For each positive integer r the class of all nilpotent r-generator groups in $\mathcal{V}$ is r-bounded.

A2) Every finitely generated group G in $\mathcal{V}$ that is residually nilpotent is nilpotent.

A3) The locally nilpotent groups in $\mathcal{V}$ form a subvariety.

For example the Burnside variety $\mathcal{B}_n$ of groups of exponent n and the n-Engel variety $\mathcal{E}_n$ of n-Engel groups satisfy these conditions.

Let us call a variety that satisfies A1)–A3) a strong generalised Burnside variety. Let C be the infinite cyclic group and for any positive integer n, let C_n be the cyclic group of order n. That the wreath products $C \operatorname{wr} C_n$ and $C_n \operatorname{wr} C$ play a crucial role for describing generalised Burnside varieties is already apparent in the work of J. Groves [20] that was later taken on by G. Endimioni [12,13] and F. Point [43] who studied what the latter refers to as Milnor identities and correspond to the generalised Burnside varieties. Building on the work of F. Point and G. Endimioni the following transparent description of these varieties was obtained in [51] (see also [14] for another proof) using again Zel'manov's results one Engel Lie-rings.

Theorem 7.1 *Let $\mathcal{V}$ be a variety. The following are equivalent.*

(1) $\mathcal{V}$ is a strong generalised Burnside variety.

(2) The groups $C_p \, wr \; C$ and $C \, wr \, C_p$ do not belong to $\mathcal{V}$ for any prime p.

The reason why this criteria is transparent is that the wreath products in (2) are metabelian groups with a simple structure. If one has a clear description of the laws that the variety $\mathcal{V}$ satisfies then it is quite straightforward to check if these groups satisfy these laws. For example, as none of these groups is periodic the Burnside varieties are generalised Burnside varieties and as none of these groups satisfy an Engel identity the n-Engel varieties are also included.

Many of the results for n-Engel groups described above have analogs for generalised Burnside varieties. For example we have the following result of G. Endimioni [13] that gives another characterisation of generalised Burnside varieties. In the following, $\bar{\mathcal{B}}_e$ is the variety of locally nilpotent groups of exponent e and $\mathcal{N}_c$ is the variety of groups that are nilpotent of class at most c.

Theorem 7.2 *Let $\mathcal{V}$ be a variety. The following are equivalent.*

(1) $\mathcal{V}$ is a strong generalised Burnside variety.

(2) There exist positive integers c, e such that all locally nilpotent groups in $\mathcal{V}$ are both in $\mathcal{N}_c\bar{\mathcal{B}}_e$ and $\bar{\mathcal{B}}_e\mathcal{N}_c$.

Remark. In view of Therorem 4.4 one can choose the integers c, e such that the second variety in (2) can be replaced by the variety of groups satisfying $[x^e, x_1, \ldots, x_c]$.

This theorem preceded Theorem 7.1 but can also be derived from it and the some of the structure theorems on n-Engel groups.

Proof (2)$\Rightarrow$(1). As none of the wreath products $C_p \, wr \, C$ are nilpotent-by-torsion and none of the wreath products $C \, wr \, C_p$ are torsion-by-nilpotent it follows from Theorem 7.1 that the variety $\mathcal{W} = \mathcal{V} \cap \mathcal{N}_c\bar{\mathcal{B}}_e \cap \bar{\mathcal{B}}_e\mathcal{N}_c$ is a generalised Burnside variety. So the locally nilpotent groups of $\mathcal{W}$ form a variety. But by our assumption these are also the locally nilpotent groups of $\mathcal{V}$. We thus conclude from A3) that $\mathcal{V}$ is a generalised Burnside variety.

(1)$\Rightarrow$(2). Let $\mathcal{W}$ be the variety consisting of the locally nilpotent groups of $\mathcal{V}$ and let F be the free 2-generator group in $\mathcal{W}$. Then F is nilpotent and thus in particular a n-Engel group for some positive integer n. It follows that all the locally nilpotent groups in $\mathcal{V}$ are n-Engel and (2) thus follows from Theorems 4.3 and 4.4. $\square$

There are related types of varieties satisfying a weaker condition that have been studied by several authors. Let us call a variety $\mathcal{V}$ a *weak generalised Burnside variety* if it satisfies the following equivalent conditions.

B1) For each positive integer r there exist positive integers $c(r)$ and $e(r)$ so that all finite r-generator groups in $\mathcal{V}$ are in $\mathcal{N}_{c(r)}\bar{\mathcal{B}}_{e(r)}$.

B2) Every finitely generated group G in $\mathcal{V}$ that is residually finite is nilpotent-by-finite.

B3) The locally nilpotent-by-finite groups in $\mathcal{V}$ form a subvariety.

The following Theorem is due to G. Endimioni [13] although not stated explicitly there in this form.

Theorem 7.3 (Endimioni) *Let $\mathcal{V}$ be a variety. The following are equivalent.*
(1) $\mathcal{V}$ is a weak generalised Burnside variety.
(2) The group C_p wr C does not belong to $\mathcal{V}$ for any prime p.

We also have the analogous criteria to Theorem 7.2 also due to G. Endimioni. It should be noted that Burns and Medvedev have independently arrived at similar results in [6]. Notice that the following theorem tells us that the $c = c(r)$ and $e = e(r)$ in B1) above can be chosen independently of r.

Theorem 7.4 *Let $\mathcal{V}$ be a variety. The following are equivalent.*
(1) $\mathcal{V}$ is a weak generalised Burnside variety.
(2) There exist positive integers c, e such that all locally nilpotent groups in $\mathcal{V}$ are in $\mathcal{N}_c \bar{\mathcal{B}}_e$.

There is an interesting open question related to strong generalised varieties. This is motivated by a result of Zel'manov [62] who has shown that for any given prime p the variety $\bar{\mathcal{B}}_p$, consisting of the locally finite groups of exponent p, is finitely based. It has been observed by G. Endimioni (written correspondence) that the following are equivalent:

(1) The variety $\bar{\mathcal{B}}_n$ of all locally nilpotent groups of exponent n is finitely based for all positive integers n.

(2) The variety $\bar{\mathcal{E}}_n$ of all locally nilpotent n-Engel groups is finitely based for all positive integers n.

(3) If a strong generalised Burnside variety $\mathcal{V}$ is finitely based then so is the variety $\bar{\mathcal{V}}$ consisting of all locally nilpotent groups in $\mathcal{V}$.

Proof $(2) \Rightarrow (3)$. Suppose that the free 2-generator group of $\bar{\mathcal{V}}$ is n-Engel. Then $\bar{\mathcal{V}} = \mathcal{V} \cap \bar{\mathcal{E}}_n$ and thus finitely based as both $\mathcal{V}$ and $\bar{\mathcal{E}}_n$ are finitely based.

$(3) \Rightarrow (1)$. Clear since $\bar{\mathcal{B}}_n$ is a strong generalised Burnside variety.

$(1) \Rightarrow (2)$. From the work of Burns and Medvedev we have that there are positive integers $e(n)$ and $c(n)$ such that any group G in $\bar{\mathcal{E}}n$ satisfies $\gamma_{c(n)}(G^{e(n)}) = \{1\}$. As the variety $\bar{\mathcal{B}}_{e(n)}$ is finitely based the same is true of the variety of all groups that are (class $c(n)$)-by-$\bar{\mathcal{B}}_{e(n)}$. As $\bar{\mathcal{E}}_n$ is the intersection of this variety and $\mathcal{E}_n$ it is finitely based. $\square$

Notice that the equivalence above also demonstrates again how the Burnside varieties and the Engel varieties are closely linked.

Problem 8 Is it true that for every strong generalised Burnside variety $\mathcal{V}$ that is finitely based, we have that the variety $\bar{\mathcal{V}}$ of all the locally nilpotent groups in $\mathcal{V}$ is also finitely based?

Remark. In particular if all n-Engel groups are locally nilpotent then the answer is yes.

7.2 Generalised Engel groups

Some of the results discussed in section 7.1 can be generalised even further. First we need some definitions and notations.

Let G be any group. For $a, t \in G$, let $H = H(a, t) = \langle a \rangle^{\langle t \rangle}$ and

$$A(a, t) = H/[H, H].$$

Then $A(a, t)$ is an abelian section of G. Let $E(a, t)$ be the ring of all endomorphisms of $A(a, t)$. Notice that t induces an endomorphism on $A(a, t)$ by conjugation.

Definition Let $I \trianglelefteq \mathbb{Z}[x]$. We say that G is an I-group if

$$a^{f(t)} = 0$$

in $A(a, t)$ for all $a, t \in G$ and for all $f \in I$.

For example any n-Engel group is an $\mathbb{Z}[x](x-1)^n$-group.

If G is any group then the set of polynomials f, such that $a^{f(t)} = 0$ in $A(a, t)$ for all $a, t \in G$, form an ideal $I(G)$. There is therefore a unique maximal ideal I such that G is an I-group. We say that two groups H and G are $\mathbb{Z}[x]$-equivalent if $I(H) = I(G)$. We now turn to the generalisations of some of the results of section 7.1. For each prime number p let f_p be the irreducible polynomial $f_p = x^{p-1} + x^{p-2} + \cdots + 1$. The following theorems come from [51]. The first one builds on a work of A. Shalev [45] on collapsing groups (see also [4]).

Theorem 7.5 *Let $f \in \mathbb{Z}[x]$ such that f is not divisible by any prime p. Then there exists positive integers $c(f)$ and $e(f)$ such that*

$$\gamma_{c(f)}(G^{e(f)}) = \{1\}$$

for any $\mathbb{Z}[x]f$-group that is nilpotent-by-finite.

Theorem 7.6 *Let $f \in \mathbb{Z}[x]$ such that f is neither divisible by p nor f_p for all primes p. For each positive integer r there exists a positive integer $c(r, f)$ such that*

$$\gamma_{c(r,f)}(G) = \{1\}$$

for any nilpotent r-generator $\mathbb{Z}[x]f$-group in G.

Theorem 7.7 *Let $f \in \mathbb{Z}[x]$ such that f is neither divisible by p nor f_p for all primes p. There exist positive integers $c(f)$ and $e(f)$ such that*

$$[G^{e(f)}, _c G] = (\gamma_{c(f)}(G))^{e(f)} = \{1\}$$

for any nilpotent $\mathbb{Z}[x]f$-group in G.

Proof It follows from Theorem 7.6 that all the nilpotent $\mathbb{Z}[x]f$-group G are $c(2, f)$-Engel and the rest now follows from Theorems 4.3 and 4.4. $\square$

References

[1] A. Abdollahi and G. Traustason, On locally finite p-groups satisfying an Engel condition, *Proc. Amer. Math. Soc.* **130** (2002), 2827–2836.

[2] S. Bachmuth and H. Y. Mochizuki, Third Engel groups and the Macdonald–Neumann conjecture, *Bull. Austral. Math. Soc.* **5** (1971), 379–385.

[3] R. Baer, Engelsche Elemente Noetherscher Gruppen, *Math. Ann.* **133** (1957), 256–270.

[4] R. G. Burns, O. Madcedoćnska and Y. Medvedev, Groups satisfying semigroup laws and nilpotent-by-Burnside varieties, *J. Algebra* **195** (1997), 510–525.

[5] R. G. Burns and Y. Medvedev, A note on Engel groups and local nilpotence, *J. Austral. Math. Soc. (Series A)* **64** (1998), 92–100.

[6] R. G. Burns and Y. Medvedev, Group laws implying virtual nilpotence, *J. Austral. Math. Soc.* **74** (2003), 295–312.

[7] W. Burnside, On an unsettled question in the theory of discontinous groups, *Quart. J. Pure Appl. Math.* **37** (1901), 230–238.

[8] W. Burnside, On groups in which every two conjugate operations are permutable, *Proc. London Math. Soc.* **35** (1902), 28–37.

[9] P. G. Crosby and G. Traustason, A remark on the structure of n-Engel groups, submitted.

[10] G. Endimioni, Une condition suffisante pour qu'n groupe d'Engel soit nilpotent, *C. R. Acad. Sci. Paris* **308** (1989), 75–78.

[11] G. Endimioni, Groupes d'Engel avec la condition maximale sur les p-sous-groupes finis abéliens, *Arch. Math.* **55** (1990), 313–316.

[12] G. Endimioni, On the locally finite p-groups in certain varieties of groups, *Quart. J. Math.* **48** (1997), 169–178.

[13] G. Endimioni, Bounds for nilpotent-by-finite groups in certain varieties, *J. Austral. Math. Soc.* **73** (2002), 393–404.

[14] G. Endimioni and G. Traustason, On varieties in which soluble groups are torsion-by-nilpotent, *Internat. J. Algebra Comput.* **15** (2005), 537–545.

[15] M. S. Garaščuk and D. A. Suprunenko, Linear groups with Engel's condition, *Dokl. Akad. Nauk BSSR* **6** (1962), 277–280.

[16] E. S. Golod, Some problems of Burnside type, Internat. Congress Math. Moscow (1966), 284–298 = *Amer. Math. Soc. Translations (2)* **84** (1969), 83–88.

[17] M. I. Golovanov, Nilpotency class of 4-Engel Lie rings, *Algebra Logika* **25** (1986), 508–532.

[18] K. W. Gruenberg, Two theorems on Engel groups, *Proc. Cambridge Philos. Soc.* **49** (1953), 377–380.

[19] K. W. Gruenberg, The upper central series in soluble groups, *Illinois J. Math.* **5** (1961), 436–466.

[20] J. R. J. Groves, Varieties of soluble groups and a dichotomy of P. Hall, *Bull Austral. Math. Soc.* **5** (1971), 394–410.

[21] N. D. Gupta, Third-Engel 2-groups are soluble, *Canad. Math. Bull.* **15** (1972), 523–524.

[22] N. D. Gupta and F. Levin, On soluble Engel groups and Lie algebras, *Arch. Math.* **34** (1980), 289–295.

[23] N. D. Gupta and M. F. Newman, On metabelian groups, *J. Austral. Math. Soc.* **6** (1966), 362–368.

[24] N. D. Gupta and M. F. Newman, Third Engel groups, *Bull. Austral. Math. Soc.* **40** (1989), 215–230.

[25] N. D. Gupta and K. W. Weston, On groups of exponent four, *J. Algebra* **17** (1971), 59–66.

[26] P. Hall, Some word-problems, *J. London Math. Soc.* **33** (1958), 482–496.

[27] G. Havas and M. R. Vaughan-Lee, 4-Engel groups are locally nilpotent, *Internat. J. Algebra Comput.* **15** (2005), 649–682.

[28] H. Heineken, Engelsche Elemente der Länge drei, *Illinois J. Math.* **5** (1961), 681–707.

[29] C. Hopkins, Finite groups in which conjuate operations are commutative, *Amer. J. Math.* **51**, (1929), 35–41

[30] L. C. Kappe and W. P. Kappe, On three-Engel groups, *Bull. Austral. Math. Soc.* **7** (1972), 391–405.

[31] Y. K. Kim and A. H. Rhemtulla, Orderable groups satisfying an Engel condition, in *Ordered algebraic structures (Gainesville, FL, 1991)*, 73–79, Kluwer Acad. Publ., Dordrecht, 1993.

[32] Y. K. Kim and A. H. Rhemtulla, Weak maximality condition and polycyclic groups, *Proc. Amer. Math. Soc.* **123**, (1995), 711–714.

[33] F. W. Levi, Groups in which the commutator operations satisfies certain algebraic conditions, *J. Indian Math. Soc.* **6** (1942), 87–97.

[34] P. Longobardi and M. Maj, Semigroup identities and Engel groups, in *Groups St Andrews 1997 in Bath II*, 527–531, London Math. Soc. Lecture Note Ser. **261**, Cambridge Univ. Pess, Cambridge 1999.

[35] P. Longobardi and M. Maj, On some classes of orderable groups, *Rend. Sem. Mat. Fis. Milano* **68** (2001), 203–216.

[36] I. D. Macdonald and B. H. Neumann, A third-Engel 5-group, *J. Austral. Math. Soc.* **7** (1967), 555–569.

[37] V. D. Mazurov and E. I. Khukhro (eds.), *The Kourovka Notebook*, Sixteenth Edition, Russian Academy of Sciences, Siberian Division Novosibirsk 2006, Unsolved problems in group theory.

[38] Y. Medvedev, On compact Engel groups, *Israel J. Math.* **135** (2003), 147–156.

[39] M. Newell, On right-Engel elements of length three, *Proc. Royal Irish Academy* **96A** (1996), 17–24.

[40] M. F. Newman and M. R. Vaughan-Lee, Engel 4 groups of exponent 5. II. Orders, *Proc. London Math. Soc. (3)* **79** (1999), 283–317.

[41] W. Nickel, Computation of nilpotent Engel groups, *J. Austral. Math. Soc.* **67** (1999), 214-222.

[42] T. A. Peng, Engel elements of groups with maximal condition on abelian subgroups, *Nanta Math.* **1** (1966), 23–28.

[43] F. Point, Milnor identities, *Comm. Algebra* **24** (1996), 3725–3744.

[44] Ju. P. Razmyslov, The Hall-Higman problem, *Izv. Akad. Nauk SSSR Ser. Mat.* **42** (1978), 833–847.

[45] A. Shalev, Combinatorial conditions in residually finite groups II, *J. Algebra* **157** (1993), 51–62.

[46] H. Smith, Bounded Engel groups with all subgroups subnormal, *Comm. Algebra* **30** (2002), no. 2, 907–909.

[47] G. Traustason, Engel Lie-algebras, *Quart. J. Math. Oxford (2)* **44** (1993), 355–384.

[48] G. Traustason, On 4-Engel groups, *J. Algebra* **178** (1995), 414–429.

[49] G. Traustason, Semigroup identities in 4-Engel groups, *J. Group Theory* **2** (1999), 39–46.

[50] G. Traustason, Locally nilpotent 4-Engel groups are Fitting groups, *J. Algebra* **270** (2003), 7–27.

[51] G. Traustason, Milnor groups and (virtual) nilpotence, *J. Group Theory* **8** (2005), 203–221.

[52] G. Traustason, Two generator 4-Engel groups, *Internat. J. Algebra Comput.* **15** (2005), 309–316.

[53] G. Traustason, A note on the local nilpotence of 4-Engel groups, *Internat. J. Algebra Comput.* **15** (2005), 757–764.

[54] M. R. Vaughan-Lee, Engel-4 groups of exponent 5, *Proc. London Math. Soc. (3)* **74** (1997), 306–334.

[55] M. R. Vaughan-Lee, On 4-Engel groups, *LMS J. Comput. Math.* **10** (2007), 341–353.

[56] J. S. Wilson, Two-generator conditions for residually finite groups, *Bull. London Math. Soc.* **23** (1991), 239–248.

[57] J. S. Wilson and E. I. Zel'manov, Identities for Lie algebras of pro-p groups, *J. Pure Appl. Algebra* **81** (1992), 103–109.

[58] E. I. Zel'manov, Engel Lie-algebras, *Dokl. Akad. Nauk SSSR* **292** (1987), 265–268.

[59] E. I. Zel'manov, Some problems in the theory of groups and Lie algebras, *Mat. Sb.* **180** (1989), 159–167.

[60] E. I. Zel'manov, The solution of the restricted Burnside problem for groups of odd exponent, *Math. USSR Izv.* **36** (1991), 41–60.

[61] E. I. Zel'manov, The solution of the restricted Burnside problem for 2-groups, *Mat. Sb.* **182** (1991), 568–592.

[62] E. I. Zel'manov, On additional laws in the Burnside problem on periodic groups, *Internat. J. Algebra Comput.* **3** (1993), 583–600.

[63] M. Zorn, Nilpotency of finite groups, *Bull. Amer. Math. Soc.* **42** (1936), 485–486.

LIE METHODS IN ENGEL GROUPS

MICHAEL VAUGHAN-LEE

Christ Church, Oxford, OX1 1DP, England
Email: vlee@maths.ox.ac.uk

1 Introduction

A group G is said to be an n-Engel group if

$$[x, \underbrace{y, y, \ldots, y}_{n}] = 1$$

for all $x, y \in G$. (Here $[x, y]$ denotes $x^{-1}y^{-1}xy$, and $[x, y, \ldots, y]$ denotes the left-normed commutator $[[\ldots[x, y], \ldots], y]$.) The big question is:

Problem 1 Are n-Engel groups locally nilpotent?

My own guess is that for large n the answer is almost certainly "no", but I have no evidence on which to base this guess. For very small n we do have some positive answers. Of course 1-Engel groups are abelian, and 2-Engel groups are nilpotent of class at most 3 — see Levi [9]. Heineken [5] proved that 3-Engel groups are locally nilpotent, and in 2005 Havas and Vaughan-Lee [4] proved that 4-Engel groups are locally nilpotent. However the problem is wide open for $n > 4$.

We know quite a bit about the structure of n-Engel groups for $n \leq 4$ in addition to the fact that they are locally nilpotent.

If x and y are elements in a 2-Engel group then the relation $[x, y, y] = 1$ implies that y commutes with $[x, y]$, and this in turn implies that y commutes with y^x. So in a 2-Engel group the normal closure of an element is abelian.

Heineken [5] proved that a 3-Engel group is nilpotent of class at most 4 if it has no elements of order 2 or 5. We will give a general construction below for non-nilpotent $(p + 1)$-Engel p-groups, so 3-Engel 2-groups do not have to be nilpotent. Bachmuth and Mochizuki [1] constructed insoluble, locally finite, 3-Engel groups of exponent 5. Kappe and Kappe [8] proved that a group is a 3-Engel group if and only if the normal closure of every element in the group is nilpotent of class at most 2.

Traustason [14] proved that if G is a locally nilpotent 4-Engel group, and if $g \in G$, then the normal closure of g in G is nilpotent of class at most 4. Havas and Vaughan-Lee's theorem that 4-Engel groups are locally nilpotent implies that Traustason's result applies to all 4-Engel groups. Traustason's bound is sharp — Nickel [11] showed that if G is the class 6 quotient of the free 4-Engel group of rank 3, and if a, b, c generate G, then the commutator $[a, b, b, b, c, b]$ has order 10. So the normal closure of an element in a 4-Engel group is nilpotent, but the nilpotency class can be as big as 4. Traustason proved that if G is a 4-Engel p-group, with $p \neq 2, 5$ then the normal closure of an element is nilpotent of class at most 3. In [15] I proved that a group is a 4-Engel group if and only if the normal closure of each element is a 3-Engel group. The examples of non-nilpotent 3-Engel 2-groups and 5-groups also provide examples of non-nilpotent 4-Engel groups. The general construction of non-nilpotent $(p+1)$-Engel p-groups gives examples of non-nilpotent 4-Engel 3-groups (see below), but Traustason [13] has shown that if G is a 4-Engel p-group with $p \neq 2, 3, 5$ then G is nilpotent of class at most 7.

As mentioned above, there is a simple general method for constructing non-nilpotent $(p+1)$-Engel p-groups. First we let A be an elementary abelian p-group of countably infinite rank, and then we form $G = C_p \text{ Wr } A$. If we let C_p be generated by b, then G has an elementary abelian normal subgroup B with basis $\{b^a \mid a \in A\}$, and G is a split extension of B by A. If we let $a_1, a_2, \ldots$ be a basis for A, then $[b, a_1, a_2, \ldots, a_n] \neq 1$ (for all n), and so G is not nilpotent. Now let $x, y \in G$. Then $[x, y], y^p \in B$, so that

$$[x, \underbrace{y, y, \ldots, y}_{p+1}] = [x, y, y^p] = 1.$$

As mentioned above, Bachmuth and Mochizuki [1] gave an example of an insoluble, locally finite, 3-Engel 5-group. Razmyslov [12] constructed insoluble $(p-2)$-Engel Lie algebras of characteristic p for all $p \geq 5$, and then used these Lie algebras to construct insoluble groups of exponent p. It turns out that these insoluble groups satisfy the $(p-2)$-Engel condition, though Razmyslov does not make this claim in his original paper. In fact, Razmyslov's construction can be used to show that for each prime $p \geq 5$ there exists an insoluble group G of exponent p with the property that if $g \in G$ then the normal closure of g is nilpotent of class most $p-3$. (See [17].)

It is still an open question whether or not 5-Engel groups are locally nilpotent.

Problem 2 Are 5-Engel groups locally nilpotent?

Although this problem is open, it is clear that 5-Engel groups do not satisfy some properties analogous to the properties of 3-Engel and 4-Engel groups. For example, the normal closure of an element in a 5-Engel group need not be 4-Engel. If we let G be the largest nilpotent 5-Engel group generated by two elements a, b of order 3, then G is nilpotent of class 9 and has order 3^{17}. And in G

$$[a, [b, a], [b, a], [b, a], [b, a]] \neq 1.$$

Gupta and Levin [2] showed that the normal closure of an element in a 5-Engel group need not be nilpotent. Actually, this is quite easy to see. Let B be the free group of countably infinite rank in the variety of groups which are nilpotent of class 2 and of exponent 3, and let A be the cyclic group of order 3. Let the free generators of B be $b_1, b_2, \ldots$, and let A be generated by a. Let G be the wreath product of A with B. Then

$$[a, [b_1, b_2], [b_1, b_3], \ldots, [b_1, b_k]]$$

is a non-trivial element of G for all $k \geq 2$. So the normal closure of b_1 in G is not nilpotent. But it is quite easy to see that G is 5-Engel — in fact G satisfies the identity

$$[w, x, y, z, z, z] = 1.$$

To see this, let N be the normal closure of A in G. Then N is an elementary abelian 3-group, and $G/N \cong B$, which is nilpotent of class 2 and has exponent 3. So if $w, x, y, z \in G$ then $[w, x, y]$ and z^3 both lie in N, and

$$1 = \left[[w, x, y], z^3\right] = [w, x, y, z]^3 [w, x, y, z, z]^3 [w, x, y, z, z, z] = [w, x, y, z, z, z].$$

As we shall see, if $p > 3$ then the normal closure of an element in a finite 5-Engel p-group is nilpotent of bounded class. But the situation in 5-Engel 2-groups is still unknown.

Problem 3 Is the normal closure of an element in a finite 5-Engel 2-group nilpotent of bounded class?

Many of the detailed structural results for 4-Engel groups were originally proved to hold in locally nilpotent 4-Engel groups using Lie methods, and in fact most of these results were established for locally nilpotent groups before it was known that *all* 4-Engel groups are locally nilpotent. In this note we show how Lie methods can be applied to locally nilpotent n-Engel groups, and we describe some recent applications of the methods to the study of locally nilpotent 5-Engel groups.

2 The associated Lie ring of an n-Engel group

The associated Lie ring of a group G is defined as follows. First we form the lower central series of G,

$$\gamma_1 \geq \gamma_2 \geq \ldots \geq \gamma_i \geq \ldots,$$

by setting $\gamma_1 = G$, and $\gamma_{i+1} = [\gamma_i, G]$ for $i = 1, 2, \ldots$. For $i = 1, 2, \ldots$ we let $L_i = \gamma_i / \gamma_{i+1}$ and we think of L_i as a $\mathbb{Z}$-module. Then we let

$$L(G) = L_1 \oplus L_2 \oplus \ldots \oplus L_i \oplus \ldots,$$

as a direct sum of $\mathbb{Z}$-modules. If $a \in L_i$ and $b \in L_j$ then we can write $a = g\gamma_{i+1}$, $b = h\gamma_{j+1}$ for some $g \in \gamma_i$, $h \in \gamma_j$. We define the Lie product $[a, b]$ to be $[g, h]\gamma_{i+j+1} \in L_{i+j}$, and we extend this Lie product to the whole of $L(G)$ by linearity. This turns $L(G)$ into a Lie ring, the associated Lie ring of G.

If G satisfies the n-Engel group identity

$$[x, \underbrace{y, y, \ldots, y}_{n}] = 1$$

then it follows immediately from the definition of $L(G)$ that

$$[a, \underbrace{b, b, \ldots, b}_{n}] = 0 \tag{1}$$

whenever a and b are *homogeneous* elements of $L(G)$ (i.e. $a \in L_i$ and $b \in L_j$ for some i, j). But it is not true in general that

$$[a, \underbrace{b, b, \ldots, b}_{n}] = 0$$

when a and b are *arbitrary* elements of $L(G)$. (It turns out that the associated Lie rings of n-Engel groups are n-Engel Lie rings for $n = 2, 3, 4$. But there are examples of 5-Engel groups with associated Lie rings which are not 5-Engel Lie rings.)

However we are able to find multilinear Lie identities which hold in $L(G)$. The first of these is usually referred to as the *multilinearized n-Engel identity*. If we substitute $y_1 y_2 \ldots y_n$ for y in the relation $[x, \underbrace{y, y, \ldots, y}_{n}] = 1$ and expand, then we obtain

$$\prod_{\sigma} [x, y_{1\sigma}, y_{2\sigma}, \ldots, y_{n\sigma}] \in \gamma_{n+2}(G),$$

where the product is taken over all permutations σ of the set $\{1, 2, \ldots, n\}$. In fact we can say more than this. We can express $\prod [x, y_{1\sigma}, y_{2\sigma}, \ldots, y_{n\sigma}]$ as a product of commutators each of which has weight at least $n + 2$, and each of which involves all the generators $x, y_1, y_2, \ldots, y_n$. This gives the identical relation

$$\sum_{\sigma} [a, b_{1\sigma}, b_{2\sigma}, \ldots, b_{n\sigma}] = 0 \qquad (2)$$

for all $a, b_1, b_2, \ldots, b_n \in L(G)$. There are other useful multilinear Lie identities which hold in $L(G)$. For example, if we substitute $x_1 x_2$ for x in the relation

$$[x, \underbrace{y, y, \ldots, y}_{n}] = 1,$$

and expand, then we obtain

$$[x_1, \underbrace{y, y, \ldots, y}_{n}][x_1, y, x_2, y, \ldots, y][x_2, \underbrace{y, y, \ldots, y}_{n}] \in \gamma_{n+3}(G),$$

which gives

$$[x_1, y, x_2, y, \ldots, y] \in \gamma_{n+3}(G).$$

It follows that $L(G)$ satisfies the multilinear Lie identity

$$\sum_{\sigma} [a_1, b_{1\sigma}, a_2, b_{2\sigma}, \ldots, b_{n\sigma}] = 0. \qquad (3)$$

There is a beautiful theory of multilinear Lie relators in varieties of groups due to Wall [18]. His theory gives a doubly infinite family of multilinear Lie relators

$$t^{(r,s)}(a_1, a_2, \ldots, a_r; b_1, b_2, \ldots b_s) = 0$$

($r \geq 1$, $s \geq n$) which hold in the associated Lie rings of n-Engel groups. Identities (2) and (3) above are $t^{(1,n)} = 0$ and $t^{(2,n)} = 0$. Wall uses a technique introduced by Higman [7], which Wall calls *smoothing*. Let F be the free group with free generating set $\{x_1, x_2, \ldots, x_r\}$, and let $u(x_1, x_2, \ldots, x_r) \in F$. The element u is said to be smooth, with support $\{x_1, x_2, \ldots, x_r\}$, if

$$u(1, x_2, x_3, \ldots, x_r) = u(x_1, 1, x_3, \ldots, x_r) = \ldots = u(x_1, x_2, \ldots, x_{r-1}, 1) = 1.$$

Now let G be a group and let $w = 1$ be an identical relation of G where $w = w(x_1, x_2, \ldots, x_r) \in F$. We *smooth* w as follows. First we smooth with respect to x_1 by setting

$$w_1(x_1, x_2, \ldots, x_r) = w(x_1, x_2, \ldots, x_r)w(1, x_2, \ldots, x_r)^{-1}.$$

Note that $w_1 = 1$ is also an identical relation of G, and that $w_1(1, x_2, \ldots, x_r) = 1$ (as an element of F). Next we smooth with respect to x_2 by setting

$$w_2(x_1, x_2, \ldots, x_r) = w_1(x_1, x_2, \ldots, x_r)w_1(x_1, 1, x_3, \ldots, x_r)^{-1}.$$

Again, $w_2 = 1$ is an identical relation of G, and now

$$w_2(1, x_2, \ldots, x_r) = w_2(x_1, 1, x_3, \ldots, x_r) = 1.$$

We continue this smoothing process setting

$$w_3(x_1, x_2, \ldots, x_r) = w_2(x_1, x_2, \ldots, x_r)w_2(x_1, x_2, 1, x_4, \ldots, x_r)^{-1},$$

and so on, systematically smoothing with respect to all the variables $x_1, x_2, \ldots, x_r$. Eventually we obtain an identical relation $w_r = 1$ of G, where $w_r(x_1, x_2, \ldots, x_r) \in F$ is smooth with support $\{x_1, x_2, \ldots, x_r\}$. By Lemma 3.1 of [18], $w_r \in \gamma_r(F)$. (See also Higman [7].) So by Lemma 3.4 of [18] we can express w_r in the form

$$w_r = \prod_\sigma [x_{1\sigma}, x_{2\sigma}, \ldots, x_{r\sigma}]^{\alpha_\sigma} \cdot v$$

where the product is taken over some permutations σ of $\{1, 2, \ldots, r\}$, where $\alpha_\sigma \in \mathbb{Z}$, and where v is a smooth word in $\gamma_{r+1}(F)$. Since $w_r = 1$ is an identical relation of G, this implies that the associated Lie ring of G satisfies the identical Lie relation

$$\sum_\sigma \alpha_\sigma [a_{1\sigma}, a_{2\sigma}, \ldots, a_{r\sigma}] = 0.$$

The identical relations $t^{(r,s)} = 0$ ($r \geq 1$, $s \geq n$) are obtained by substituting $x_1 x_2 \ldots x_r$ for x and substituting $y_1 y_2 \ldots y_s$ for y in the identical relation

$$[x, \underbrace{y, y, \ldots, y}_{n}] = 1,$$

and then smoothing the resulting word with respect to the variables x_1, x_2, $\ldots$, x_r, y_1, y_2, $\ldots$, y_s. (The identical relations $t^{(r,s)} = 0$ are defined for all $r, s \geq 1$, but $t^{(r,s)}$ is identically zero if $s < n$.)

So Wall's theory implies that if L is the associated Lie ring of an n-Engel group then $t^{(r,s)} = 0$ is an identical Lie relation in L. Wall's theory also implies that any multilinear identical relation which holds in the associated Lie rings of n-Engel groups is a consequence of the relations $t^{(r,s)} = 0$.

3 Applications of the Lie method

If G is an n-Engel group with associated Lie ring L, then (as we showed in Section 2) L satisfies the following relations:

1. $[a, \underbrace{b, b, \ldots, b}_{n}] = 0$ whenever a and b are homogeneous elements of L,

2. $\sum_{\sigma}[a, b_{1\sigma}, b_{2\sigma}, \ldots, b_{n\sigma}] = 0$ for all $a, b_1, b_2, \ldots, b_n \in L$.

From Zelmanov's solution of the restricted Burnside problem [21], [22] we see that these relations imply that L is locally nilpotent. In fact we can say more — for each $m > 0$ the nilpotency class of m generator subalgebras of L can be bounded in terms of m and n. In particular, if G is an m generator n-Engel group then the nilpotency class of L can be bounded in terms of m and n. This means that the class of nilpotent quotients of G can be bounded in terms of m and n, but this cannot be used directly to show that G is nilpotent. See also Wilson [19].

The Lie method is particularly useful when G is a finite n-Engel p-group. Then the associated Lie ring L has the same order and nilpotency class as G. Furthermore L/pL has the same nilpotency class as G, and we can view L/pL as a Lie algebra over the field $\text{GF}(p)$. In addition we can use relation (1) above, together with the identical relations $t^{(r,s)} = 0$, to obtain information about L/pL.

For example Newman and Vaughan-Lee [10] proved that the associated Lie rings of free 4-Engel groups of exponent 5 are free Lie rings in the variety of Lie rings determined by the identical relations $5x = 0$, $t^{(r,4)} = 0$ $(r = 1, 2, 3)$. They then used the nilpotent quotient algorithm for graded Lie algebras ([3]) to compute the orders of the free Lie rings in this variety. I had earlier shown in [16] that 4-Engel groups of exponent 5 are locally nilpotent, so that this Lie algebra calculation enabled us to obtain the order and nilpotency classes of free 4-Engel groups of exponent 5.

There have been other applications of this method to 4-Engel 2-groups. In his proof that the normal closure of an element in a 4-Engel 2-group is nilpotent of class 4, Traustason [14] made very subtle use of various identities of the form $t^{(1,s)} = 0$ and $t^{(r,4)} = 0$ in the associated Lie rings of 4-Engel 2-groups. And in my proof that a group is 4-Engel if and only if the normal closure of each element is 3-Engel ([15]), I worked extensively with Lie algebras over $\text{GF}(2)$ satisfying the identities $t^{(r,s)} = 0$ for $r + s \leq 7$.

4 The Higman algebra

In his solution of the restricted Burnside problem for exponent 5, Higman [6] used a beautiful reduction theorem to prove that if L is the associated Lie ring of a group of exponent 5 then the the ideal generated by an element of L is nilpotent of bounded class. Higman's reduction is as follows. It has been known since the 1950s that the associated Lie rings of groups of exponent p satisfy the identical relations

1. $px = 0$,

2. $\sum_\sigma [x, y_{1\sigma}, y_{2\sigma}, \ldots, y_{(p-1)\sigma}] = 0$, where the summation is taken over all permutations σ of $\{1, 2, \ldots, p-1\}$.

The second of these relations is the multilinearized $(p-1)$-Engel identity, which is equivalent to the $(p-1)$-Engel identity in characteristic p. So the associated Lie rings of groups of exponent 5 can be thought of as 4-Engel Lie algebras over GF(5). Now let F be the free Lie algebra over GF(5), freely generated by $x, a_1, a_2, \ldots$, and let I be the ideal of F generated by values of

$$\sum_\sigma [x, y_{1\sigma}, y_{2\sigma}, y_{3\sigma}, y_{4\sigma}]$$

in F. Let $L = F/I$, so that L is a free Lie algebra in the variety of 4-Engel Lie algebras. With some abuse of notation, we denote the free generators of L by $x, a_1, a_2, \ldots$. Let J be the ideal of L generated by $\{[a_i, a_j] \mid i, j \geq 1\}$. Higman's reduction was to prove that $\mathrm{Id}_L(x)$ is nilpotent if and only if $\mathrm{Id}_{L/J}(x + J)$ is nilpotent. He proved moreover that if $\mathrm{Id}_{L/J}(x + J)$ is nilpotent of class c then so is $\mathrm{Id}_L(x)$. Higman then went on to show that L/J is nilpotent, and hence that $\mathrm{Id}_{L/J}(x + J)$ is nilpotent. Higman's proof used hand computation, and he did not obtain explicit bounds on these nilpotency classes. Havas, Newman and Vaughan-Lee [3] used the nilpotent quotient algorithm for graded Lie rings to prove that L/J is nilpotent of class 12, and that $\mathrm{Id}_{L/J}(x + J)$ is nilpotent of class 6. So the ideal generated by an element in a 4-Engel Lie algebra of characteristic 5 is nilpotent of class at most 6. This implies that an m generator 4-Engel Lie algebra of characteristic 5 is nilpotent of class at most $6m$. (Higman's reduction relies critically on the fact that the ideal I above is multigraded, so that L is multigraded.)

Higman's reduction has been used to study 4-Engel p-groups. For any prime p, we let L be the (relatively) free Lie algebra over GF(p) generated by $x, a_1, a_2, \ldots$ in the variety of Lie algebras determined by the identical relations $t^{(r,4)} = 0$ for $r = 1, 2, 3$. Again we let J be the ideal of L generated by $\{[a_i, a_j] \mid i, j \geq 1\}$, and we form the quotient L/J. The nilpotent quotient algorithm for graded Lie rings can be used to show that when $p = 5$, L/J is nilpotent of class 6 and $\mathrm{Id}_{L/J}(x + J)$ is nilpotent of class 4. When $p > 5$ L/J is nilpotent of class 6 and $\mathrm{Id}_{L/J}(x + J)$ is nilpotent of class 3. It follows from these results that in a 4-Engel 5-group the normal closure of an element is nilpotent of class at most 4, and in a 4-Engel p-group for $p > 5$ the normal closure of an element is nilpotent of class at most 3.

The situation is slightly trickier for 4-Engel 2-groups and 4-Engel 3-groups since L/J is not nilpotent for $p = 2, 3$. Traustason [14] used Higman's reduction in his proof that in a 4-Engel 2-group the normal closure of an element is nilpotent of class at most 4, and in his proof that in a 4-Engel 3-group the normal closure of an element is nilpotent of class at most 3. He worked in Lie rings satisfying the identical relations $t^{(r,4)} = 0$ and $t^{(1,s)}$ for various r and s, but in addition he imposed some non-identical relations involving the generator x. His paper contains a careful proof that Higman's reduction is justified in this slightly different setting.

As mentioned above, in 5-Engel 3-groups the normal closure of an element does not have to be nilpotent. Dagmara Milian has recently obtained some (as yet unpublished) results about the normal closure of elements in finite 5-Engel p-groups for $p > 3$. For $p = 5$ she has shown that the normal closure of an element is nilpotent of class at most 6, for $p = 7$ it is nilpotent of class at most 5, and for $p > 7$ it is nilpotent of class at most 4. These results are best possible.

The following table gives bounds on the nilpotency class of the normal closure of an element in a finite n-Engel p-group for $n \leq 5$.

	$p = 2$	$p = 3$	$p = 5$	$p = 7$	$p > 7$
$n = 2$	1	1	1	1	1
$n = 3$	2	2	2	2	2
$n = 4$	4	3	4	3	3
$n = 5$	?	∞	6	5	4

5 Global nilpotence

There is a famous theorem of Zelmanov [20] that if L is an n-Engel Lie algebra over a field of characteristic zero, then L is globally nilpotent. If the nilpotency class is c then this means that in the free Lie ring F over the integers $\mathbb{Z}$, with free generators $x_1, x_2, \ldots, x_{c+1}$ we have

$$m[x_1, x_2, \ldots, x_{c+1}] = \sum_{i=1}^{k} \left(\sum_{\sigma \in \mathrm{Sym}(n)} [a_i, b_{i1\sigma}, b_{i2\sigma}, \ldots, b_{in\sigma}] \right)$$

for some integer m and some elements $a_i, b_{ij} \in F$. If p is a prime and if p does not divide m, then this implies that every n-Engel Lie algebra of characteristic p is nilpotent of class at most c. So there are constants P and c such that if $p > P$ then any finite n-Engel p-group is nilpotent of class at most c. For small n (at least) the numbers c and P are surprisingly small.

n	P	c
2	1	3
3	5	4
4	5	7
5	7	10

For $n = 2$ these figures come from Levi [9]. For $n = 3$ they come from Heineken [5], and for $n = 4$ they come from Traustason [13]. For $n = 5$ the figures come from recent unpublished results of Dagmara Milian. The largest finite 3-generator 5-Engel group of exponent 7 has class 11 and order 7^{560}, so the figure $P = 7$ when $n = 5$ is best possible.

References

[1] S. Bachmuth and H.Y. Mochizuki, Third Engel groups and the Macdonald–Neumann conjecture, *Bull. Austral. Math. Soc.* **5** (1971), 379–386.

[2] N.D. Gupta and F. Levin, On soluble Engel groups and Lie algebras, *Arch. Math.* **34** (1980), 289–295.

[3] G. Havas, M.F. Newman, and M.R. Vaughan-Lee, A nilpotent quotient algorithm for
     graded Lie rings, *J. Symbolic Comput.* **9** (1990), 653–664.

[4] George Havas and M.R. Vaughan-Lee, 4-Engel groups are locally nilpotent, *Internat.
     J. Algebra Comput.* **15** (2005), 649–682.

[5] H. Heineken, Engelsche Elemente der Länge drei, *Illinois J. Math.* **5** (1961), 681–707.

[6] G. Higman, On finite groups of exponent five, *Proc. Camb. Phil. Soc.* **52** (1956),
     381–390.

[7] G. Higman, Some remarks on varieties of groups, *Quart. J. Math. Oxford (2)* **10**
     (1959), 165–178.

[8] L.C. Kappe and W.P. Kappe, On three Engel groups, *Bull. Austral. Math. Soc.* **7**
     (1972), 391–405.

[9] F.W. Levi, Groups in which the commutator operation satisfies certain algebraic con-
     ditions, *J. Indian Math. Soc.* **6** (1942), 87–97.

[10] M.F. Newman and Michael Vaughan-Lee, Engel-4 groups of exponent 5 II. Orders,
     Proc. London Math. Soc. (3) **79** (1999), 283–317.

[11] W. Nickel, Computation of nilpotent Engel groups, *J. Austral. Math. Soc. Ser. A* **67**
     (1999), 214–222.

[12] Ju. P. Razmyslov, On Engel Lie algebras, *Algebra i Logika* **10** (1971), 33–44.

[13] G. Traustason, On 4-Engel groups, *J. Algebra* **178** (1995), 414–429.

[14] Gunnar Traustason, Locally nilpotent 4-Engel groups are Fitting groups, *J. Alge-
     bra* **270** (2003), 7–27.

[15] Michael Vaughan-Lee, On 4-Engel groups, *L.M.S. J. Comput. Math.* **10** (2007), 341–
     353.

[16] M.R. Vaughan-Lee, Engel-4 groups of exponent 5, *Proc. London Math. Soc.* **74** (1997),
     306–334.

[17] M.R. Vaughan-Lee, An insoluble $(p - 2)$-Engel group of exponent p, *J. Algebra* to
     appear.

[18] G.E. Wall, Multilinear Lie relators for varieties of groups, *J. Algebra* **157** (1993),
     341–393.

[19] J.S. Wilson, Two-generator conditions for residually finite groups, *Bull. London Math.
     Soc.* **23** (1991), 239–248.

[20] E.I. Zel'manov, Engel Lie algebras, *Dokl. Akad. Nauk SSSR* **292** (1987), 265–268.

[21] E.I. Zel'manov, The solution of the restricted Burnside problem for groups of odd
     exponent, *Izv. Math. USSR* **36** (1991), 41–60.

[22] E.I. Zel'manov, The solution of the restricted Burnside problem for 2-groups, *Mat.
     Sb.* **182** (1991), 568–592.

ON THE DEGREE OF COMMUTATIVITY OF p-GROUPS OF MAXIMAL CLASS

A. VERA-LÓPEZ and M.A. GARCÍA-SÁNCHEZ

Dpto. Matemáticas, Fc Ciencia y Tecnología, University of the Basque Country, Aptdo. 644, 48080, Bilbao, Spain

Email: antonio.vera@ehu.es, mariasun.garcia@ehu.es

Abstract

The first major study of p-groups of maximal class was made by Blackburn in 1958. He showed that an important invariant of these groups is the 'degree of commutativity', which is a measure of the commutativity among the members of the lower central series of G.

In this paper, we survey the main lower bounds for the degree of commutativity of p-groups of maximal class. Firstly, we study the case of p-groups of maximal class with Y_1 of class 2. Next, we extend the results for any p-group of maximal class. Finally, we apply these lower bounds in order to obtain new conclusions about p-groups of maximal class such that the nilpotency class of Y_1 is at most 3, the number of non-isomorphic p-groups of maximal class of order p^m and the order of the subgroups of $\mathrm{Aut}(G)$.

1 Introduction

A group G of order p^m, with $m \geq 4$, is said to be a p-group of maximal class if $Y_{m-1} \neq 1$, where $Y_0 = G$, $Y_i = \underbrace{[G, \ldots, G]}_{i}$ for every $i \geq 2$ and Y_1 satisfies $Y_1/Y_4 = C_{G/Y_4}(Y_2/Y_4)$.

The most important invariant parameter of a p-group of maximal class G is its degree of commutativity. It was introduced by Blackburn (cf. [1]) and it is defined by

$$c = \max\{k \leq m - 2 \mid [Y_i, Y_j] \leq Y_{i+j+k}, \forall i, j \geq 1\}.$$

It is interesting to obtain lower bounds for the degree of commutativity of p-groups of maximal class because they can be translated into structural properties of G. For example, if we try to obtain the defining relations of a p-group of maximal class, an improvement of only one unit in the lower bound for the degree of commutativity allows us to eliminate a lot of variables in the commutator structure of the defining relations.

Because of this, lower bounds of type $2c \geq m - h(p)$, where $h(p)$ is a polynomial in p and c is the degree of commutativity of a p-group of maximal class of order p^m, has been searched. Blackburn (cf. [1]) proved that $c = m - 2$, if $p = 2$, whereas $c \geq m - 4$, for $p = 3$ and $2c \geq m - 6$ for $p = 5$. Later, Shepherd (cf. [9]) found out that $2c \geq m - 9$ for $p = 7$. These results for $p \leq 7$ were independently generalized to arbitrary primes by Shepherd (cf. [9]) and Leedham-Green and McKay (cf. [6]) who proved that $2c \geq m - 3p + 6$. This last lower bound was improved by Fernández-Alcober (cf. [3]) who showed that $2c \geq m - 2p + 5$, for $p > 7$. Moreover, there exist p-groups of maximal class such that they match this bound (cf. [7]). So, $h(p) = 2p - 5$ is the best possible function for this type of bound.

However, if we consider also the invariant parameters v (cf. [10]), given by

$$v = \min\{k \in [2, m - c - 2] \mid [Y_1, Y_k] = Y_{1+k+c}\}$$

and c_0, residual class of c modulo $p - 1$, sharper inequalities of type $2c \geq m - g(p, v, c_0)$, where $g(p, v, c_0)$ is a polynomial in the variables v, p and c_0, can be obtained. Indeed, in Section 2 we give the expression of $g(p, v, c_0)$ for a p-group G of maximal class of order p^m with Y_1 of nilpotency class 2. In Section 3, we explain the fundamental steps of an algorithm that returns us a value $g(p, v, c_0)$ such that $2c \geq m - g(p, v, c_0)$, being p, v and c_0 initial data, for any p-group G of maximal class with these invariant parameters. In Section 4, we generalize the results of Section 2 for a p-group G of maximal class of order p^m satisfying certain conditions on v and c_0. Furthermore, we collect results about p-groups of maximal class such that the nilpotency class of Y_1 is at most 3, the number of non-isomorphic p-groups of maximal class of order p^m and the order of the subgroups of $\mathrm{Aut}(G)$.

In order to finish this section, we introduce some notation and preliminary results. As we have said, in [10] the invariant parameter v for a p-group of maximal class is defined. Furthermore, they proved that v is an even integer number $2l$ satisfying $2 \leq v = 2l \leq p - 1$ and if $v = 2l = p - 1$, then $m = c + p$ (cf. [10]). So, throughout this paper, we consider the invariant parameter $l = v/2$ for a p-group of maximal class instead of v and we take $1 \leq l \leq (p - 3)/2$. Consequently, $(l, c_0) \in \mathcal{I}$, where $\mathcal{I} = \{1, \ldots, (p-3)/2\} \times \{0, \ldots, p-2\}$ and the parameters of the function $g(p, v, c_0)$ will be $g(p, l, c_0)$ hereafter.

Following N. Blackburn's ideas, if G is a p-group of maximal class, we can choose a pair of elements $s \in G \setminus (Y_1 \cup C_G(Y_{m-2}))$ and $s_1 \in Y_1 \setminus Y_2$, and define recursively $s_i = [s_{i-1}, s] \in Y_i \setminus Y_{i+1}$, for $i = 2, \ldots, m - 1$. For $i + j \leq m - c - 1$, let $\alpha_{i,j} \in \mathbb{F}_p$ be determined by the congruence

$$[s_i, s_j] \equiv s_{i+j+c}^{\alpha_{i,j}} \pmod{Y_{i+j+c+1}}.$$

It is well known that $\alpha_{i,j}$ satisfies the following properties:

(P1) $\alpha_{i,j} = -\alpha_{j,i}$.

(P2) $\alpha_{i,i} = 0$, if $2i \leq m - c - 1$.

(P3) $\alpha_{i,j} = \alpha_{i+1,j} + \alpha_{i,j+1}$, if $i + j + 1 \leq m - c - 1$ (Bernoulli's property).

(P4) $\alpha_{i,j} = \alpha_{i+p-1,j} = \alpha_{i,j+p-1}$, if $i + j + p - 1 \leq m - c - 1$ (periodicity modulo $p - 1$).

(P5) $f(i, j, k) = \alpha_{i,j}\alpha_{i+j+c,k} + \alpha_{j,k}\alpha_{j+k+c,i} + \alpha_{k,i}\alpha_{k+i+c,j} = 0$ for any positive integers i, j, k satisfying $i + j + k \leq m - 2c - 1$ (Jacobi's identity).

Let t be a positive integer number such that $t \leq m - 2c - 1$. We denote

$$\mathcal{S}(t) = \{f(i, j, k) \mid i + j + k \leq t\}$$

and
$$x_\lambda = \alpha_{\lambda,\lambda+1}.$$
It is known (cf. [10]) that $x_i = 0$ for $i = 1, \ldots, l-1$ and $x_l \neq 0$. From the properties
(P1)–(P3), it is easy to check that

$$\alpha_{i,j} = \sum_{k=i}^{[(i+j-1)/2]} (-1)^{k-i} \binom{j-k-1}{k-i} x_k, \quad \forall i+j \leq m-c-1,$$

and $f(i,j,k)$ is a quadratic form in the variables x_λ. Moreover, (P4) implies that
x_λ, with $\lambda > (p-3)/2$, is a linear combination of $x_1, \ldots, x_{(p-3)/2}$.

2 p-groups of maximal class with Y_1 of class 2

The introduction of the invariant parameters l and c_0 is specially useful, if G is a
p-group of maximal class with Y_1 of nilpotency class 2. In [5], the following result
is shown:

Theorem 2.1 *If G is a p-group of maximal class of order p^m with Y_1 of class 2
and c, l and c_0 are its invariant parameters, then $2c \geq m - g(p,l,c_0)$, where*

$$g(p,l,c_0) = \begin{cases} 2l + c_0 + 2, & \text{if } l + c_0 \leq (p-3)/2; \\ 2l + c_0 + 2, & \text{if } l + c_0 = (p-1)/2 \text{ and } c = c_0; \\ 2l + c_0 + 1, & \text{if } l + c_0 = (p-1)/2 \text{ and } c > c_0; \\ 2l + p - c_0, & \text{if } l + c_0 \geq (p-1)/2. \end{cases}$$

Moreover, there exist p-groups maximal class with Y_1 of class 2 such that $2c = m - g(p,l,c_0)$, for all $(l,c_0) \in \mathcal{I}$.

We notice that if G is a p-group of maximal class of order p^m with Y_1 of class 2,
the value $g(p,l,c_0) = 2p - 5$ only appears for $(l,c_0) = ((p-3)/2, 2)$ and for the
rest of the pairs $(l,c_0) \in \mathcal{I} - \{((p-3)/2,2)\}$, $g(p,l,c_0)$ is less than $2p - 5$. So
it is worth obtaining similar bounds for any p-group of maximal class without the
restriction over the class of Y_1.

To illustrate Theorem 2.1, the division according to the expression of $g(p,l,c_0)$
for all 23-groups of maximal class with Y_1 of class 2, $(l,c_0) \in \mathcal{I}$ and $c > c_0$ appears
in the left table of Figure 1. The particular value of c_0 appears in the first column
and the value of l in the first row.

3 The Algorithm

When we work on p-groups of maximal class for a fixed prime p, it is easy to design
an algorithm that returns us a lower bound for c. The main idea of the algorithm
is quite simple. Fixed the parameters l, c_0 and p, we suppose, by contradiction,
that $a \leq m - 2c - 1$. Then, the conditions (P1)–(P5) hold. If we are able to

derive a contradiction from these conditions, the initial hypothesis must be false, so $2c \geq m - a$ and this number a is the value returned by the program.

The scheme of the algorithm is the following one:

1. We fix p, l, c_0 and define $n = 6$. We define $x_\nu = 0$, for $\nu = 1, \ldots, l - 1$ and x_l non-zero.

2. For fixed level n, we compute all quadratic forms $f(i, j, k)$, written in terms of x_λ, such that $i < j < k$ and $i + j + k = n$. Among these $f(i, j, k)$, we choose only those ones that factorizes as two linear forms over $\mathbb{F}_p$. If there is not any, we increase the level n by 1 until we find one. It is well known that the existence of a quadratic form $f(i, j, k)$ that factorizes over $\mathbb{F}_p$ is guaranteed.

3. If $f(i, j, k)$ is one of the quadratic forms that factorizes over $\mathbb{F}_p$, we write $f(i, j, k) = s_1(x_1, \ldots, x_{(p-3)/2}) \, s_2(x_1, \ldots, x_{(p-3)/2})$. But $f(i, j, k) = 0$ implies $s_1(x_1, \ldots, x_{(p-3)/2}) = 0$ or $s_2(x_1, \ldots, x_{(p-3)/2}) = 0$. We make one of the factor equal to 0 and keep the other one for a posterior analysis.

4. The factor $s_r(x_1, \ldots, x_{(p-3)/2})$ chosen allows us to write one x_λ as linear combination of the remaining x_ν. This value of x_λ is substituted in the other $f(i, j, k) \in \mathcal{S}(n)$. It can occur that $\mathcal{S}(n) = \{0\}$ or not. If so, we proceed to next step. If not, we check whether there exists a $f(i, j, k) \in \mathcal{S}(n) - \{0\}$ that factorizes over $\mathbb{F}_p$. If so, we return to the previous step. If there is not a $f(i, j, k) \in \mathcal{S}(n) - \{0\}$ that factorizes over $\mathbb{F}_p$, we increases by 1 the level n and return to the second step, unless we derive a contradiction.

5. When $\mathcal{S}(n) = \{0\}$, we take $n = n + 1$ and repeat the process, until we derive a contradiction with the previous relationships.

6. If a contradiction appears in a branch, we check if the level n is greater than the maximum level of the other branches. If so, we change the previous maximum level by n and we check whether the assignations of the x_ν satisfy $\mathcal{S}(n - 1)$ or not. If so, we print the level n and the solution of $\mathcal{S}(n - 1)$. If not, we print only the maximum level.

7. We undo the last assignation of x_λ and we take the other linear factor of the last factorization of the corresponding quadratic form not studied in order to repeat the process until all the possibilities have been checked. The maximum level $n = a$ obtained in the program satisfies $2c \geq m - a$. Furthermore, the assignation of the variables x_ν satisfying $\mathcal{S}(a - 1)$, if there exists, define a Lie algebra of maximal class.

There are two keys that are worth emphasizing: we only use the quadratic forms $f(i, j, k)$ that factorize as two linear forms over $\mathbb{F}_p$ and the assignations that we do along the process give us a tree with different branches, which allow us to obtain the Lie algebra in many cases. Related to the first fact, we could think about taking all $f(i, j, k)$ for a particular level and try solving the associated quadratic systems in $\mathbb{F}_p$. But the process would be more difficult to program, it would take longer and we could not improve the bound for the most of (l, c_0), because the found bounds by our algorithm are exact in the most of cases since we can obtain a Lie algebra, or a p-group of maximal class, that attains the bound, as we see in next section.

We have implemented this algorithm in Maple. The most difficult step is to make a procedure that checks whether a quadratic form factorizes as a product of two linear forms in $\mathbb{F}_p$ or not. The main idea of this procedure is the following one: if the expression of the quadratic form is $\sum a_{i,j} x_i x_j$, we look for the first $a_{i,i}$ non-zero, if there exists. Suppose that a_{i_1,i_1} is not zero. Then if there exists factorization in linear forms it must be appear $a_{i_1,i_1}(x_{i_1} + h_1)(x_{i_1} + h_2)$, where x_{i_1} does not appear in neither h_1 nor h_2. Then, we look for the first non-zero a_{i_1,j_1} such that $i_1 \neq j_1$. So, the factors are of type $(x_{i_1} + b_1 x_{j_1} + h_3)(x_{i_1} + b_2 x_{j_1} + h_4)$, being $-b_1$ and $-b_2$ the roots of the equation $y^2 + a_{i_1 j_1} y + a_{j_1,j_1}$. If the last equation has not roots over $\mathbb{F}_p$, the factorization is not possible. In other case, we repeat the process looking for the first j_2 such that $j_2 \neq i_1, j_1$ and a_{i_1,j_2} is non zero. This continues until we get the factors or we conclude that the quadratic form does not factor. If $a_{i,i}$ is zero for all i and the factorization exists, it will be of the type $(\sum b_i x_i)(\sum c_i x_i)$, with $b_i = 0$, if $c_i \neq 0$. Looking for the first $a_{i,j}$ non-zero, we can search easily the relation between the coefficients of the factors in order to obtain the factors or to conclude that it does not exist.

We have run our program for $p \leq 43$ and $(l, c_0) \in \mathcal{I}$. The running time depends on the prime p and the particular (l, c_0) chosen. The greater the prime p, the longer it takes. Moreover, for a fixed prime p, the lower l, the running time is longer. Finally, for fixed p and l, the lower c_0, the running time is usually longer.

4 Main results

The data obtained for the program, if $p \leq 43$ and $(l, c_0) \in \mathcal{I}$, can been collected in tables $\mathcal{T}_p$ that contains in the cell (i, j) the corresponding value $a_{i,j}$ returned by the program when the initial data are p, $c_0 = i - 1$ and $l = j$.

We have seen uniformities in these tables $\mathcal{T}_p$, that allow us to divide each table in regions not depending on p. So, we have conjectured the existence of a function $g(p, l, c_0)$ such that $2c \geq m - g(p, l, c_0)$. Unfortunately, the expression of $g(p, l, c_0)$ obtained in Theorem 2.1 does not match in the obtained $\mathcal{T}_p$, with $p \leq 43$, if G is a p-group of maximal class with Y_1 not necessarily of nilpotency class 2. However, the values that appear in $\mathcal{T}_p$ are lower than $2p - 5$, except for $(l, c_0) \in \left\{ \big((p-3)/2, 2\big), \big((p-3)/2, 3\big) \right\}$. Consequently, it is worth extending Theorem 2.1 to any p-group of maximal class. Indeed, we have proved the following result in [11], [12], [13] and [14]:

Theorem 4.1 *Let G be a p-group of maximal class of order p^m, degree of commutativity c and invariant parameters l and c_0. Then, $2c \geq m - g(p, l, c_0)$ for the values of $g(p, l, c_0)$ given by Table 1.*

The regions defined in Theorem 4.1 cover almost all values of c_0 and l. For example, for $p = 23$, the regions A–G appear shaded in the right table of Figure 1 and the corresponding bounds for c are valid for all 23-groups of maximal class with the invariant parameters l, c_0 and c.

Region	Conditions on l and c_0	$g(p, l, c_0)$
A	$l + c_0 = \frac{p-1}{2}$, $l \geq \frac{p+5}{6}$, $c \geq p$	$2l + c_0 + 1$
B_1	$p - 2 \geq c_0 \geq p - l$, $l \geq 2$	
B_2	$c_0 \geq 2p - 4l + 1$	
B_3	$2p - 4l - 1 \leq c_0 \leq 2p - 4l + 2$ and $l > \frac{p-1}{3}$	$p + 2l - c_0$
B_4	$c_0 \in \{2p - 4l - 2, 2p - 4l - 4\}$, $l + c_0 > \frac{p+1}{2}$, and $p \leq 3l - 1$	
B_5	$l \geq \frac{p+5}{6}$ and $l + c_0 = \frac{p+1}{2}$	
C	$l + c_0 = p - 1$, $l < p/3$	$p + 1$
D	$l = 1$, $\frac{4p-5}{5} \leq c_0 \leq p - 3$, $c \geq p$	$2c_0 - p + 7$
E	$l \geq 3$, $l + c_0 \leq \frac{p-3}{2}$, $1 \leq c_0 \leq 2l - 1$	$2l + c_0 + 2$
F	$c_0 = p - 4l + 3\lambda + \mu$ with $\lambda \geq 1$ and $\mu = 1, 3, 5$	$p + 2l - c_0 + 1$
G	$l \geq 3$, $l + c_0 \leq \frac{p-3}{2}$, $c_0 > 2l - 1$	$2l + c_0 + 3$

Table 1. Values of $g(p, l, c_0)$ for Theorem 4.1.

Besides, Theorem 5.5 of [5] gives us examples of p-groups of maximal class of order p^m, such that $2c = m - g(p, l, c_0)$ for the regions A, B and E. So, the given bound is exact in these regions. Consequently, in these zones, we have extended Theorem 2.1 to any p-group of maximal class. On the other hand, using techniques of Lie algebras, it is easy to check that the bound for regions C and D can not be improved (cf. [13]).

On the other hand, in [3], by using the bound $2c \geq m - 2p + 5$, it is shown that the nilpotency class of Y_1, $cl(Y_1)$, is at most 3, if $m \geq 6p - 25$. However, by applying Theorem 4.1, the value $6p$ can be reduced to $3p$. Indeed, it is easy to prove:

Corollary 4.2 *Let G be a p-group of maximal class of order p^m, degree of commutativity c and invariant parameters l and c_0. If $m \geq 3p - 7$ and the pair (l, c_0) satisfies the conditions to belong to region A, C, D, E or G, then $cl(Y_1) \leq 3$.*

Corollary 4.3 *Let G be a p-group of maximal class of order p^m, degree of commutativity c and invariant parameters l and c_0 such that $m \geq 3p - 7$ and $c_0 \geq 2l$. If one of the following conditions holds*
1. $2 \leq l$ and $p - l \leq c_0 \leq p - 2$,
2. $2p - 4l + 1 \leq c_0$,
3. $c_0 = p - 4l + 3\lambda + \mu$ with $\lambda \geq 1$ and $\mu \in \{1, 3, 5\}$,
then $cl(Y_1) \leq 3$.

Corollary 4.4 *Let G be a p-group of maximal class of order p^m, degree of commutativity c and invariant parameters l and c_0. If $m \geq 3p - 7 + 3(2l - c_0)$, $c_0 \leq 2l - 1$ and one of the following conditions holds:*
1. $c_0 = 2p - 4l - 4$,

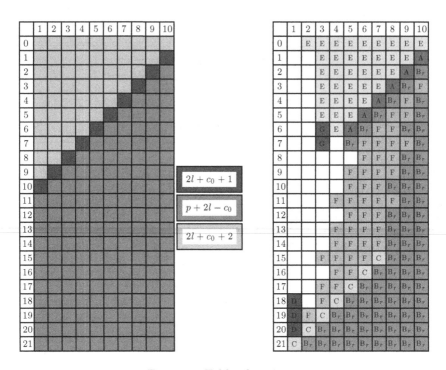

Figure 1. Tables for $p = 23$

2. $2p - 4l - 2 \leq c_0$,

3. $l \geq \frac{p+5}{6}$, $l + c_0 = \frac{p+1}{2}$,

4. $c_0 = p - 4l + 3\lambda + \mu$ with $\lambda \geq 1$, $\mu \in \{1, 3, 5\}$,

then $cl(Y_1) \leq 3$.

On the other hand, in Corollary 3.5, [4], an upper bound for the number of p-groups of maximal class of order p^m, with $p > 5$, has been given. In fact, if we denote by $a(p, m) = \frac{1}{4}(p - 3)(m + p - 6) + 3$ and $\mathbf{t}(p, m)$ is the number of non-isomorphic p-groups of maximal class of order p^m, then $\mathbf{t}(p, m) \leq p^{a(p,m)}$ for $p > 5$. However, if we fix the invariant parameters c_0 and l this bound can be improved for the regions of Theorem 4.1.

Corollary 4.5 *Let* $\mathbf{t}(p, m, c_0, l)$ *be the number of non-isomorphic p-groups of maximal class of order* p^m *with the invariant parameters* c_0 *and* l, $p \geq 7$. *Then, for* $m \geq p + c - 1$, $\mathbf{t}(p, m, c_0, l) \leq p^{a(p,m) - b(p, c_0, l)}$, *where* $a(p, m) = \frac{1}{4}(p - 3)(m + p - 6) + 3$

and $b(p, c_0, l)$ is given by

$$b(p, c_0, l) = \begin{cases} \dfrac{p-3}{4}\left(\dfrac{3p-11}{2} - l\right) - l & \text{in region } A, E, \text{ or } G \\[2mm] \dfrac{p-3}{4}(p + c_0 - 2l - 5) - l + 1 & \text{in region } B \\[2mm] \dfrac{p-3}{4}(p - 6) - l + 1 & \text{in region } C \\[2mm] \dfrac{p-3}{4}(3p - 2c_0 - 12) & \text{in region } D \\[2mm] \dfrac{p-3}{4}(p - 6 + c_0 - 2l) - l + 1 & \text{in region } F. \end{cases}$$

Finally, we improve Juhasz (cf. [8]) and Caranti and Mattarei (cf. [2]) result about $\mathrm{Aut}(G)$. In fact, by using different methods, Juhasz, Caranti and Mattarei proved that $\mathrm{Aut}(G)$ contains a subgroup of order $p^{\lfloor(3m-2p+5)/2\rfloor}$, if G is a p-group of maximal class of order p^m, with $m > p + 1$ and $p \geq 5$. However, an argument like the one used by Caranti and Mattarei but considering the lower bound of Theorem 4.1 for c is enough to prove:

Corollary 4.6 *Let G be a p-group of maximal class of order p^m, with $m > p + 1$ and $p \geq 7$, degree of commutativity c and invariant parameters l and c_0. Then, $\mathrm{Aut}(G)$ contains a subgroup of order at least $p^{[m_1]}$, where m_1 is defined by*

$$m_1 = \begin{cases} \dfrac{3m - p + 1}{2}, & \text{in Region } A \text{ or } E; \\[2mm] \dfrac{6m - 3p + 9}{4}, & \text{in Region } B_1; \\[2mm] \dfrac{3m - 2p + 10}{2}, & \text{in Region } B_2; \\[2mm] \dfrac{3m - 2p + 8}{2}, & \text{in Region } B_3; \\[2mm] \dfrac{3m - 2p + 7}{2}, & \text{in Region } B_4; \\[2mm] \dfrac{3m - 2p + 5}{2}, & \text{in Region } B_5 \text{ or } F; \\[2mm] \dfrac{3m - p - 1}{2}, & \text{in Region } C \text{ or } D; \\[2mm] \dfrac{3m - p}{2}, & \text{in Region } G. \end{cases}$$

References

[1] N. Blackburn, On a special class of p-groups, *Acta Math.* **100** (1958), 45–92
[2] A. Caranti and S. Mattarei, Automorphisms of p-groups of maximal class, *Rend. Sem. Mat. Univ. Padova.* **115** (2006), 189–198
[3] G. A. Fernández-Alcober, The exact lower bound for the degree of commutativity of a p-group of maximal class, *J. Algebra* **174** (1995), 523–530.

[4] A. Jaikin-Zapirain, A. Vera-López, On the use of Lazard's correspondence in the classification of p-groups of maximal class, *J. Algebra* **228** (2000), 477–490.

[5] A. Jaikin-Zapirain, A. Vera-López, On p-groups of maximal class, *Math. Nachr.*, to appear.

[6] C.R. Leedham-Green and S. McKay, On p-groups of maximal class I, *Quart. J. Math. Oxford Ser. (2)* **37** (1976), 297–311.

[7] C.R. Leedham-Green and S. McKay, On p-groups of maximal class II, *Quart. J. Math. Oxford Ser. (2)* **29** (1978), 175–186.

[8] A. Juhasz, The group of automorphisms of a class of finite p-groups, *Trans. Amer. Mat. Soc.* **270** (1982), no. 2, 469–481.

[9] R. Shepherd, *p-Groups of Maximal Class*, Ph.D. thesis, University of Chicago, 1970.

[10] A. Vera-López, J. M. Arregi, and F. J. Vera-López, Some bounds on the degree of commutativity of a p-group of maximal class, II, *Comm. in Algebra* **23** (1995), 2765–2795.

[11] A. Vera-López, J. M. Arregi, and F. J. Vera-López, Some bounds on the degree of commutativity of a p-group of maximal class, III, *Math. Proc. Cambridge Philos. Soc.* **122** (1997), 251–260.

[12] A. Vera-López, J. M. Arregi, M.A. García-Sánchez, R. Esteban-Romero and F. J. Vera-López, The exact bound for the degree of commutativity of a p-group of maximal class, I, *J. Algebra* **256** (2002), 375–401.

[13] A. Vera-López, J. M. Arregi, M.A. García-Sánchez, R. Esteban-Romero and F. J. Vera-López, The exact bound for the degree of commutativity of a p-group of maximal class, II, *J. Algebra* **273** (2004), 806–853.

[14] A. Vera-López, J. M. Arregi, M.A. García-Sánchez and L. Ormaetxea, On the degree of commutativity of a p-group of maximal class, submitted to *J. Pure Appl. Algebra*.

CLASS PRESERVING AUTOMORPHISMS OF FINITE p-GROUPS: A SURVEY

MANOJ K. YADAV

School of Mathematics, Harish-Chandra Research Institute, Chhatnag Road, Jhunsi,
Allahabad – 211 019, India
Email: myadav@hri.res.in

Abstract

In this short survey article, we list a number of known results on class preserving
automorphisms of finite p-groups.

1 Introduction

Let us start with a finite group G. For $x \in G$, x^G denotes the conjugacy class
of x in G. By $\mathrm{Aut}(G)$ we denote the group of all automorphisms of G. An automorphism α of G is called *class preserving* if $\alpha(x) \in x^G$ for all $x \in G$. The
set of all class preserving automorphisms of G, denoted by $\mathrm{Aut}_c(G)$, is a normal
subgroup of $\mathrm{Aut}(G)$. Notice that $\mathrm{Inn}(G)$, the group of all inner automorphisms
of G, is a normal subgroup of $\mathrm{Aut}_c(G)$. The group of all class preserving outer
automorphisms of G, i.e., $\mathrm{Aut}_c(G)/\mathrm{Inn}(G)$ is denoted by $\mathrm{Out}_c(G)$.

The story begins with the following question of W. Burnside [2, Note B]: Does
there exist any finite group G such that G has a non-inner class preserving automorphism? In 1913, Burnside [3] himself gave an affirmative answer to this question.
He constructed a group G of order p^6 isomorphic to the group W consisting of all
3×3 matrices

$$M = \begin{pmatrix} 1 & 0 & 0 \\ x & 1 & 0 \\ z & y & 1 \end{pmatrix} \qquad \cdot$$

with x, y, z in the field $\mathbb{F}_{p^2}$ of p^2 elements, where p is an odd prime. For this
group G, $\mathrm{Out}_c(G) \neq 1$. He also proved that $\mathrm{Aut}_c(G)$ is an elementary abelian
p-group of order p^8.

In 1947, G. E. Wall [30] constructed examples of arbitrary finite groups G such
that $\mathrm{Out}_c(G) \neq 1$. Interestingly his examples contain 2-group having class preserving outer automorphisms. The smallest of these groups is a group of order 32.
The members of the class of groups, constructed by Wall, appear as a semidirect
product of a cyclic group by an abelian group.

At that stage we had examples of arbitrary finite groups having class preserving
outer automorphisms. But, unfortunately, this problem was not taken up (to my
knowledge) for the next 20 years. However in this span of time many results were
proved on automorphisms of finite groups, and more precisely on automorphisms
of finite p-groups. But all of these results were centered around finding a non-
inner p-automorphism (an automorphism having order a power of a prime p) of

finite p-groups with order at least p^2. This was finally achieved by W. Gaschutz in 1966 [14] by using cohomology of finite groups.

In 1968, C. H. Sah [25] studied first some properties of $\text{Aut}_c(G)$ for an arbitrary group G. He used the concepts of c-closed and strongly c-closed subgroups (cf. [25, pg 48] for definition), along with cohomological techniques to explore many nice basic properties of $\text{Aut}_c(G)$ and $\text{Out}_c(G)$ for a given finite group G. We record a few of his results in Section 4.

After this, in 1980, H. Heineken [9], on the way to producing examples of finite groups in which all normal subgroups are characteristic, constructed finite p-groups G with $\text{Aut}(G) = \text{Aut}_c(G)$. In particular, $\text{Out}_c(G) \neq 1$ for all of these group G. His groups are p-groups of nilpotency class 2. These groups also satisfy the property that $x\gamma_2(G) = x^G$ for all $x \in G - \gamma_2(G)$.

Continuing in this direction, I. Malinowska [19], in 1992, constructed finite p-groups G of nilpotency class 3 and order p^6 such that $\text{Aut}(G) = \text{Aut}_c(G)$. In the same paper, she also constructed p-groups G of nilpotency class r, for any prime $p > 5$ and any integer $r > 2$ such that $\text{Out}_c(G) \neq 1$. In 1988, W. Feit and G. M. Seitz [6, Section C] proved that $\text{Out}_c(G) = 1$ for all finite simple groups G.

Motivation for studying class preserving automorphisms of finite groups also arises from other branches of mathematics. T. Ono and H. Wada proved that $\text{Out}_c(G) = 1$ for the cases when G is a free group, $SL_n(D)$, $GL_n(D)$ (where D is an Euclidean domain), S_n and A_n (the symmetric and alternating groups on n symbols) in [23, 24, 28, 29]. Their motivation for studying these things arose from the "Hasse principle" for smooth curves on number field. They associated a Shafarevich–Tate set to the given curve. The curve is said to enjoy "Hasse principle" if the corresponding Shafarevich–Tate set is trivial. If the group G, involved in defining Shafarevich-Tate set, is finite then this set enjoys the group structure, which is isomorphic to $\text{Out}_c(G)$. The interested reader can find all the details in [21, 22].

The other motivation comes from integral groups rings. The interested reader can refer to [12]. A. Hermann and Y. Li [10] and M. Hertweck and E. Jespers [13] proved that $\text{Out}_c(G) = 1$ for all Blackburn group. These groups were classified by N. Blackburn and satisfy the property that the intersection of all its non-normal subgroups is non-trivial. M. Hertweck [11] constructed a class of Frobenius group G such that $\text{Out}_c(G) \neq 1$. He also proved that $\text{Out}_c(G) = 1$ for all finite metabelian A-groups and for all A-groups having all Sylow subgroups elementary abelian. By an A-group, we here mean a solvable group with abelian Sylow subgroups.

Other interesting examples of finite groups G such that $\text{Out}_c(G) \neq 1$ are given by F. Szechtman [27]. He constructed a very concrete method to find such examples. The interesting thing here is that his method produces examples in a variety of ways arising from different kind of Lie algebras, regular representations of finite fields and linear algebra. We would like to remark that his method produced examples of group having outer n-inner automorphims. An automorphism α of a group G is said to be n-inner if given any subset S of G with cardinality less than n, there exists an inner automorphism of G agreeing with α on S. Notice that 2-inner automorphisms are class preserving.

The present survey intersects with Section 5 of a survey article by I. Malinowska [20].

Our notation for objects associated with a finite multiplicative group G is mostly standard. The abelian group of all homomorphisms from an abelian group H to an abelian group K is denoted by $\mathrm{Hom}(H, K)$. To say that some H is a subset or a subgroup of G we write $H \subseteq G$ or $H \leq G$ respectively. To indicate, in addition, that H is properly contained in G, we write $H \subset G$, $H < G$, respectively. If $x, y \in G$, then x^y denotes the conjugate element $y^{-1}xy \in G$ and $[x, y]$ denotes the commutator $x^{-1}y^{-1}xy = x^{-1}x^y \in G$. Here $[x, G]$ denotes the set of all $[x, w]$ where w runs over every element of G. Since $x^w = x[x, w]$, for all $w \in G$, we have $x^G = x[x, G]$. For $x \in G$, $\mathrm{C}_H(x)$ denotes the centralizer of x in H, where $H \leq G$. The center of G will be denoted by $\mathrm{Z}(G)$. By $\gamma_2(G)$ we denote the commutator subgroup of a group G which is generated by all the commutators $[x, y]$ in G.

2 Some defintions and basic results

We start with the concept of isoclinism of finite groups which was introduced by P. Hall [8] (also see [4, pg. 39–40] for details).

Let X be a finite group and $\bar{X} = X/\mathrm{Z}(X)$. Then commutation in X gives a well defined map $a_X : \bar{X} \times \bar{X} \mapsto \gamma_2(X)$ such that $a_X(x\,\mathrm{Z}(X), y\,\mathrm{Z}(X)) = [x, y]$ for $(x, y) \in X \times X$. Two finite groups G and H are called *isoclinic* if there exists an isomorphism ϕ of the factor group $\bar{G} = G/\mathrm{Z}(G)$ onto $\bar{H} = H/\mathrm{Z}(H)$, and an isomorphism θ of the subgroup $\gamma_2(G)$ onto $\gamma_2(H)$ such that the following diagram is commutative

$$\begin{array}{ccc} \bar{G} \times \bar{G} & \xrightarrow{\ a_G\ } & \gamma_2(G) \\ {\scriptstyle \phi \times \phi}\Big\downarrow & & \Big\downarrow{\scriptstyle \theta} \\ \bar{H} \times \bar{H} & \xrightarrow{\ a_H\ } & \gamma_2(H). \end{array}$$

The resulting pair (ϕ, θ) is called an *isoclinism* of G onto H. Notice that isoclinism is an equivalence relation among finite groups. Each isoclinism class has a subgroup G such that $\mathrm{Z}(G) \leq \gamma_2(G)$, which is called a stem group of the family.

Let G and H be two isoclinic groups. Then the nilpotency class of G is equal to the nilpotency class of H. Also the terms in the lower central series of G are isomorphic to the respective terms of the lower central series of H. More precisely, one can say that the commutator structure of G is similar to the commutator structure of H. The consequence of these simple results is the following theorem [33, Theorem 4.1], which essentially says that the group $\mathrm{Aut}_c(G)$ is independent of the choice of the group G in its isoclinism class.

Theorem 2.1 *Let G and H be two finite non-abelian isoclinic groups. Then* $\mathrm{Aut}_c(G) \cong \mathrm{Aut}_c(H)$.

Since every isoclinism class has a stem group, we readily get the following result.

Corollary 2.2 *It is sufficient to study* $\mathrm{Aut}_c(G)$ *for all finite groups G such that* $\mathrm{Z}(G) \leq \gamma_2(G)$.

Let G be a finite group and N a non-trivial proper normal subgroup of G. The pair (G, N) is called a *Camina pair* if $xN \subseteq x^G$ for all $x \in G - N$. A group G is called a *Camina group* if $(G, \gamma_2(G))$ is a Camina pair. A basic result for Camina p-groups, which is due to R. Dark and C. M. Scoppola [5] is the following.

Theorem 2.3 *Let G be a finite Camina p-group. Then the nilpotency class of G is at most* 3.

A finite group G is called *purely non-abelian* if it does not have a non-trivial abelian direct factor. An automorphism α of a group G is called central if it commutes with every inner automorphism of G. The group of all central automorphisms of a group G is denoted by $\text{Autcent}(G)$. The following nice result is due to J. E. Adney and T. Yen [1, Corollary 2].

Theorem 2.4 *If G is a purely non-abelian p-group, then* $\text{Autcent}(G)$ *is also a p-group.*

3 $\text{Out}_c(G)$ for finite p-groups

In this section we record several classes of finite p-groups G, for which it is decidable whether $\text{Out}_c(G) = 1$ or not. For the convenience of the reader we divide this section into several parts.

3.1 p-groups of small orders.

In this part we consider p-groups of small orders and mention results on class-preserving automorphisms of p-groups of order at most p^5. Since $\text{Aut}_c(G) = \text{Inn}(G)$ for all abelian groups G, it is sufficient to consider the groups of order at least p^3. For the proof of the following well known result, one can see [26, pg 69].

Proposition 3.1 *Let G be an extraspecial p-group. Then* $\text{Out}_c(G) = 1$.

Since every non-abelian group of order p^3 is extraspecial, we get the following.

Corollary 3.2 *Let G be a group of order p^3. Then* $\text{Out}_c(G) = 1$.

We (jointly with L. R. Vermani) proved the following result in [17].

Theorem 3.3 *Let G be a group of order p^4. Then* $\text{Out}_c(G) = 1$.

Now we consider the groups of order p^5, where p is an odd prime. Set

$$G_7 = \langle a, b, x, y, z \mid \mathcal{R}, [b, a] = 1 \rangle \text{ for } p \geq 3 \tag{1}$$

and

$$G_{10} = \langle a, b, x, y, z \mid \mathcal{R}, [b, a] = x \rangle \text{ for } p \geq 5, \tag{2}$$

where $\mathcal{R}$ is defined as follows:

$$\mathcal{R} = \{a^p = b^p = x^p = y^p = z^p = 1\} \cup \{[x,y] = [x,z] = [y,z] = 1\}$$
$$\cup \{x^b = xz, y^b = y, z^b = z\} \cup \{x^a = xy, y^a = yz, z^a = z\}.$$

For $p = 3$ define a group H of order 3^5 by

$$H = \langle a, b, c \mid a^3 = b^9 = c^9 = 1, [b,c] = c^3, [a,c] = b^3, [b,a] = c \rangle \qquad (3)$$
$$= \langle a \rangle \ltimes (\langle b \rangle \ltimes \langle c \rangle).$$

Remark 3.4 The group $\phi_7(1^5)$ in the isoclinism family (7) of [15, Section 4.5] is isomorphic to G_7. The group $\phi_{10}(1^5)$ in the isoclinism family (10) of [15, Section 4.5] is isomorphic to G_{10} for $p \geq 5$ and is isomorphic to H for $p = 3$.

The proof of the following theorem [33, Theorem 5.5] uses the classification of groups of order p^5 from [15] and Theorem 2.1 above:

Theorem 3.5 *Let G be a finite group of order p^5, where p is an odd prime. Then $\mathrm{Out}_c(G) \neq 1$ if and only if G is isoclinic to one of the groups G_7, G_{10} and H.*

This theorem, along with the examples of Wall [30], proves that 4 is the smallest value of an integer n such that all the groups of order less than or equal to p^n has the property that $\mathrm{Out}_c(G) = 1$.

3.2 Groups with large cyclic subgroups.

In this part we study class preserving automorphisms of finite p-groups of order p^n having cyclic subgroups of order p^{n-1} or p^{n-2}.

The following result is Theorem 3.1 in [16]

Theorem 3.6 *Let G be a finite p-group having a maximal cyclic subgroup. Then $\mathrm{Out}_c(G) = 1$.*

For odd primes p, we have a similar result for finite p-groups G having a cyclic subgroup of index p^2, [18] and [7].

Theorem 3.7 *Let G be a group of order p^n having a cyclic subgroup of order p^{n-2}, where p is an odd prime. Then $\mathrm{Out}_c(G) = 1$.*

But the situation is different for $p = 2$. Let us set

$$G_1 = \langle x, y, z \mid x^{2^{m-2}} = 1 = y^2 = z^2, \; yxy = x^{1+2^{m-3}}, \; zyz = y,$$
$$zxz = x^{-1+2^{m-3}} \rangle;$$
$$G_2 = \langle x, y, z \mid x^{2^{m-2}} = 1 = y^2 = z^2, \; yxy = x^{1+2^{m-3}}, \; zxz = x^{-1+2^{m-3}},$$
$$zyz = yx^{2^{m-3}} \rangle.$$

Then we get the following result [18]:

Theorem 3.8 *Let G be a group of order 2^n, $n > 4$, such that G has a cyclic normal subgroup of order 2^{n-2}, but does not have any element of order 2^{n-1}. Then $\mathrm{Out}_c(G) \neq 1$ if and only if G is isomorphic to G_1 or G_2.*

3.3 Groups of nilpotency class 2.

Let G be a finite nilpotent group of class 2. Let $\phi \in \mathrm{Aut}_c(G)$. Then the map $g \mapsto g^{-1}\phi(g)$ is a homomorphism of G into $\gamma_2(G)$. This homomorphism sends $Z(G)$ to 1. So it induces a homomorphism $f_\phi \colon G/Z(G) \to \gamma_2(G)$, sending $gZ(G)$ to $g^{-1}\phi(g)$, for any $g \in G$. It is easily seen that the map $\phi \mapsto f_\phi$ is a monomorphism of the group $\mathrm{Aut}_c(G)$ into $\mathrm{Hom}(G/Z(G), \gamma_2(G))$.

Any $\phi \in \mathrm{Aut}_c(G)$ sends any $g \in G$ to some $\phi(g) \in g^G$. Then $f_\phi(gZ(G)) = g^{-1}\phi(g)$ lies in $g^{-1}g^G = [g,G]$. Denote

$$\{ f \in \mathrm{Hom}(G/Z(G), \gamma_2(G)) \mid f(gZ(G)) \in [g,G] \text{ for all } g \in G \}$$

by $\mathrm{Hom}_c(G/Z(G), \gamma_2(G))$. Thus $f_\phi \in \mathrm{Hom}_c(G/Z(G), \gamma_2(G))$ for all $\phi \in \mathrm{Aut}_c(G)$. On the other hand, if $f \in \mathrm{Hom}_c(G/Z(G), \gamma_2(G))$, then the map sending any $g \in G$ to $gf(gZ(G))$ is an automorphism $\phi \in \mathrm{Aut}_c(G)$ such that $f_\phi = f$. Thus we have

Proposition 3.9 *Let G be a finite nilpotent group of class 2. Then the above map $\phi \mapsto f_\phi$ is an isomorphism of the group $\mathrm{Aut}_c(G)$ onto $\mathrm{Hom}_c(G/Z(G), \gamma_2(G))$.*

This correspondence gives the following result [33, Theorem 3.5]:

Theorem 3.10 *Let G be a finite p-group of class 2. Let $\{x_1, x_2, \ldots, x_d\}$ be a minimal generating set for G such that $[x_i, G]$ is cyclic, $1 \leq i \leq d$. Then $\mathrm{Out}_c(G) = 1$.*

In particular, we get

Corollary 3.11 *Let G be a finite p-group of class 2 such that $\gamma_2(G)$ is cyclic. Then $\mathrm{Out}_c(G) = 1$.*

3.4 Camina p-groups.

Notice that extraspecial p-groups are Camina groups and $\mathrm{Out}_c(G) = 1$ for every extraspecial p-group G. The following result, which is proved in [31, Theorem 5.4], shows that these are the only Camina p-groups of class 2 for which $\mathrm{Out}_c(G) = 1$.

Theorem 3.12 *Let G be a finite Camina p-group of nilpotency class 2. Then $|\mathrm{Aut}_c(G)| = |\gamma_2(G)|^d$, where d is the number of elements in a minimal generating set for G.*

The following result [32, Corollary 4.4] deals with finite p-groups G such that $(G, Z(G))$ is a Camina pair.

Theorem 3.13 *Let G be a finite p-group of nilpotency class at least 3 such that $(G, Z(G))$ is a Camina pair and $|Z(G)| \geq p^2$. Then $|\mathrm{Aut}_c(G)| \geq |G|$.*

Since $(G, Z(G))$ is a Camina pair for every Camina p-group G, we have the following immediate corollary:

Corollary 3.14 *Let G be a finite Camina p-group of nilpotency class 3 and $|Z(G)| \geq p^2$. Then $|\text{Aut}_c(G)| \geq |G|$.*

It is not diffucult to prove that $\text{Out}_c(G) \neq 1$ for every Camina p-group of nilpotency class 3.

4 Results of Sah

In this section, we record a few results of C. H. Sah [25]. This paper contains lots of reduction techniques and nice results, but it is very difficult to mention all the results here. For the readers, who are interested in class preserving automorphisms of group, we strongly recommend this beautiful article.

Theorem 4.1 (Proposition 1.8, [25]) *Let G be a nilpotent group of class c. Then $\text{Aut}_c(G)$ is a nilpotent group of class $c - 1$.*

As an immediate consequence of this result one readily gets the following corollary.

Corollary 4.2 *Let G be a nilpotent group of class c. Then $\text{Out}_c(G)$ is a nilpotent group of class at most $c - 1$.*

Theorem 4.3 (Theorem 2.9, [25]) *Let G be a finite solvable group. Then $\text{Aut}_c(G)$ is solvable.*

Theorem 4.4 (Theorem 2.10, [25]) *Let G be a group admitting a composition series. Suppose that for each composition factor F of G, the group $\text{Aut}(F)/\text{Inn}(F)$ is solvable. Then $\text{Out}_c(G)$ is solvable.*

Since every finite group admits a composition series and the Schreier's conjecture, i.e., the group of outer automorphisms of a finite simple group is solvable, holds true for all finite simple groups, the following result is a consequence of the above theorem.

Corollary 4.5 *Let G be a finite group. Then $\text{Out}_c(G)$ is solvable.*

Sah also constructed examples of finite groups G of order p^{5n}, $n \geq 3$, such that $\text{Out}_c(G)$ is non-abelian. Thus, these examples contradict the intuitions of W. Burnside [2, Note B] that $\text{Out}_c(G)$ should be abelian for all finite groups G.

5   An upper bound for $\mathrm{Aut}_c(G)$ for a finite p-group G

In this section, we list results from [31]. Let G be a finite p-group of order p^n. Let $\{x_1, \ldots, x_d\}$ be any minimal generating set for G. Let $\alpha \in \mathrm{Aut}_c(G)$. Since $\alpha(x_i) \in x_i^G$ for $1 \le i \le d$, there are at the most $|x_i^G|$ choices for the image of x_i under α. Thus it follows that

$$|\mathrm{Aut}_c(G)| \le \prod_{i=1}^{d} |x_i^G|. \tag{4}$$

Let $|\gamma_2(G)| = p^m$. Let $\Phi(G)$ denote the Frattini subgroup of G. Since $\gamma_2(G)$ is contained in $\Phi(G)$, by the Burnside basis theorem we have $d \le n - m$. Notice that $|x_i^G| \le |\gamma_2(G)| = p^m$ for all $i = 1, 2, \ldots, d$. So from (4) we get

$$|\mathrm{Aut}_c(G)| \le p^{md} \le (p^m)^{n-m} = p^{m(n-m)}. \tag{5}$$

Theorem 5.1 (Theorem 5.1, [31]) *Let G be a finite p-group. Equality holds in (5) if and only if G is either an abelian p-group or a non-abelian Camina special p-group.*

Theorem 5.2 (Theorem 5.5, [31]) *Let G be a non-trivial p-group having order p^n. Then*

$$|\mathrm{Aut}_c(G)| \le \begin{cases} p^{(n^2-4)/4}, & \text{if } n \text{ is even;} \\ p^{(n^2-1)/4}, & \text{if } n \text{ is odd.} \end{cases} \tag{6}$$

Since the bound in Theorem 5.2 is attained by all non-abelian groups of order p^3 (n odd) and the group constructed by W. Burnside (n even), it follows that the bound in the theorem is optimal.

In the next theorem we are going to classify all finite p-groups which attain the bound in Theorem 5.2. For the statement of the next result we need the following group of order p^6, which is the group $\phi_{21}(1^6)$ in the isoclinism family (21) of [15]:

$$R = \langle \alpha, \alpha_1, \alpha_2, \beta, \beta_1, \beta_2 \mid [\alpha_1, \alpha_2] = \beta, [\beta, \alpha_i] = \beta_i, [\alpha, \alpha_1] = \beta_2,$$

$$[\alpha, \alpha_2] = \beta_1^\nu, \alpha^p = \beta^p = \beta_i^p = 1, \alpha_1^p = \beta_1^{\binom{p}{3}}, \alpha_2^p = \beta_1^{-\binom{p}{3}}, \ i = 1, 2 \rangle, \tag{7}$$

where ν is the smallest positive integer which is a non-quadratic residue mod p and β_1 and β_2 are central elements. We would like to remark here that this is not a minimal presentation of the group G. But it is sufficient for our purpose.

The following theorem gives a classification of all finite p-groups G for which the upper bound in Theorem 5.2 is achieved by $|\mathrm{Aut}_c(G)|$. Recall that W is the group of order p^6, defined in the second paragraph of the introduction.

Theorem 5.3 (Theorem 5.12, [31]) *Let G be a non-abelian finite p-group of order p^n. Then equality holds in* (6) *if and only if one of the following holds:*

G *is an extra-special p-group of order p^3;* $\hfill$ (8a)

G *is a group of nilpotency class 3 and order p^4;* $\hfill$ (8b)

G *is a Camina special p-group isoclinic to the group W and $|G| = p^6$;* $\hfill$ (8c)

G *is isoclinic to R and $|G| = p^6$.* $\hfill$ (8d)

6 Some problems

In this section, we formulate some research problems for finite p-groups, which certainly make sense for arbitrary finite groups.

We know that $\text{Out}_c(G) = 1$ for all finite simple group [6]. But the proof uses classification of finite simple groups. We think that there should be a direct proof.

Problem 6.1 Let G be a finite simple group. Without using classification of finite simple groups, prove that $\text{Out}_c(G) = 1$.

Problem 6.2 Let G be a finite p-group of nilpotency class 2. Find necessary and sufficient conditions on G such that $\text{Out}_c(G) = 1$.

Let G be a finite p-group of nilpotency class 2. Then $\text{Aut}_c(G) \leq \text{Autcent}(G)$.

Problem 6.3 Classify all finite p-groups G of class 2 such that $\text{Autcent}(G) = \text{Aut}_c(G)$.

As we mentioned in the introduction that H. Heineken [9] constructed examples of finite p-groups G of nilpotency class 2 such that $\text{Aut}(G) = \text{Aut}_c(G)$. This gives rise to the following natural problem.

Problem 6.4 Classify all finite p-groups G of class 2 such that $\text{Aut}(G) = \text{Aut}_c(G)$.

The following problem arises from the work of Sah [25, pg 61], which is also given in Malinowska's survey [20] as Question 12.

Problem 6.5 Let G be a finite p-group of nilpotency class $c > 2$. Give a sharp upper bound for the nilpotency class of $\text{Out}_c(G)$.

Let G be a finite p-group with a minimal generating set $\{x_1, x_2, \ldots, x_d\}$. Then $|\text{Aut}_c(G)| \leq \prod_{i=1}^{d} |x_i^G|$.

Problem 6.6 Classify all finite p-groups G with a minimal generating set $\{x_1, x_2, \ldots, x_d\}$ such that $|\text{Aut}_c(G)| = \prod_{i=1}^{d} |x_i^G|$.

Since $|x^G| \leq |\gamma_2(G)|$ for all $x \in G$, we can even formulate the following particular case of the preceeding problem.

Problem 6.7 Classify all finite p-groups G such that $|\text{Aut}_c(G)| = |\gamma_2(G)|^d$, where d is the number of elements in a minimal generating set for G.

Notice that the Problem 6.7 has a solution (Theorem 5.1) when $\gamma_2(G) = \Phi(G)$. Let G be a finite p-group such that $Z(G) \leq \gamma_2(G)$, then it follows that G is purely non-abelian and therefore $\text{Autcent}(G)$ is a p-group (Theorem 2.4).

Problem 6.8 Let G be a finite p-group such that $\text{Out}_c(G) \neq 1$ and $Z(G) \leq \gamma_2(G)$. Find a sharp lower bound for $|\text{Aut}_c(G)\,\text{Autcent}(G)|$.

References

[1] J. E. Adney and T. Yen, Automorphisms of a p-group, *Illinois J. Math.* **9** (1965), 137–143.

[2] W. Burnside, *Theory of groups of finite order, 2nd Ed.*, Dover Publications, Inc., 1955. Reprint of the 2nd edition (Cambridge, 1911).

[3] W. Burnside, On the outer automorphisms of a group, *Proc. London Math. Soc. (2)* **11** (1913), 40–42.

[4] E. C. Dade and M. K. Yadav, Finite groups with many product conjugacy classes, *Israel J. Math.* **154** (2006), 29–49.

[5] R. Dark and C. M. Scoppola, On Camina groups of prime power order, *J. Algebra* **181** (1996), 787–802.

[6] W. Feit and G. M. Seitz, On finite rational groups and related topics, *Illinois J. Math.* **33** (1988), 103–131.

[7] M. Fuma and Y. Ninomiya, "Hasse principle" for finite p-groups with cyclic subgroups of index p^2, *Math. J. Okayama Univ.* **46** (2004), 31–38.

[8] P. Hall, The classification of prime power groups, *Journal für die reine und angewandte Mathematik* **182** (1940), 130–141.

[9] H. Heineken, Nilpotente Gruppen, deren sämtliche Normalteiler charakteristisch sind, *Arch. Math. (Basel)* **33** (1980), no. 6, 497–503.

[10] A. Herman and Y. Li, Class preserving automorphisms of Blackburn groups, *J. Austral. Math. Soc.* **80** (2006), 351–358.

[11] M. Hertweck, Class-preserving automorphisms of finite groups, *J. Algebra* **241** (2001), 1–26.

[12] M. Hertweck, *Contributions to the integral representation theory of groups*, Habilitationsschrift, University of Stuttgart (2004). Available at
http://elib.uni-stuttgart.de/opus/volltexte/2004/1638

[13] M. Hertweck and E. Jespers, Class-preserving automorphisms and normalizer property for Blackburn groups, *J. Group Theory* **12** (2009), 157–169.

[14] W. Gaschütz, Nichtabelsche p-gruppen besitzen äussere p-automorphismen, *J. Algebra* **4** (1966), 1–2.

[15] R. James, The groups of order p^6 (p an odd prime), *Math. Comp.* **34** (1980), 613–637.

[16] M. Kumar and L. R. Vermani, "Hasse principle" for extraspecial p-groups, *Proc. Japan Acad.* **76**, Ser. A, no. 8 (2000), 123–125.

[17] M. Kumar and L. R. Vermani, "Hasse principle" for groups of order p^4, *Proc. Japan Acad.* **77**, Ser. A, no. 6 (2001), 95–98.

[18] M. Kumar and L. R. Vermani, On automorphisms of some p-groups, *Proc. Japan Acad.* **78**, Ser. A, no. 4 (2002), 46–50.

[19] I. Malinowska, On quasi-inner automorphisms of a finite p-group, *Publ. Math. Debrecen* **41** (1992), no. 1–2, 73–77.

[20] I. Malinowska, p-automorphisms of finite p-groups: problems and questions, *Advances in Group Theory* (1992), 111–127.

[21] T. Ono, A note on Shafarevich–Tate sets for finite groups, *Proc. Japan Acad.* **74**, Ser. A (1998), 77–79.

[22] T. Ono, On Shafarevich-Tate sets, *Advanced Studies in Pure Mathematics: Class Field Theory — Its Centenary and prospect* **30** (2001), 537–547.

[23] T. Ono and H. Wada, "Hasse principle" for free groups, *Proc. Japan Acad.* **75**, Ser. A (1999), 1–2.

[24] T. Ono and H. Wada, "Hasse principle" for symmetric and alternating groups, *Proc. Japan Acad.* **75**, Ser. A (1999), 61–62.

[25] C. H. Sah, Automorphisms of finite groups, *J. Algebra* **10** (1968), 47–68.

[26] M. Suzuki, *Group Theory II*, Springer, New York — Berlin — Heidelberg — Tokyo (1986).

[27] F. Szechtman, n-inner automorphisms of finite groups, *Proc. Amer. Math. Soc.* **131** (2003), 3657–3664.

[28] H. Wada, "Hasse principle" for $SL_n(D)$, *Proc. Japan Acad.* **75**, Ser. A (1999), 67–69.

[29] H. Wada, "Hasse principle" for $GL_n(D)$, *Proc. Japan Acad.* **76**, Ser. A (2000), 44–46.

[30] G. E. Wall, Finite groups with class preserving outer automorphisms, *J. London Math. Soc.* **22** (1947), 315–320.

[31] M. K. Yadav, Class preserving automorphisms of finite p-groups, *J. London Math. Soc.* **75** (2007), no. 3, 755–772.

[32] M. K. Yadav, On automorphisms of finite p-groups, *J. Group Theory* **10** (2007), 859–866.

[33] M. K. Yadav, On automorphisms of some finite p-groups, *Proc. Indian Acad. Sci. (Math. Sci.)* **118** (2008), no. 1, 1–11.

SYMMETRIC COLORINGS OF FINITE GROUPS

YULIYA ZELENYUK

School of Mathematics, University of the Witwatersrand, Private Bag 3, Wits 2050, South Africa
Email: yuliya.zelenyuk@wits.ac.za

Abstract

Let G be a finite group and let $r \in \mathbb{N}$. An *r-coloring* of G is any mapping $\chi \colon G \to \{1, \ldots, r\}$. Colorings χ and φ are *equivalent* if there exists $g \in G$ such that $\chi(xg^{-1}) = \varphi(x)$ for all $x \in G$. A coloring χ is *symmetric* if there exists $g \in G$ such that $\chi(gx^{-1}g) = \chi(x)$ for all $x \in G$. We derive formulae for computing the number of symmetric r-colorings of G and the number of equivalence classes of symmetric r-colorings of G.

1 Introduction

Let G be a finite group and let $r \in \mathbb{N}$. A *coloring* (an *r-coloring*) of G is any mapping $\chi : G \to \mathbb{N}$ ($\chi : G \to \{1, \ldots, r\}$). The group G naturally acts on the colorings. For every coloring χ and $g \in G$, the coloring χg is defined by

$$\chi g(x) = \chi(xg^{-1}).$$

Let $[\chi]$ and $St(\chi)$ denote the orbit and the stabilizer of a coloring χ, that is,

$$[\chi] = \{\chi g : g \in G\} \text{ and } St(\chi) = \{g \in G : \chi g = \chi\}.$$

As in the general case of an action, we have that

$$|[\chi]| = |G : St(\chi)| \text{ and } St(\chi g) = g^{-1} St(\chi) g.$$

Let $\sim$ denote the equivalence on the colorings corresponding to the partition into orbits, that is, $\chi \sim \varphi$ if and only if there exists $g \in G$ such that $\chi(xg^{-1}) = \varphi(x)$ for all $x \in G$.

Obviously, the number of all r-colorings of G is $r^{|G|}$. Applying Burnside's Lemma [1, I, §3] shows that the number of equivalence classes of r-colorings of G is equal to

$$\frac{1}{|G|} \sum_{g \in G} r^{|G : \langle g \rangle|},$$

where $\langle g \rangle$ is the subgroup generated by g.

A coloring χ of G is *symmetric* if there exists $g \in G$ such that

$$\chi(gx^{-1}g) = \chi(x)$$

Supported by NRF grant IFR2008041600015 and The John Knopfmacher Centre for Applicable Analysis and Number Theory.

for all $x \in G$. That is, a coloring is symmetric if it is invariant under some symmetry, and a symmetry is any mapping of the form

$$G \to Gx \qquad\qquad \mapsto gx^{-1}g,$$

where $g \in G$. A coloring equivalent to a symmetric one is also symmetric (see Lemma 2.1). Let $S_r(G)$ denote the set of all symmetric r-colorings of G.

Theorem 1.1 ([2, Theorem 1]) *Let G be a finite Abelian group. Then*

$$|S_r(G)| = \sum_{X \leq G} \sum_{Y \leq X} \frac{\mu(Y,X)|G/Y|}{|B(G/Y)|} r^{(|G/X|+|B(G/X)|)/2},$$

$$|S_r(G)/\sim| = \sum_{X \leq G} \sum_{Y \leq X} \frac{\mu(Y,X)}{|B(G/Y)|} r^{(|G/X|+|B(G/X)|)/2}.$$

Here, X runs over subgroups of G, Y over subgroups of X, $\mu(Y,X)$ is the Möbius function on the lattice of subgroups of G, and $B(G) = \{x \in G : x^2 = e\}$. Given a finite partially ordered set, the Möbius function is defined as follows:

$$\mu(a,b) = \begin{cases} 1 & \text{if } a = b \\ -\sum_{a < z \leq b} \mu(z,b) & \text{if } a < b \\ 0 & \text{otherwise .} \end{cases}$$

See [1, IV] for more information about the Möbius function.

In this paper we generalize Theorem 1.1 to an arbitrary finite group G. Our approach is based on constructing a partially ordered set of so called optimal partitions of G.

For every partition π of G, define the *stabilizer* and the *center* of π by

$$St(\pi) = \{g \in G : \text{ for every } x \in G, x \text{ and } xg^{-1} \text{ belong to the same cell of } \pi\}$$

and

$$Z(\pi) = \{g \in G : \text{ for every } x \in G, x \text{ and } gx^{-1}g \text{ belong to the same cell of } \pi\}.$$

We shall show that $St(\pi)$ is a subgroup of G and $Z(\pi)$ is a union of left cosets of G modulo $St(\pi)$. Furthermore, if $e \in Z(\pi)$, then $Z(\pi)$ is also a union of right cosets of G modulo $St(\pi)$ and for every $a \in Z(\pi)$, $\langle a \rangle \subseteq Z(\pi)$. We say that a partition π of G is *optimal* if $e \in Z(\pi)$ and for every partition π' of G with $St(\pi') = St(\pi)$ and $Z(\pi') = Z(\pi)$, one has $\pi \leq \pi'$. The latter means that every cell of π is contained in some cell of π', or equivalently, the equivalence corresponding to π is contained in that of π'. The partially ordered set of optimal partitions of G can be naturally identified with the partially ordered set of pairs (A,B) of subsets of G such that $A = St(\pi)$ and $B = Z(\pi)$ for some partition π of G with $e \in Z(\pi)$. For every partition π, we write $|\pi|$ to denote the number of cells of π.

In Section 2 we prove the following result.

Theorem 1.2 *Let P be the partially ordered set of optimal partitions of G. Then*

$$|S_r(G)| = |G| \sum_{x \in P} \sum_{y \leq x} \frac{\mu(y, x)}{|Z(y)|} r^{|x|},$$

$$|S_r(G)/\sim| = \sum_{x \in P} \sum_{y \leq x} \frac{\mu(y, x)|St(y)|}{|Z(y)|} r^{|x|}.$$

In Section 3 we illustrate Theorem 1.2 by computing the numbers $|S_r(G)|$ and $|S_r(G)/\sim|$ for the alternating group A_4.

2 Proof of Theorem 1.2

For every coloring χ of G, let

$$Z(\chi) = \{g \in G : \chi(gx^{-1}g) = \chi(x) \text{ for all } x \in G\}.$$

Note that a coloring χ is symmetric if and only if $Z(\chi) \neq \emptyset$.

Lemma 2.1 $Z(\chi h) = Z(\chi)h$ *for all* $h \in G$.

Proof Let $g \in G$. Then

$$
\begin{aligned}
g \in Z(\chi h) \quad &\Leftrightarrow \quad \chi h(gx^{-1}g) = \chi h(x) \text{ for all } x \in G \\
&\Leftrightarrow \quad \chi(gx^{-1}gh^{-1}) = \chi(xh^{-1}) \text{ for all } x \in G \\
&\Leftrightarrow \quad \chi(gh^{-1}(xh^{-1})^{-1}gh^{-1}) = \chi(xh^{-1}) \text{ for all } x \in G \\
&\Leftrightarrow \quad \chi(gh^{-1}x^{-1}gh^{-1}) = \chi(x) \text{ for all } x \in G \\
&\Leftrightarrow \quad gh^{-1} \in Z(\chi) \\
&\Leftrightarrow \quad g \in Z(\chi)h.
\end{aligned}
$$

$\square$

For every coloring χ of G and $g \in G$, let

$$[\chi]_g = \{\varphi \in [\chi] : g \in Z(\varphi)\}.$$

It is clear that if a coloring χ is not symmetric, then $[\chi]_g = \emptyset$ for all $g \in G$. It follows from the next lemma that if χ is symmetric, then $[\chi]_g \neq \emptyset$ for all $g \in G$.

Lemma 2.2 *For every* $g, h \in G$, $[\chi]_g h = [\chi]_{gh}$.

Proof Let $\varphi \in [\chi]$. Then

$$\varphi \in [\chi]_g h \quad \Leftrightarrow \quad \varphi h^{-1} \in [\chi]_g \quad \Leftrightarrow \quad g \in Z(\varphi h^{-1}).$$

By Lemma 2.1, $Z(\varphi h^{-1}) = Z(\varphi)h^{-1}$. Hence

$$\varphi \in [\chi]_g h \quad \Leftrightarrow \quad g \in Z(\varphi)h^{-1} \quad \Leftrightarrow \quad gh \in Z(\varphi) \quad \Leftrightarrow \quad \varphi \in [\chi]_{gh}.$$

$\square$

Corollary 2.3 *For every $g \in G$, $|[\chi]_g| = |[\chi]_e|$.*

Proof By Lemma 2.2, $[\chi]_e g = [\chi]_g$ and $[\chi]_g g^{-1} = [\chi]_e$. It follows that $|[\chi]_g| \leq |[\chi]_e|$ and $|[\chi]_e| \leq |[\chi]_g|$. Hence, $|[\chi]_g| = |[\chi]_e|$.   □

Lemma 2.4 $Z(\chi) = \{g \in G : e \in Z(\chi g^{-1})\} = \{g \in G : \chi g^{-1} \in [\chi]_e\}$.

Proof Let $g \in G$. Then

$$
\begin{aligned}
g \in Z(\chi) \quad &\Leftrightarrow \quad \chi(gx^{-1}g) = \chi(x) \text{ for all } x \in G \\
&\Leftrightarrow \quad \chi(g(xg)^{-1}g) = \chi(xg) \text{ for all } x \in G \\
&\Leftrightarrow \quad \chi(x^{-1}g) = \chi(xg) \text{ for all } x \in G \\
&\Leftrightarrow \quad \chi g^{-1}(x^{-1}) = \chi g^{-1}(x) \text{ for all } x \in G \\
&\Leftrightarrow \quad e \in Z(\chi g^{-1}).
\end{aligned}
$$

□

Lemma 2.5 $Z(\chi) \cdot St(\chi) = Z(\chi)$ *and* $|Z(\chi)| = |St(\chi)| \cdot |[\chi]_e|$.

Proof Consider the mapping

$$
\begin{aligned}
f : G &\to [\chi] \\
g &\mapsto \chi g^{-1}.
\end{aligned}
$$

Clearly f is surjective. Also for every $\varphi \in [\chi]$, $f^{-1}(\varphi)$ is a left coset of $St(\chi)$ in G. Indeed, for every $g, h \in G$,

$$
\chi g^{-1} = \chi h^{-1} \quad \Leftrightarrow \quad \chi g^{-1} h = \chi \quad \Leftrightarrow \quad g^{-1} h \in St(\chi) \quad \Leftrightarrow \quad h \in g \cdot St(\chi).
$$

Finally, by Lemma 2.4, we have that $f^{-1}([\chi]_e) = Z(\chi)$.   □

Lemma 2.6 *If $e \in Z(\chi)$, then $St(\chi) \cdot Z(\chi) = Z(\chi)$.*

Proof Clearly $Z(\chi) \subseteq St(\chi) \cdot Z(\chi)$. To show the converse inclusion, let $h \in St(\chi)$ and $g \in Z(\chi)$. For every $x \in G$, we have that

$$
\begin{aligned}
\chi(hgx^{-1}hg) = \chi(g^{-1}h^{-1}xg^{-1}h^{-1}) &= \chi(g^{-1}h^{-1}xg^{-1}) = \chi(gx^{-1}hg) \\
&= \chi(h^{-1}x) = \chi(x^{-1}h) = \chi(x^{-1}) = \chi(x).
\end{aligned}
$$

Hence, $hg \in Z(\chi)$.   □

Lemma 2.7 *If $e \in Z(\chi)$, then*

$$
Z(\chi) = \{g \in G : \chi(gxg) = \chi(x) \text{ for all } x \in G\}.
$$

Proof Let $g \in G$. Then

$$
\begin{aligned}
g \in Z(\chi) \quad &\Leftrightarrow \quad \chi(gx^{-1}g) = \chi(x) \text{ for all } x \in G \\
&\Leftrightarrow \quad \chi(g(gxg)^{-1}g) = \chi(gxg) \text{ for all } x \in G \\
&\Leftrightarrow \quad \chi(x^{-1}) = \chi(gxg) \text{ for all } x \in G \\
&\Leftrightarrow \quad \chi(x) = \chi(gxg) \text{ for all } x \in G.
\end{aligned}
$$

$\square$

Corollary 2.8 *If $e \in Z(\chi)$, then for every $a \in Z(\chi)$, one has $\langle a \rangle \subseteq Z(\chi)$.*

Proof Indeed, let $a \in Z(\chi)$ and $n > 1$. Then for every $x \in G$, we have that

$$
\chi(a^n x a^n) = \chi(a(a^{n-1}xa^{n-1})a) = \chi(a^{n-1}xa^{n-1}) = \cdots = \chi(axa) = \chi(x).
$$

$\square$

Corollary 2.9 *If $e \in Z(\chi)$ and G is Abelian, then $Z(\chi) = \{g \in G : g^2 \in St(\chi)\}$.*

Proof Indeed,

$$
\begin{aligned}
Z(\chi) &= \{g \in G : \chi(gxg) = \chi(x) \text{ for all } x \in G\} \\
&= \{g \in G : \chi(xg^2) = \chi(x) \text{ for all } x \in G\} \\
&= \{g \in G : (g^2)^{-1} \in St(\chi)\} \\
&= \{g \in G : g^2 \in St(\chi)\}.
\end{aligned}
$$

$\square$

Let P be the set of all pairs $x = (A_x, B_x)$ such that $A_x = St(\chi)$ and $B_x = Z(\chi)$ for some coloring χ of G with $e \in Z(\chi)$. Define the order $\leq$ on P by

$$
x \leq y \quad \Leftrightarrow \quad A_x \subseteq A_y \text{ and } B_x \subseteq B_y.
$$

For every $x \in P$, let π_x denote the finest partition of G with $St(\pi) = A_x$ and $Z(\pi) = B_x$. Recall that a partition π of G is *optimal* if $e \in Z(\pi)$ and for every partition π' of G with $St(\pi') = St(\pi)$ and $Z(\pi') = Z(\pi)$, one has $\pi \leq \pi'$. Note that $\{\pi_x : x \in P\}$ is the set of all optimal partitions of G. The next lemma says that as a partially ordered set it can be identified with P.

Lemma 2.10 $\pi_x \leq \pi_y \Leftrightarrow x \leq y$.

Proof Let π, π' be partitions of G, let ρ, ρ' be the corresponding equivalences on G, and let $\pi \leq \pi'$. We have that

$$
St(\pi) = \{h \in G : (g, gh^{-1}) \in \rho \text{ for all } g \in G\},
$$

$$
St(\pi') = \{h \in G : (g, gh^{-1}) \in \rho' \text{ for all } g \in G\}.
$$

Since $\rho \subseteq \rho'$, it follows that $St(\pi) \subseteq St(\pi')$. Similarly,

$$Z(\pi) = \{a \in G : (g, ag^{-1}a) \in \rho \text{ for all } g \in G\},$$

$$Z(\pi') = \{a \in G : (g, ag^{-1}a) \in \rho' \text{ for all } g \in G\},$$

and so $Z(\pi) \subseteq Z(\pi')$.

Now let $x \leq y$ and let ρ_x, ρ_y be the equivalences on G corresponding to the partitions π_x, π_y. Note that ρ_x, ρ_y are the transitive closures of the relations

$$R_x = \{(g, g^{-1}) : g \in G\} \cup \{(g, gh^{-1}) : g \in G, h \in A_x\} \cup \{(g, ag^{-1}a) : g \in G, a \in B_x\},$$

$$R_y = \{(g, g^{-1}) : g \in G\} \cup \{(g, gh^{-1}) : g \in G, h \in A_y\} \cup \{(g, ag^{-1}a) : g \in G, a \in B_y\}.$$

Since $A_x \subseteq A_y$ and $B_x \subseteq B_y$, it follows that $R_x \subseteq R_y$, and so $\rho_x \subseteq \rho_y$. Hence $\pi_x \leq \pi_y$. $\qquad\square$

Now we are in a position to prove Theorem 1.2.

Proof of Theorem 1.2 Let $C = \{\chi \in S_r(G) : e \in Z(\chi)\}$ and for every $x \in P$, let

$$C(x) = \{\chi \in C : St(\chi) = St(x) \text{ and } Z(\chi) = Z(x)\}.$$

Then $\{C(x) : x \in P\}$ is a partition of C and for every $x \in P$ and $\chi \in C(x)$, one has $|[\chi]_e| = |Z(x)|/|St(x)|$ and $|[\chi]| = |G|/|St(x)|$, but in general, $[\chi]_e \nsubseteq C(x)$. To correct this situation, define the equivalence $\equiv$ on P by

$$x \equiv y \quad \Leftrightarrow \quad St(y) = aSt(x)a^{-1} \text{ and } Z(y) = Z(x)a^{-1} \text{ for some } a \in Z(x).$$

Clearly $\equiv$ is induced by $\sim$. For every $x \in P$, let $\bar{x}$ denote the $\equiv$-class containing x and let $C(\bar{x}) = \bigcup_{y \in \bar{x}} C(y)$. Then whenever $y \in \bar{x}$ and $\chi \in C(y)$,

$$[\chi]_e \subseteq C(\bar{x}), \quad |[\chi]_e| = \frac{|Z(x)|}{|St(x)|} \quad \text{and} \quad |[\chi]| = \frac{|G|}{|St(x)|}.$$

It follows that

$$|C(\bar{x})/\sim| = \frac{|St(x)||C(\bar{x})|}{|Z(x)|} = \sum_{y \in \bar{x}} \frac{|St(y)||C(y)|}{|Z(y)|}$$

and the number of all colorings equivalent to colorings from $C(\bar{x})$ is

$$|C(\bar{x})/\sim| \cdot \frac{|G|}{|St(x)|} = |G| \sum_{y \in \bar{x}} \frac{|C(y)|}{|Z(y)|}.$$

Consequently,

$$|S_r(G)/\sim| = |C(G)/\sim| = \sum_{y \in P} \frac{|St(y)||C(y)|}{|Z(y)|},$$

$$|S_r(G)| = |G| \sum_{y \in P} \frac{|C(y)|}{|Z(y)|}.$$

Now to compute $|C(y)|$, note that

$$\sum_{y \leq x} |C(x)| = r^{|y|}.$$

Then applying Möbius inversion (see [1, IV, §2]) gives us that

$$|C(y)| = \sum_{y \leq x} \mu(y,x) r^{|x|}.$$

Finally, we obtain that

$$|S_r(G)/\sim| = \sum_{y \in P} \sum_{y \leq x} \frac{\mu(y,x)|St(y)|}{|Z(y)|} r^{|x|} = \sum_{x \in P} \sum_{y \leq x} \frac{\mu(y,x)|St(y)|}{|Z(y)|} r^{|x|},$$

$$|S_r(G)| = |G| \sum_{y \in P} \sum_{y \leq x} \frac{\mu(y,x)}{|Z(y)|} r^{|x|} = |G| \sum_{x \in P} \sum_{y \leq x} \frac{\mu(y,x)}{|Z(y)|} r^{|x|}.$$

We conclude this section by suggesting how to construct the partially ordered set of optimal partitions π of G together with parameters $|St(\pi)|$, $|Z(\pi)|$ and $|\pi|$. This can be done by starting with the finest optimal partition $\{\{x, x^{-1}\} : x \in G\}$ and using the following simple fact.

Lemma 2.11 Let π be an optimal partition of G and let $A \subseteq G$. Let π_1 be the finest partition of G such that $\pi \leq \pi_1$ and $A \subseteq St(\pi_1)$, and let π_2 be the finest partition of G such that $\pi \leq \pi_2$ and $A \subseteq Z(\pi_2)$. Then the partitions π_1 and π_2 are also optimal.

Proof Let π_1' be any partition of G such that $St(\pi_1') = St(\pi_1)$ and $Z(\pi_1') = Z(\pi_1)$. Since π is optimal and

$$St(\pi) \subseteq St(\pi_1) = St(\pi_1') \text{ and } Z(\pi) \subseteq Z(\pi_1) = Z(\pi_1'),$$

we obtain that $\pi \leq \pi_1'$. It follows that $\pi_1 \leq \pi_1'$. Hence, π_1 is optimal. The proof that π_2 is optimal is the same. $\square$

Remark 2.12 If G is Abelian, then for every coloring χ of G with $e \in Z(\chi)$, $Z(\chi) = \{g \in G : g^2 \in St(\chi)\}$ (Corollary 2.9), so $Z(\chi)$ is completely determined by $St(\chi)$. It follows that the partially ordered set of optimal partitions of G can be identified with the lattice of subgroups of G and Theorem 1.2 becomes Theorem 1.1.

3 Example

In this section we illustrate Theorem 1.2 by computing the numbers $|S_r(G)|$ and $|S_r(G)/\sim|$ for the alternating group

$$A_4 = \{e, (12)(34), (13)(24), (14)(23),$$
$$(123), (132), (134), (143), (124), (142), (234), (243)\}.$$

It has three subgroups of order 2:

$$\{e, (12)(34)\}, \ \{e, (13)(24)\}, \ \{e, (14)(23)\},$$

four subgroups of order 3:

$$\{e, (123), (132)\}, \ \{e, (134), (143)\}, \ \{e, (124), (142)\}, \ \{e, (234), (243)\}$$

and one subgroup of order 4:

$$\{e, (12)(34), (13)(24), (14)(23)\}.$$

So the lattice of subgroups of A_4 is the following:

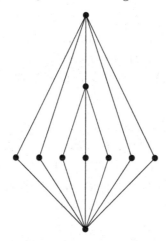

Now we list all optimal partitions π of A_4 together with parameters $|St(\pi)|$, $|Z(\pi)|$ and $|\pi|$.
The finest partition

$$\pi : \{e\}, \{(12)(34)\}, \{(13)(24)\}, \{(14)(23)\},$$
$$\{(123), (132)\}, \{(134), (143)\}, \{(124), (142)\}, \{(234), (243)\},$$
$$St(\pi) = \{e\}, Z(\pi) = \{e\},$$
$$|St(\pi)| = 1, |Z(\pi)| = 1, |\pi| = 8.$$

Three partitions of the form

$$\pi : \{e\}, \{(12)(34)\}, \{(13)(24)\}, \{(14)(23)\},$$
$$\{(123), (132), (124), (142)\}, \{(134), (143), (234), (243)\},$$
$$St(\pi) = \{e\}, Z(\pi) = \{e, (12)(34)\},$$
$$|St(\pi)| = 1, |Z(\pi)| = 2, |\pi| = 6.$$

One partition

$$\pi : \{e\}, \{(12)(34)\}, \{(13)(24)\}, \{(14)(23)\},$$

$$\{(123), (132), (124), (142), (134), (143), (234), (243)\},$$
$$St(\pi) = \{e\}, Z(\pi) = \{e, (12)(34), (13)(24), (14)(23)\},$$
$$|St(\pi)| = 1, |Z(\pi)| = 4, |\pi| = 5.$$

Three partitions of the form

$$\pi : \{e, (12)(34)\}, \{(13)(24), (14)(23)\},$$
$$\{(123), (132), (234), (243), (124), (142), (134), (143)\},$$
$$St(\pi) = \{e, (12)(34)\}, Z(\pi) = \{e, (12)(34), (13)(24), (14)(23)\},$$
$$|St(\pi)| = 2, |Z(\pi)| = 4, |\pi| = 3.$$

One partition

$$\pi : \{e, (12)(34), (13)(24), (14)(23)\},$$
$$\{(123), (132), (234), (243), (124), (142), (134), (143)\},$$
$$St(\pi) = \{e, (12)(34), (13)(24), (14)(23)\},$$
$$Z(\pi) = \{e, (12)(34), (13)(24), (14)(23)\},$$
$$|St(\pi)| = 4, |Z(\pi)| = 4, |\pi| = 2.$$

Four partitions of the form

$$\pi : \{e, (123), (132)\}, \{(12)(34), (124), (142)\},$$
$$\{(13)(24), (134), (143)\}, \{(14)(23), (234), (243)\},$$
$$St(\pi) = \{e\}, Z(\pi) = \{e, (123), (132)\},$$
$$|St(\pi)| = 1, |Z(\pi)| = 3, |\pi| = 4.$$

Six partitions of the form

$$\pi : \{e, (123), (132), (12)(34), (124), (142)\},$$
$$\{(13)(24), (134), (143), (14)(23), (234), (243)\},$$
$$St(\pi) = \{e\}, Z(\pi) = \{e, (12)(34), (123), (132), (124), (142)\},$$
$$|St(\pi)| = 1, |Z(\pi)| = 6, |\pi| = 2.$$

Four partitions of the form

$$\pi : \{e, (123), (132)\},$$
$$\{(12)(34), (13)(24), (14)(23), (124), (142), (134), (143), (234), (243)\},$$
$$St(\pi) = \{e, (123), (132)\}, Z(\pi) = \{e, (123), (132)\},$$
$$|St(\pi)| = 3, |Z(\pi)| = 3, |\pi| = 2.$$

And the coarsest partition

$$\pi : \{A_4\},$$
$$|St(\pi)| = 12, |Z(\pi)| = 12, |\pi| = 1.$$

Next, we draw the partially ordered set of optimal partitions π together with parameters $|St(\pi)|$, $|Z(\pi)|$ and $|\pi|$. The picture below shows also the values of the Möbius function of the form $\mu(a, 1)$.

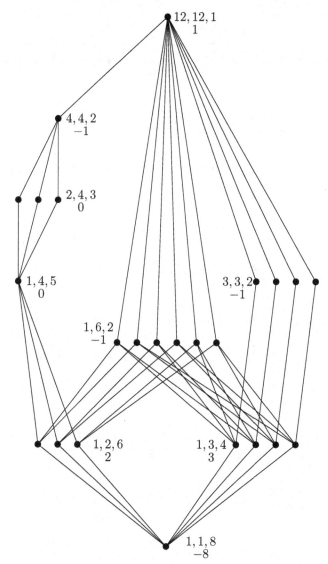

Finally, by the formulae from Theorem 1.2, we obtain that

$$|S_r(A_4)| = |A_4| \sum_{x \in P} \sum_{y \le x} \frac{\mu(y,x)}{|Z(y)|} r^{|x|}$$

$$= 12 \left(r^8 + 3r^6 \left(\frac{1}{2} - 1 \right) + r^5 \left(\frac{1}{4} - \frac{3}{2} + 2 \right) + 3r^3 \left(\frac{1}{4} - \frac{1}{4} \right) \right.$$

$$+ r^2 \left(\frac{1}{4} - \frac{3}{4} + \frac{2}{4} \right) + 4r^4 \left(\frac{1}{3} - 1 \right) + 6r^2 \left(\frac{1}{6} - \frac{1}{2} - \frac{2}{3} + 2 \right)$$

$$\left. + 4r^2 \left(\frac{1}{3} - \frac{1}{3} \right) + r \left(\frac{1}{12} - \frac{1}{4} - \frac{6}{6} - \frac{4}{3} + 3 + 4 - 8 \right) \right)$$

$$= 12\left(r^8 - \frac{3}{2}r^6 + \frac{3}{4}r^5 - \frac{8}{3}r^4 + 6r^2 - \frac{7}{2}r\right)$$

$$= 12r^8 - 18r^6 + 9r^5 - 32r^4 + 72r^2 - 42r,$$

$$|S_r(A_4)/\sim| = \sum_{x\in P}\sum_{y\leq x}\frac{\mu(y,x)|St(y)|}{|Z(y)|}r^{|x|}$$

$$= r^8 + 3r^6\left(\frac{1}{2}-1\right) + r^5\left(\frac{1}{4}-\frac{3}{2}+2\right) + 3r^3\left(\frac{2}{4}-\frac{1}{4}\right)$$

$$+ r^2\left(1-\frac{3}{2}+\frac{2}{4}\right) + 4r^4\left(\frac{1}{3}-1\right) + 6r^2\left(\frac{1}{6}-\frac{1}{2}-\frac{2}{3}+2\right)$$

$$+ 4r^2\left(1-\frac{1}{3}\right) + r\left(1-1-\frac{6}{6}-4+3+4-8\right)$$

$$= r^8 - \frac{3}{2}r^6 + \frac{3}{4}r^5 - \frac{8}{3}r^4 + \frac{3}{4}r^3 + \frac{26}{3}r^2 - 6r.$$

In particular, $|S_2(A_4)| = 1900$ and $|S_2(A_4)/\sim| = 170$, while the number of all 2-colorings of A_4 is $2^{12} = 4096$ and the number of equivalence classes of all 2-colorings of A_4 is

$$\frac{1}{|A_4|}\sum_{g\in A_4} 2^{|A_4/\langle g\rangle|} = \frac{1}{12}(2^{12} + 3\cdot 2^6 + 8\cdot 2^4) = 368.$$

References

[1] M. Aigner, *Combinatorial Theory*, Springer-Verlag, Berlin-Heidelberg-New York, 1979.
[2] Y. Gryshko (Zelenyuk), Symmetric colorings of regular polygons, *Ars Combin.* **78** (2006), 277–281.

Printed in the United States
by Baker & Taylor Publisher Services